SPON'S HOUSE IMPROVEMENT PRICE BOOK

Other books by Bryan Spain
also available from Taylor & Francis

Spon's First Stage Estimating Price Book (2010 edition)
978-0-415-54715-4

Spon's Estimating Costs Guide to Electrical Works (2009 edition)
978-0-415-46904-3

Spon's Estimating Costs Guide to Plumbing and Heating (2009 edition)
978-0-415-46905-0

Spon's Estimating Costs Guide to Minor Works, Alterations and Repairs to Fire, Flood, Gate and Theft Damage (2009 edition)
978-0-415-46906-7

Spon's Estimating Costs Guide to Small Groundworks, Landscaping and Gardening (2007 edition)
978-0-415-43442-3

Spon's Estimating Costs Guide to Finishings: painting and decorating, plastering and tiling (2008 edition)
978-0-415-43443-0

Spon's Estimating Costs Guide to Roofing (2005 edition)
978-0-415-34412-8

Spon's Construction Resource Handbook (1998 edition)
978-0-419-23680-1

Information and ordering details

For price, availability and ordering visit our website **www.sponpress.com**

Alternatively our books are available from all good bookshops.

SPON'S HOUSE IMPROVEMENT PRICE BOOK

Fourth Edition

House extensions, storm damage work, alterations, loft conversions, insulation and kitchens

Edited by

Bryan Spain

Spon Press
an imprint of Taylor & Francis
LONDON AND NEW YORK

First published 1998 by E&FN Spon
Second edition published 2003 by Spon Press
Third edition published 2005 by Taylor & Francis

This edition published 2010
by Spon Press
2 Park Square, Milton Park, Abingdon, Oxon OX14 4RN

Simultaneously published in the USA and Canada
by Spon Press
270 Madison Avenue, New York, NY 10016, USA

Spon Press is an imprint of the Taylor & Francis Group, an informa business

Publisher's Note
This book has been prepared from camera-ready copy supplied by the author.

Printed and bound in Great Britain by
TJ International Ltd, Padstow, Cornwall

British Library Cataloguing in Publication Data
A catalogue record for this book is available from the British Library

ISBN10: 0–415–54716–4 (pbk)
ISBN10: 0–203–87594–0 (ebk)

ISBN13: 978–0–415–54716–1 (pbk)
ISBN13: 978–0–203–87594–0 (ebk)

Contents

One storey, flat roof (cont'd)

One storey, flat roof (cont'd)

One storey, flat roof (cont'd)

One storey, pitched roof

One storey, pitched roof (cont'd)

One storey, pitched roof (cont'd)

One storey, pitched roof (cont'd)

Two storey, flat roof

Two storey, flat roof (cont'd)

Two storey, flat roof (cont'd)

Two storey, flat roof (cont'd)

Two storey, pitched roof

Two storey, pitched roof (cont'd)

Two storey, pitched roof (cont'd)

Two storey, pitched roof (cont'd)

Part Two: Loft Conversions

Preface

This is the fourth edition of Spon's House Improvement Price Book and is published to meet the demands of readers who found the previous editions a useful tool in preparing estimates in the domestic construction market. The uncertainty in the housing market continues but the interest in house extensions continues as house sellers see them as the answer to the problems of space caused by increasing family sizes.

New feedback from contractors using previous editions of this book confirm that they have continued to receive substantial benefits by using the book by saving time in the preparation of quotations and estimates and it is expected that this benefit will continue with this new edition.

The sections covering work in kitchens, loft conversions, insulation, damage repairs and alterations are included again together with summaries of the costs at the end of their respective sections and are presented in a style to show the percentage value of each element together with a cost per square metre for each project for comparison purposes.

It will be noted that an index has not been included again in the book because it is felt that providing an index that repeatedly referred to the same items in the 40 house extensions and 3 loft conversions would not be helpful and a comprehensive contents list is included instead.

I have again received a great deal of support in the research necessary for this book and I am grateful to those individuals and firms who helped me by providing cost information. Although every care has been taken in the preparation of the book's contents, neither the publishers nor I can accept any responsibility for the use of the information made by any person or firm. Finally, I would welcome constructive criticism of the book together with suggestions on how it could be improved for future editions.

Bryan Spain
April 2009

Introduction

The contents of this book cover the cost of domestic construction work in the following areas:

  extensions
  loft conversions
  insulation work
  damage repairs
  kitchens
  alterations
  plant and tool hire

The house extensions section comprises 40 different sizes and types of extensions ranging from 2 x 3m to 2 x 6m, flat roofs and pitched roofs, and one storey and two storey in height. Each extension is presented in a bill of quantities format and individual rates are broken down into labour hours, labour costs, material costs, overheads and profit.
  Each extension contains a maximum of thirteen elements:

  preliminaries
  substructure
  external walls
  flat roof*
  pitched roof*
  windows and external doors
  internal partitions and doors*
  wall finishes
  floor finishes
  ceiling finishes
  electrics
  heating work
  alteration work.

* where applicable

Each element has separate totals for labour hours, labour costs, material costs, overheads and profit. These elemental totals are carried to a general summary at the end of each extension to produce final totals.
  For example, on page 36 the total cost of a one storey extension size 2 x 3m with a flat roof is £8,558.95. This is made up of £1,132.00 for preliminaries, £3,030.86 for labour, £3,427.36 for materials, and £968.73 for overheads and profit. It should be noted that the horizontal and vertical totals do not always coincide due to rounding off.
  The total time is shown as 178.80 hours but it should not be assumed that by dividing this figure by 37.5 it would produce the number of weeks needed to construct the work. This is due to many factors including some trades working simultaneously and the inevitable gaps that occur on small projects between the completion of one package of work and the commencement of the next.
  The hourly labour rates are based upon current wage awards and are £19.00 for electricians, £18.00 for plumbers, £17 for other craftsmen and £13 for unskilled workers.

The purchase of materials at competitive rates can be difficult for small contractors. They cannot obtain large discounts because the size of their orders is generally too small and they often experience a higher than average percentage of waste by not always using all the materials in large packs. Nevertheless, it has been assumed that the contractor has access to facilities to store surplus materials and this should reduce the incidence of high wastage.

The total cost of the one storey extension size 2 x 3m with a flat roof on page 36 is transferred to the Summary of Extension Costs on page 353 and entered as £8,559. Each element in the extension is expressed as a percentage of the total cost. The floor area is stated as 4.08 square metres (that is the area inside the external walls) and the cost per square metre is shown as £2,098. This high figure is caused by the small floor area – in larger extensions this figure is reduced per square metre.

Access to square metre prices for a wide range of sizes and types of extensions can be invaluable to a small contractor, particularly in dealing with casual enquiries where the preparation of time-consuming quotations can be expensive. In general terms, the standard of finish is basic but a list of alternative items is included on page 361 so that adjustments can be made to standard extension as set out in the book.

There are eleven elements in the loft conversion section:

> preliminaries
> preparation
> dormer window
> roof window
> stairs
> flooring
> partitions and doors
> ceilings and soffits
> wall finishes
> electrics
> heating work.

The costs of these elements are transferred to a summary on page 409. Square metre prices are shown in the summary in the same way as in the extensions section. Drawings are included for both the extensions and the loft conversions but they are not drawn to scale and are for illustrative purposes only.

Changes to the building regulations came into effect on 1 April 2002 and three methods of complying with U-values were introduced: Elemental Method, Target U-value Method and the Carbon Index Method.

The Elemental Method is the most relevant to house extension and loft conversion work and materials to meet the targets are included in this book but contractors should seek specialist advice if necessary.

The new section included in the last edition proved very popular and covers the cost of damage to property caused by fire, flood, gale and theft including some costs of emergency measures. A combination of variations in extreme weather conditions, rising crime levels and an increase in insurance claims has placed demands on small contractors undertaking this type of work. It is hoped that the cost information provided in this section will help those involved in this sector of the industry.

A new section on kitchens is included and the next section deals with alteration work including the forming and filling in of openings and repair work generally.

Part One

HOUSE EXTENSIONS

Standard items

Drawings

One storey, flat roof

Two storey, flat roof

One storey, pitched roof

Two storey, pitched roof

Summary of extension costs

Alternative items

	Ref	Qty	Hours	Hours £	Mat'ls £	O & P £	Total £
PART A **PRELIMINARIES**							
Concrete mixer	A1	wk					50.00
Small tools	A2	wk					40.00
Scaffolding (m²/weeks)	A3	n2/wk					6.00
Skip	A4	wk					110.00
Clean up	A5	hour					13.00
PART B **SUBSTRUCTURE TO** **DPC LEVEL**							
Excavate topsoil 150mm thick by hand and deposit on site	B1	m²	0.30	3.90	0.00	0.59	4.49
Excavate to reduce levels by hand and deposit on site	B2	m³	2.50	32.50	0.00	4.88	37.38
Excavate for trench foundations by hand and deposit on site	B3	m³	2.60	33.80	0.00	5.07	38.87
Earthwork support to sides of trenches	B4	m³	0.40	5.20	1.85	1.06	8.11
Backfilling to sides of trenches with excavated material	B5	m²	0.60	7.80	0.00	1.17	8.97
Hardcore filling in bed 225mm thick blinded with sand to receive damp-proof membrane	B6	m²	0.20	2.60	6.29	1.33	10.22
Hardcore filling to sides of trenches	B7	m³	0.50	6.50	6.29	1.92	14.71
Concrete grade (1:3:6) in in foundations	B8	m³	1.35	17.55	85.62	15.48	118.65
Concrete grade (1:2:4) in in sub-floor 150mm thick	B9	m²	0.30	3.90	15.32	2.88	22.10
Concrete grade (1:2:4) in filling to hollow wall	B10	m²	0.20	2.60	2.38	0.75	5.73

	Ref	Qty	Hours	Hours £	Mat'ls £	O & P £	Total £
Polythene damp-proof membrane	B11	m²	0.04	0.52	2.32	0.43	3.27
Steel fabric reinforcement ref A193 in strip foundations	B12	m²	0.12	1.56	2.02	0.54	4.12
Steel fabric reinforcement ref A193 in slab	B13	m²	0.15	1.95	2.00	0.59	4.54
Solid blockwork 140mm thick in skin of cavity wall	B14	m²	1.30	22.10	16.68	5.82	44.60
Common bricks (£180 per 1000) 112.5mm thick in skin of cavity wall	B15	m²	1.70	28.90	16.45	6.80	52.15
Facing bricks (£340 per 1000) 112.5mm thick in skin of cavity wall	B16	m²	1.80	30.60	28.55	8.87	68.02
Form cavity 50mm wide with 5 nr butterfly wall ties per square metre in cavity wall	B17	m²	0.03	0.51	0.52	0.15	1.18
Pitch polymer-based damp-proof course 112mm wide	B18	m²	0.05	0.85	1.02	0.28	2.15
Pitch polymer-based damp-proof course 140mm wide	B19	m²	0.06	1.02	1.38	0.36	2.76
Cut, tooth and bond 140mm thick blockwork to existing faced wall	B20	m²	0.44	7.48	2.91	1.56	11.95
Cut, tooth and bond half brick to existing faced wall	B21	m²	0.35	5.95	2.68	1.29	9.92
50mm thick Polyform Plus insulation board or similar	B22	m²	0.30	5.10	7.20	1.85	14.15

PART C
EXTERNAL WALLS

	Ref	Qty	Hours	Hours £	Mat'ls £	O & P £	Total £
Solid blockwork 140mm thick in skin of cavity wall	C1	m²	1.30	22.10	16.68	5.82	44.60

	Ref	Qty	Hours	Hours £	Mat'ls £	O & P £	Total £
Facing bricks (£340 per 1000) 112.5mm thick in skin of cavity wall	C2	m²	1.80	30.60	28.55	8.87	68.02
75mm thick Crown Dritherm or similar as cavity filling	C3	m²	0.22	3.74	5.34	1.36	10.44
Standard galvanised steel lintel 256mm high, 2400mm long to hollow wall	C4	nr	0.25	4.25	148.29	22.88	175.42
Standard galvanised steel lintel 256mm high, 1500mm long to hollow wall	C5	nr	0.20	3.40	121.36	18.71	143.47
Standard galvanised steel lintel 256mm high, 1150mm long to hollow wall	C6	nr	0.15	2.55	98.72	15.19	116.46
Close cavity of hollow wall at jambs	C7	m	0.05	0.85	2.81	0.55	4.21
Close cavity of hollow wall at cills	C8	m	0.05	0.85	2.81	0.55	4.21
Close cavity of hollow wall at top	C9	m	0.05	0.85	2.81	0.55	4.21
Pitch polymer-based damp-proof course 112mm wide at jambs	C10	m	0.05	0.85	1.02	0.28	2.15
Pitch polymer-based damp-proof course 112mm wide at cills	C11	m	0.05	0.85	1.02	0.28	2.15

**PART D
FLAT ROOF**

	Ref	Qty	Hours	Hours £	Mat'ls £	O & P £	Total £
200 x 50mm sawn softwood joists	D1	m	0.25	4.25	3.58	1.17	9.00
200 x 50mm sawn softwood sprocket pieces 500mm long	D2	nr	0.14	2.38	1.79	0.63	4.80

	Ref	Qty	Hours	Hours £	Mat'ls £	O & P £	Total £
18mm thick WPB grade plywood roof decking fixed to roof joists	D3	m²	0.90	15.30	10.47	3.87	29.64
50mm wide sawn softwood tapered firring pieces average depth 50mm	D4	m	0.18	3.06	1.91	0.75	5.72
High density polyethylene vapour barrier 150mm thick	D5	m²	0.20	3.40	11.19	2.19	16.78
100 x 75mm sawn softwood wall plate bedded in cement mortar	D6	m	0.30	5.10	2.74	1.18	9.02
100 x 75mm sawn softwood tilt fillet	D7	m	0.25	4.25	2.61	1.03	7.89
Build in ends of 200 x 50mm softwood joists to existing faced wall	D8	nr	0.35	5.95	0.22	0.93	7.10
Rake out joint of existing faced brick wall to receive flashing and point up on completion	D9	m	0.35	5.95	0.24	0.93	7.12
6mm thick asbestos-free insulation board soffit 150mm wide	D10	m	0.40	6.80	2.11	1.34	10.25
19mm thick wrought softwood fascia board 200mm high	D11	m	0.50	8.50	2.71	1.68	12.89
Three layer polyester-based mineral-surfaced roofing felt	D12	m²	0.55	9.35	13.04	3.36	25.75
Turn down to edge of roof 100mm girth	D13	m	0.10	1.70	1.40	0.47	3.57
Flashing 150mm girth including dressing over tilt fillet	D14	m	0.10	1.70	1.72	0.51	3.93
PVC-U gutter, 112mm half round with gutter union joints, fixed to softwood fascia board with support brackets at 1m maximum centres	D15	m	0.26	4.42	3.24	1.15	8.81

	Ref	Qty	Hours	Hours £	Mat'ls £	O & P £	Total £
Extra for stop end	D16	nr	0.14	2.38	2.12	0.68	5.18
Extra for stop end outlet	D17	nr	0.25	4.25	4.19	1.27	9.71
PVC-U down pipe, 68mm diameter, loose spigot and socket joints, plugged to faced brickwork with pipe clips at 2m maximum centres	D18	m	0.25	4.25	5.70	1.49	11.44
Extra for shoe	D19	nr	0.30	5.10	3.15	1.24	9.49
Apply one coat primer, one oil-based undercoat and one coat gloss paint to fascia and soffit exceeding 300mm girth	D20	m²	0.20	3.40	1.60	0.75	5.75

PART E
PITCHED ROOF

	Ref	Qty	Hours	Hours £	Mat'ls £	O & P £	Total £
100 x 75mm sawn softwood wall plate bedded in cement mortar	E1	m	0.30	5.10	2.61	1.16	8.87
200 x 50mm sawn softwood pole plate plugged to brick wall	E2	m	0.30	5.10	3.32	1.26	9.68
100 x 50mm sawn softwood rafters	E3	m	0.20	3.40	2.48	0.88	6.76
125 x 50mm sawn softwood purlins	E4	m	0.20	3.40	2.90	0.95	7.25
150 x 50mm sawn softwood joists	E5	m	0.20	3.40	3.12	0.98	7.50
150 x 50mm sawn softwood sprocket pieces 500mm long	E6	nr	0.12	2.04	1.56	0.54	4.14
Crown Wool insulation or similar 100mm thick fixed between joists with chicken wire and 150mm thick layer laid over joists	E7	m²	0.48	8.16	11.60	2.96	22.72
6mm thick asbestos-free insulation board soffit 150mm wide	E8	m	0.40	6.80	2.11	1.34	10.25

	Ref	Qty	Hours	Hours £	Mat'ls £	O & P £	Total £
19mm thick wrought softwood fascia/barge board 200mm high	E9	m	0.50	8.50	2.71	1.68	12.89
Marley Plain roofing tiles size 278 x 165mm, 65mm lap, type 1F reinforced felt and 38 x 19mm softwood battens	E10	m²	1.90	32.30	36.24	10.28	78.82
Extra for double eaves course	E11	m	0.35	5.95	4.21	1.52	11.68
Extra for verge with plain tile undercloak	E12	m	0.25	4.25	6.61	1.63	12.49
Lead flashing, code 5, 200mm girth	E13	m	0.60	10.20	8.62	2.82	21.64
Rake out joint of existing faced brick wall to receive flashing and point up on completion	E14	m	0.35	5.95	0.24	0.93	7.12
PVC-U gutter, 112mm half round with gutter union joints, fixed to softwood fascia board with support brackets at 1m maximum centres	E15	m	0.26	4.42	3.24	1.15	8.81
Extra for stop end	E16	nr	0.14	2.38	2.12	0.68	5.18
Extra for stop end outlet	E17	nr	0.25	4.25	4.19	1.27	9.71
PVC-U down pipe, 68mm diameter, loose spigot and socket joints, plugged to faced brickwork with pipe clips at 2m maximum centres	E18	m	0.25	4.25	5.70	1.49	11.44
Extra for shoe	E19	nr	0.30	5.10	3.15	1.24	9.49
Apply one coat primer, one oil-based undercoat and one coat gloss paint to fascia and soffit exceeding 300mm girth	E20	m²	0.20	3.40	1.06	0.67	5.13

	Ref	Qty	Hours	Hours £	Mat'ls £	O & P £	Total £
PART F **WINDOWS AND** **EXTERNAL DOORS**							
PVC-U door size 840 x 1980mm complete (B)	F1	nr	2.50	42.50	278.67	48.18	369.35
PVC-U sliding patio door size 1700 x 2075mm complete (C)	F2	nr	7.00	119.00	362.10	72.17	553.27
PVC-U window size 1200 x 1200mm complete (A)	F3	nr	2.00	34.00	182.14	32.42	248.56
25 x 225mm wrought softwood window board with rounded edge	F4	m	0.30	5.10	4.02	1.37	10.49
Apply one coat primer, one oil-based undercoat and one coat gloss paint to window board not exceeding 300mm girth	F5	m	0.14	2.38	0.92	0.50	3.80
PART G **INTERNAL** **PARTITIONS AND** **DOORS**							
50 x 75mm sawn softwood sole plate	G1	m	0.22	3.74	1.34	0.76	5.84
50 x 75mm sawn softwood head	G2	m	0.22	3.74	1.34	0.76	5.84
50 x 75mm sawn softwood studs	G3	m	0.28	4.76	1.34	0.92	7.02
50 x 75mm sawn softwood noggings	G4	m	0.28	4.76	1.34	0.92	7.02
Plasterboard 9.5mm thick fixed to softwood studding, filled joints and taped to receive decoration	G5	m²	0.36	6.12	2.67	1.32	10.11
Flush door 35mm thick, size 762 x 1981mm, internal quality, half-hour fire check, veneered finish both sides	G6	nr	1.25	21.25	69.91	13.67	104.83
38 x 150mm wrought softwood lining	G7	m	0.22	3.74	4.62	1.25	9.61

	Ref	Qty	Hours	Hours £	Mat'ls £	O & P £	Total £
13 x 38mm wrought softwood door stop	G8	m	0.20	3.40	0.68	0.61	4.69
19 x 50mm wrought softwood chamfered architrave	G9	m	0.15	2.55	1.34	0.58	4.47
19 x 100mm wrought softwood chamfered skirting	G10	m	0.17	2.89	2.51	0.81	6.21
100mm rising steel butts	G11	pair	0.30	5.10	1.92	1.05	8.07
Silver anodised aluminium mortice latch with lever furniture	G12	nr	0.80	13.60	16.22	4.47	34.29
Two coats emulsion paint to plasterboard walls	G13	m²	0.26	4.42	0.96	0.81	6.19
Apply one coat primer, one oil-based undercoat and one coat gloss paint to general surfaces exceeding 300mm girth	G14	m²	0.20	3.40	1.60	0.75	5.75

PART H
WALL FINISHES

	Ref	Qty	Hours	Hours £	Mat'ls £	O & P £	Total £
19 x 100mm wrought softwood chamfered skirting	H1	m	0.17	2.89	2.51	0.81	6.21
12mm plasterboard fixed to walls with dabs	H2	m²	0.39	6.63	2.85	1.42	10.90
12mm plasterboard less than 300mm wide fixed to walls with dabs	H3	m	0.18	3.06	0.92	0.60	4.58
Two coats emulsion paint to walls	H4	m²	0.26	4.42	0.96	0.81	6.19
Apply one coat primer, one oil-based undercoat and one coat gloss paint to general surfaces surfaces not exceeding 300mm	H5	m	0.20	3.40	0.92	0.65	4.97

	Ref	Qty	Hours	Hours £	Mat'ls £	O & P £	Total £
PART J **FLOORING FINISHES**							
Cement and sand (1:3) floor screed, 40mm thick, steel trowelled to level floors	J1	m²	0.25	4.25	4.76	1.35	10.36
Vinyl floor tiles, size 300 x 300 x 2mm thick, fixing with adhesive to floor screed	J2	m²	0.17	2.89	7.76	1.60	12.25
25mm thick tongued and grooved softwood flooring	J3	m²	0.74	12.58	12.48	3.76	28.82
150 x 50mm sawn softwood joists	J4	m	0.22	3.74	3.12	1.03	7.89
Cut and pin ends of joists to existing brick wall	J5	nr	0.18	3.06	0.30	0.50	3.86
Build in ends of joists to brickwork	J6	nr	0.10	1.70	0.24	0.29	2.23
PART K **CEILING FINISHES**							
Plasterboard 9.5mm thick fixed to ceiling joists, joints filled with filler and taped to receive decoration	K1	m²	0.36	6.12	2.85	1.35	10.32
One coat skim plaster to plasterboard ceilings including scrimming joints	K2	m²	0.50	8.50	1.73	1.53	11.76
Two coats emulsion paint to plastered ceilings	K3	m²	0.26	4.42	0.96	0.81	6.19
PART L **ELECTRICAL WORK**							
13 amp double switched socket outlet with neon	L1	nr	0.80	15.20	8.81	3.60	27.61
Single lighting point	L2	nr	0.70	13.30	7.34	3.10	23.74
Single one way lighting switch	L3	nr	0.70	13.30	4.42	2.66	20.38

	Ref	Qty	Hours	Hours £	Mat'ls £	O & P £	Total £
Lighting wiring	L4	m	0.20	3.80	1.86	0.85	6.51
Power cable	L5	m	0.30	5.70	2.71	1.26	9.67

PART M
HEATING WORK

	Ref	Qty	Hours	Hours £	Mat'ls £	O & P £	Total £
Copper pipe 15mm diameter, capillary fittings, fixed with pipe clips to softwood	M1	m	0.44	7.92	2.15	1.51	11.58
Extra for elbow	M2	nr	0.56	10.08	2.44	1.88	14.40
Extra for equal tee	M3	nr	0.68	12.24	2.90	2.27	17.41
Radiator, steel panelled double convector, size 1400 x 520mm, plugged and screwed to blockwork with concealed brackets, complete with 3mm chromium-plated air valve, 15mm straight valve with union and 15mm lockshield valve with union	M4	nr	1.30	23.40	139.11	24.38	186.89
Break into 15mm diameter pipe and insert tee	M5	nr	0.75	13.50	4.42	2.69	20.61

PART N
ALTERATION WORK

	Ref	Qty	Hours	Hours £	Mat'ls £	O & P £	Total £
Take out existing window size 1500 x 1000mm and lintel over, adapt opening to receive 1770 x 2075mm patio door and insert new lintel over (both measured separately) and make good	N1	nr	20.00	340.00	32.10	55.82	427.92
Take out existing window size 1500 x 1000mm and lintel over, adapt opening to receive 840 x 1980mm door (measured separately) and make good	N2	nr	24.00	408.00	58.16	69.92	536.08

One and two storey extension size 2 × 3m
with either flat or pitched roof (not to scale)

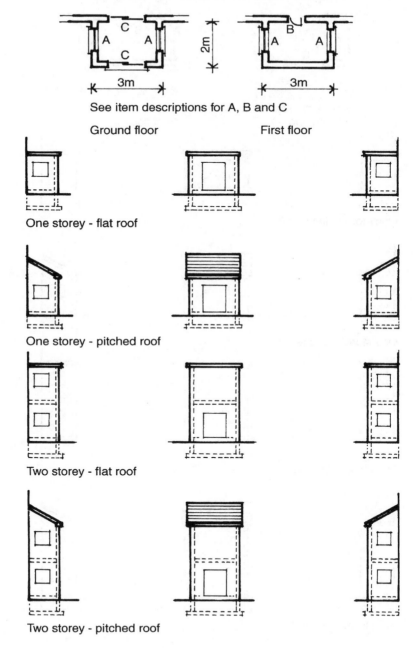

See item descriptions for A, B and C

Ground floor   First floor

One storey - flat roof

One storey - pitched roof

Two storey - flat roof

Two storey - pitched roof

One and two storey extension size 2 × 4m
with either flat or pitched roof (not to scale)

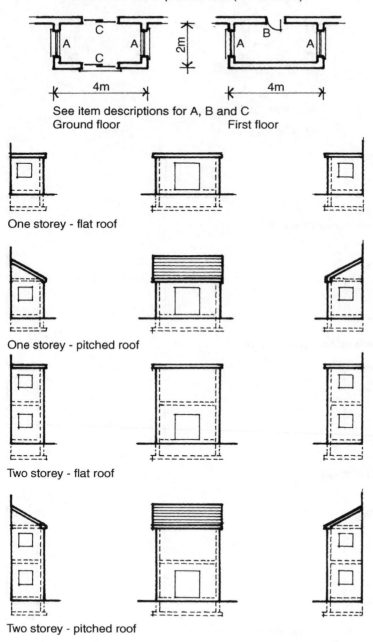

See item descriptions for A, B and C
Ground floor   First floor

One storey - flat roof

One storey - pitched roof

Two storey - flat roof

Two storey - pitched roof

One and two storey extension size 2 × 5m
with either flat or pitched roof (not to scale)

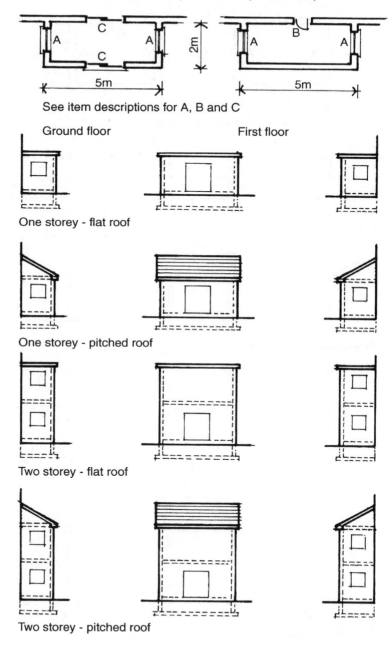

See item descriptions for A, B and C

Ground floor   First floor

One storey - flat roof

One storey - pitched roof

Two storey - flat roof

Two storey - pitched roof

One and two storey extension size 3×3m
with either flat or pitched roof (not to scale)

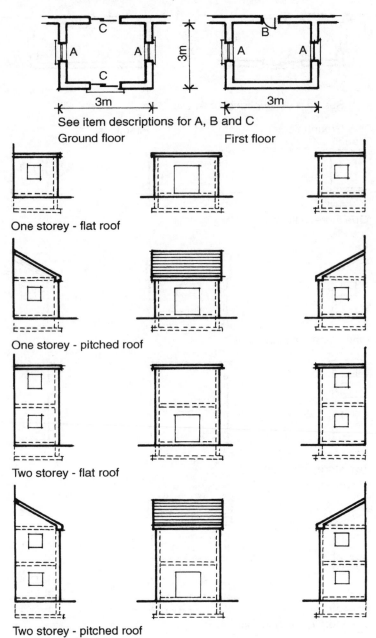

See item descriptions for A, B and C

Ground floor   First floor

One storey - flat roof

One storey - pitched roof

Two storey - flat roof

Two storey - pitched roof

One and two storey extension size 3 × 4m
with either flat or pitched roof (not to scale)

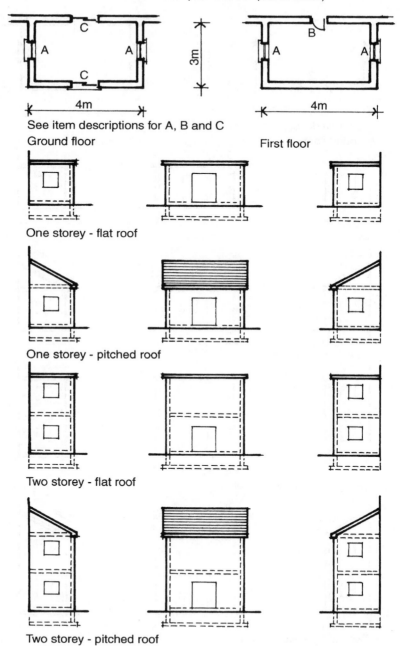

See item descriptions for A, B and C
Ground floor   First floor

One storey - flat roof

One storey - pitched roof

Two storey - flat roof

Two storey - pitched roof

One and two storey extension size 3 × 5m
with either flat or pitched roof (not to scale)

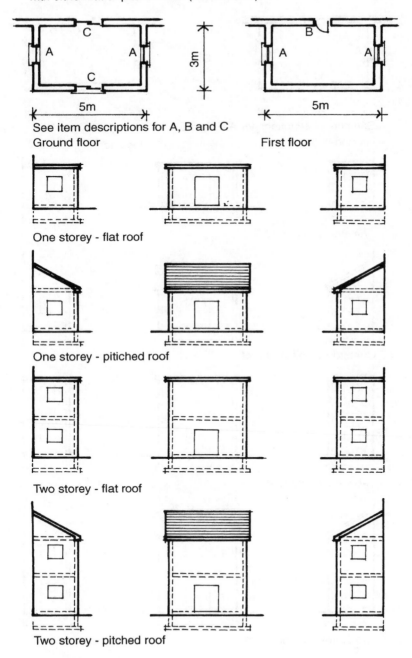

See item descriptions for A, B and C

Ground floor   First floor

One storey - flat roof

One storey - pitiched roof

Two storey - flat roof

Two storey - pitched roof

One and two storey extension size 3 × 6m
with either flat or pitched roof (not to scale)

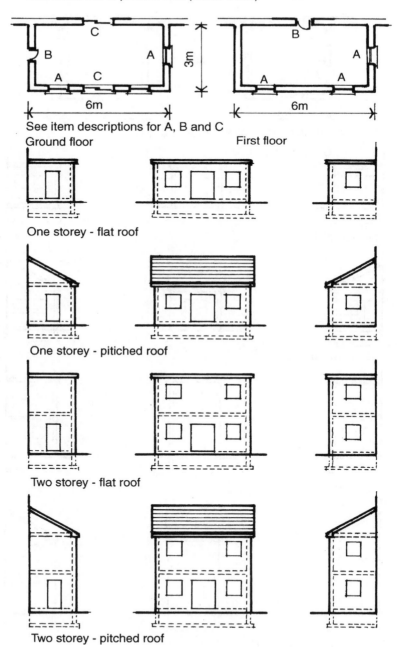

See item descriptions for A, B and C

Ground floor   First floor

One storey - flat roof

One storey - pitiched roof

Two storey - flat roof

Two storey - pitched roof

One and two storey extension size 4 × 4m
with either flat or pitched roof (not to scale)

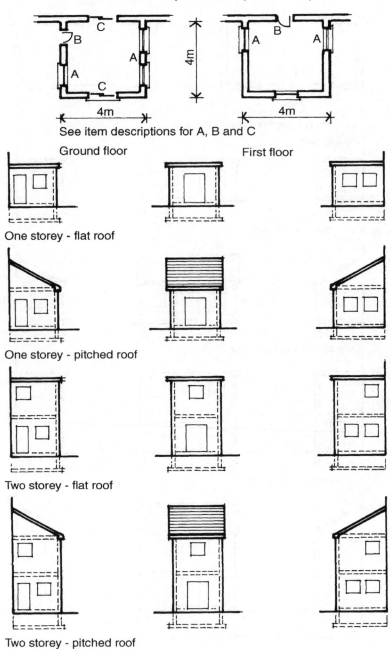

See item descriptions for A, B and C

Ground floor   First floor

One storey - flat roof

One storey - pitched roof

Two storey - flat roof

Two storey - pitched roof

One and two storey extension size 4×5m
with either flat or pitched roof (not to scale)

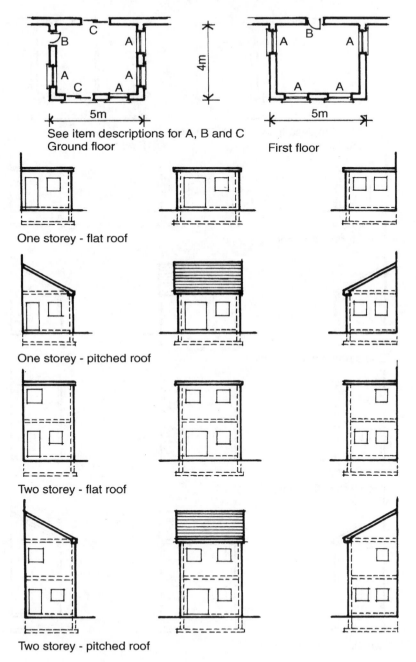

See item descriptions for A, B and C
Ground floor   First floor

One storey - flat roof

One storey - pitched roof

Two storey - flat roof

Two storey - pitched roof

One and two storey extension size 4 × 6m
with either flat or pitched roof (not to scale)

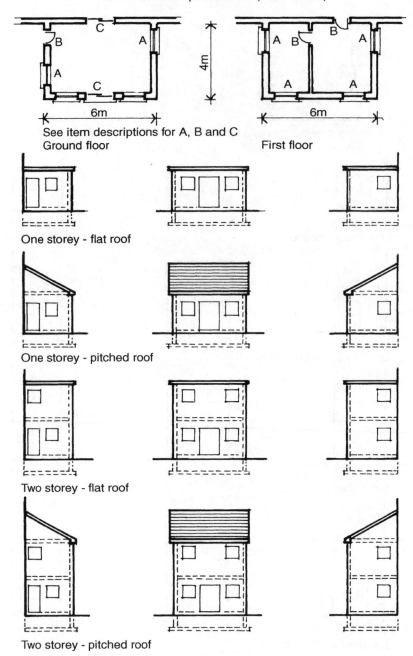

See item descriptions for A, B and C

Ground floor   First floor

One storey - flat roof

One storey - pitched roof

Two storey - flat roof

Two storey - pitched roof

One and two storey extension size 4 × 7m
with either flat or pitched roof (not to scale)

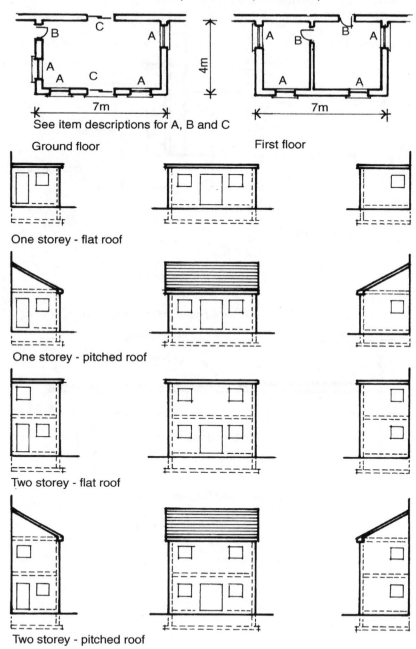

See item descriptions for A, B and C

Ground floor   First floor

One storey - flat roof

One storey - pitched roof

Two storey - flat roof

Two storey - pitched roof

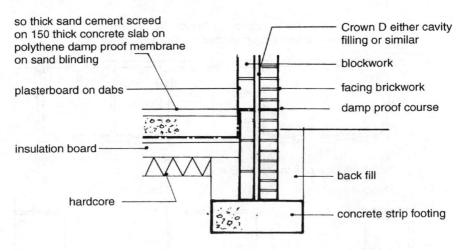

so thick sand cement screed
on 150 thick concrete slab on
polythene damp proof membrane
on sand blinding

plasterboard on dabs

insulation board

hardcore

Crown D either cavity
filling or similar

blockwork

facing brickwork

damp proof course

back fill

concrete strip footing

Foundation detail

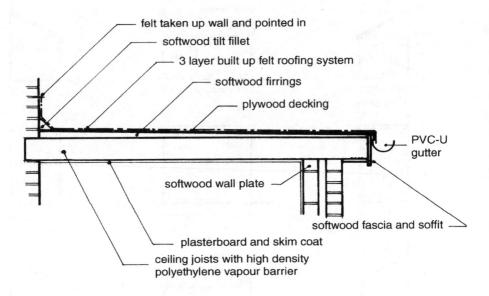

felt taken up wall and pointed in

softwood tilt fillet

3 layer built up felt roofing system

softwood firrings

plywood decking

PVC-U
gutter

softwood wall plate

softwood fascia and soffit

plasterboard and skim coat

ceiling joists with high density
polyethylene vapour barrier

Typical cross section - flat roof

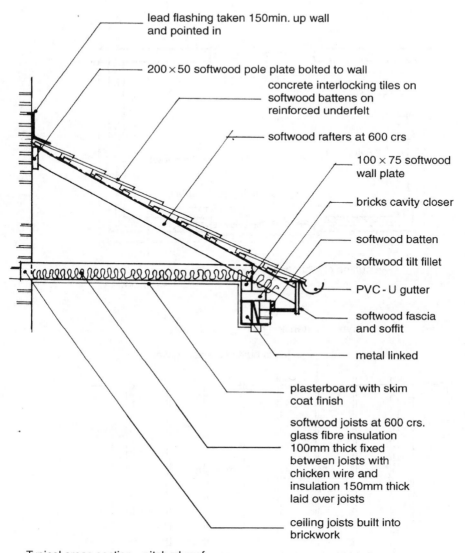

lead flashing taken 150min. up wall
and pointed in

200 × 50 softwood pole plate bolted to wall

concrete interlocking tiles on
softwood battens on
reinforced underfelt

softwood rafters at 600 crs

100 × 75 softwood
wall plate

bricks cavity closer

softwood batten

softwood tilt fillet

PVC - U gutter

softwood fascia
and soffit

metal linked

plasterboard with skim
coat finish

softwood joists at 600 crs.
glass fibre insulation
100mm thick fixed
between joists with
chicken wire and
insulation 150mm thick
laid over joists

ceiling joists built into
brickwork

Typical cross section - pitched roof

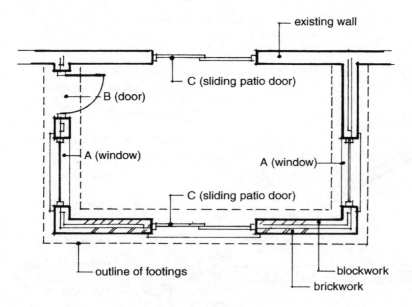

Typical extension plan (ground floor)

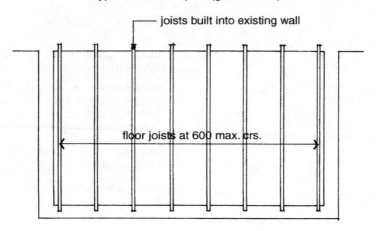

Typical floor joist layout

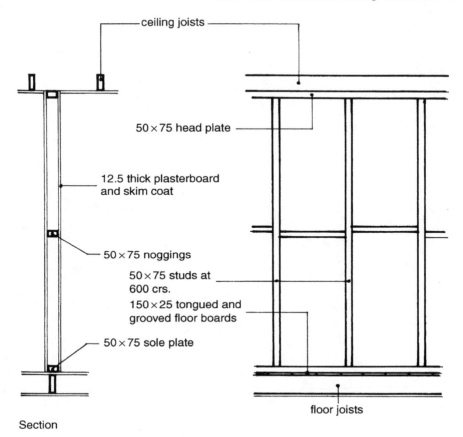

ceiling joists

50 × 75 head plate

12.5 thick plasterboard
and skim coat

50 × 75 noggings

50 × 75 studs at
600 crs.

150 × 25 tongued and
grooved floor boards

50 × 75 sole plate

floor joists

Section

Timber stud wall

Elevation

	Ref	Unit	Qty	Hours	Hours £	Mat'ls £	O & P £	Total £
PART A **PRELIMINARIES**								
Concrete mixer	A1	wks	4.00					200.00
Small tools	A2	wks	6.00					240.00
Scaffolding (m²/weeks)	A3		70					420.00
Skip	A4	wks	2.00					220.00
Clean up	A5	hrs	4.00					52.00
Carried to summary								1,132.00
PART B **SUBSTRUCTURE TO** **DPC LEVEL**								
Excavate topsoil 150mm thick by hand	B1	m²	7.10	2.13	27.69	0.00	4.15	31.84
Excavate to reduce levels	B2	m³	2.13	5.33	69.23	0.00	10.38	79.61
Excavate for trench foundations by hand	B3	m³	1.06	2.76	35.83	0.00	5.37	41.20
Earthwork support to sides of trenches	B4	m²	9.34	3.74	48.57	17.28	9.88	75.72
Backfilling with excavated material	B5	m³	0.40	0.24	3.12	0.00	0.47	3.59
Hardcore 225mm thick	B6	m²	4.08	0.82	10.61	25.66	5.44	41.71
Hardcore filling to trench	B7	m³	0.08	0.04	0.52	6.29	1.02	7.83
Concrete grade (1:3:6) in foundations	B8	m³	0.86	1.16	15.09	73.63	13.31	102.04
Concrete grade (1:2:4) in bed 150mm thick	B9	m²	4.08	1.22	15.91	62.51	11.76	90.18
Concrete (1:2:4) in cavity wall filling	B10	m²	3.20	0.64	11.97	7.62	2.94	22.52
Carried forward				18.07	238.53	192.99	64.73	496.25

	Ref	Unit	Qty	Hours	Hours £	Mat'ls £	O & P £	Total £
Brought forward				18.07	238.53	192.99	64.73	496.25
Damp-proof membrane	B11	m²	4.38	0.18	2.28	10.16	1.87	14.31
Reinforcement ref A193 in foundation	B12	m²	3.22	0.39	5.02	6.50	1.73	13.26
Steel fabric reinforcement ref A193 in slab	B13	m²	4.08	0.61	7.96	8.16	2.42	18.53
Solid blockwork 140mm thick in cavity wall	B14	m²	4.16	5.41	91.94	69.39	24.20	185.52
Common bricks 112.5mm thick in cavity wall	B15	m²	3.20	5.44	92.48	52.64	21.77	166.89
Facing bricks in 112.5mm thick in skin of cavity wall	B16	m²	0.96	1.73	29.38	28.55	8.69	66.61
Form cavity 50mm wide in cavity wall	B17	m²	4.16	0.12	2.12	6.32	1.27	9.71
DPC 112mm wide	B18	m	6.40	0.32	5.44	6.53	1.80	13.76
DPC 140mm wide	B19	m	6.40	0.38	6.53	8.83	2.30	17.66
Bond in block wall	B20	m	1.30	0.57	9.72	3.78	2.03	15.53
Bond in half brick wall	B21	m	0.30	0.11	1.79	0.80	0.39	2.98
50mm thick insulation board	B22	m²	4.08	1.22	20.81	29.38	7.53	57.71
Carried to summary				34.55	513.99	424.04	140.70	1,078.73

PART C
EXTERNAL WALLS

	Ref	Unit	Qty	Hours	Hours £	Mat'ls £	O & P £	Total £
Solid blockwork 140mm thick in cavity wall	C1	m²	10.50	13.65	232.05	175.14	61.08	468.27
Facing brickwork 112.5mm thick in cavity wall	C2	m²	10.50	18.90	321.30	299.78	93.16	714.24
75mm thick insulation in cavity wall	C3	m²	10.50	2.31	39.27	56.07	14.30	109.64
Carried forward				34.86	592.62	530.99	168.54	1,292.15

	Ref	Unit	Qty	Hours	Hours £	Mat'ls £	O & P £	Total £
Brought forward				34.86	592.62	530.99	168.54	1,292.15
Steel lintel 2400mm long	C4	nr	2.00	0.50	8.50	296.58	45.76	350.84
Steel lintel 1500mm long	C5	nr	2.00	0.40	6.80	242.72	37.43	286.95
Close cavity wall at jambs	C7	m	8.96	0.45	7.62	25.18	4.92	37.71
Close cavity wall at cills	C8	m	3.67	0.18	3.12	10.31	2.01	15.45
Close cavity wall at top	C9	m	6.40	0.32	5.44	17.98	3.51	26.94
DPC 112mm wide at jambs	C10	m	8.96	0.45	7.62	9.14	2.51	19.27
DPC 112mm wide at cills	C11	m	3.67	0.18	3.12	3.74	1.03	7.89
Carried to summary				37.34	634.83	1,136.64	265.72	2,037.19

PART D
FLAT ROOF

	Ref	Unit	Qty	Hours	Hours £	Mat'ls £	O & P £	Total £
200 x 50mm sawn softwood joists	D1	m	15.05	3.76	63.96	53.88	17.68	135.52
200 x 50mm sawn softwood sprocket pieces	D2	nr	8.00	1.12	19.04	14.32	5.00	38.36
18mm thick WPB grade decking	D3	m²	7.10	6.39	108.63	81.44	28.51	218.58
50 x 50mm (avg) wide sawn softwood firrings	D4	m	15.05	2.71	46.05	28.75	11.22	86.02
High density polyethylene vapour barrier 150mm thick	D5	m²	6.00	1.20	20.40	67.14	13.13	100.67
100 x 75mm sawn softwood wall plate	D6	m	7.00	2.10	35.70	19.18	8.23	63.11
100 x 75mm sawn softwood tilt fillet	D7	m	3.30	0.83	14.03	8.61	3.40	26.03
Build in ends of 200 x 50mm joists	D8	nr	7.00	2.45	41.65	1.54	6.48	49.67
Carried forward				20.56	349.46	274.85	93.65	717.96

	Ref	Unit	Qty	Hours	Hours £	Mat'ls £	O & P £	Total £
Brought forward				20.56	349.46	274.85	93.65	717.96
Rake out joint for flashing	D9	m	3.30	1.16	19.64	0.79	3.06	23.49
6mm thick soffit 150mm wide	D10	m	7.30	2.92	49.64	15.40	9.76	74.80
19mm wrought softwood fascia 200mm high	D11	m	7.30	3.65	62.05	19.78	12.27	94.11
Three layer fibre-based roofing felt	D12	m²	7.10	3.91	66.39	92.58	23.85	182.81
Felt turn-down 100mm girth	D13	m	7.30	0.73	12.41	10.22	3.39	26.02
Felt flashing 150mm girth	D14	m	3.30	0.33	5.61	5.68	1.69	12.98
112mm dia. PVC-U gutter	D15	m	3.30	0.86	14.59	10.69	3.79	29.07
Stop end	D16	nr	1.00	0.14	2.38	2.12	0.68	5.18
Stop end outlet	D17	nr	1.00	0.25	4.25	4.19	1.27	9.71
68mm diameter PVC-U down pipe	D18	m	2.50	0.63	10.63	14.25	3.73	28.61
Shoe	D19	nr	1.00	0.30	5.10	3.15	1.24	9.49
Paint fascia and soffit	D20	m²	2.19	0.44	7.45	3.50	1.64	12.59
Carried to summary				35.86	609.58	457.22	160.02	1,226.82
PART E **PITCHED ROOF**				N/A	N/A	N/A	N/A	N/A
PART F **WINDOWS AND** **EXTERNAL DOORS**								
PVC-U door size 840 x 1980mm complete (B)	F1	nr	0.00	0.00	0.00	0.00	0.00	0.00
PVC-U sliding patio door size 1700 x 2075mm (C)	F2	nr	2.00	14.00	238.00	616.94	128.24	983.18
Carried forward				14.00	238.00	616.94	128.24	983.18

	Ref	Unit	Qty	Hours	Hours £	Mat'ls £	O & P £	Total £
Brought forward				14.00	238.00	616.94	128.24	983.18
PVC-U window size 1200 x 1200mm complete (A)	F3	nr	2.00	4.00	68.00	364.28	64.84	497.12
25 x 225mm wrought softwood window board	F4	m	2.40	0.72	12.24	9.65	3.28	25.17
Paint window board	F5	m	2.40	0.14	2.38	2.21	0.69	5.28
Carried to summary				18.86	320.62	993.08	197.05	1,510.75
PART G INTERNAL PARTITIONS AND DOORS				N/A	N/A	N/A	N/A	N/A
PART H WALL FINISHES								
19 x 100mm wrought softwood skirting	H1	m	4.54	0.77	13.12	11.40	3.68	28.19
12mm plasterboard fixed to walls with dabs	H2	m²	9.00	3.51	59.67	25.65	12.80	98.12
12mm plasterboard fixed to walls less than 300mm wide	H3	m	12.63	2.27	38.65	11.62	7.54	57.81
Two coats emulsion paint to walls	H4	m²	10.90	2.83	48.18	10.46	8.80	67.44
Paint skirting	H5	m	4.54	0.91	15.44	4.18	2.94	22.55
Carried to summary				10.30	175.05	63.31	35.75	274.11
PART J FLOOR FINISHES								
Cement and sand floor screed 40mm thick	J1	m²	4.08	1.02	17.34	19.42	5.51	42.27
Vinyl floor tiles, size 300 x 300mm	J2	m²	4.08	0.69	11.79	31.66	6.52	49.97
Carried to summary				1.71	29.13	51.08	12.03	92.24

	Ref	Unit	Qty	Hours	Hours £	Mat'ls £	O & P £	Total £
PART K								
CEILING FINISHES								
Plasterboard with taped butt joints fixed to joists	K1	m²	4.08	1.47	24.97	11.63	5.49	42.09
5mm skim coat to plasterboard ceilings	K2	m²	4.08	2.04	34.68	7.06	6.26	48.00
Two coats emulsion paint to ceilings	K3	m²	4.08	1.06	18.03	3.92	3.29	25.24
Carried to summary				4.57	77.68	22.60	15.04	115.33
PART L								
ELECTRICAL WORK								
13 amp double switched socket outlet with neon	L1	nr	2.00	1.60	30.40	17.62	7.20	55.22
Lighting point	L2	nr	2.00	1.40	26.60	14.68	6.19	47.47
Lighting switch	L3	nr	2.00	1.40	26.60	8.84	5.32	40.76
Lighting wiring	L4	m	5.00	1.00	19.00	9.30	4.25	32.55
Power cable	L5	m	12.00	3.60	68.40	32.52	15.14	116.06
Carried to summary				9.00	171.00	82.96	38.09	292.05
PART M								
HEATING WORK								
15mm copper pipe	M1	m	4.00	1.76	31.68	8.60	6.04	46.32
Elbow	M2	nr	4.00	2.24	40.32	9.76	7.51	57.59
Tee	M3	nr	1.00	0.56	10.08	2.44	1.88	14.40
Radiator, double convector size 1400 x 520mm	M4	nr	1.00	1.30	23.40	139.11	24.38	186.89
Break into existing pipe and insert tee	M5	nr	1.00	0.75	13.50	4.42	2.69	20.61
Carried to summary				6.61	118.98	164.33	42.50	325.81

	Ref	Unit	Qty	Hours	Hours £	Mat'ls £	O & P £	Total £

PART N
ALTERATION WORK

	Ref	Unit	Qty	Hours	Hours £	Mat'ls £	O & P £	Total £
Take out existing window size 1500 x 1000mm and lintel over, adapt opening to receive 1770 x 2000mm patio door and insert new lintel over (both measured separately) and make good	N1	nr	1.00	20.00	340.00	32.10	55.82	427.92
Carried to summary				20.00	340.00	32.10	55.82	427.92

SUMMARY

	Hours	Hours £	Mat'ls £	O & P £	Total £
PART A PRELIMINARIES	0.00	0.00	0.00	0.00	1,132.00
PART B SUBSTRUCTURE TO DPC LEVEL	34.55	513.99	424.04	140.70	1,078.73
PART C EXTERNAL WALLS	37.34	634.83	1,136.64	265.72	2,037.19
PART D FLAT ROOF	35.86	609.58	457.22	160.02	1,226.82
PART E PITCHED ROOF	0.00	0.00	0.00	0.00	0.00
PART F WINDOWS AND EXTERNAL DOORS	18.86	320.62	993.08	197.05	1,510.75
PART G INTERNAL PARTITIONS AND DOORS	0.00	0.00	0.00	0.00	0.00
PART H WALL FINISHES	10.30	175.05	63.31	35.75	274.11
PART J FLOOR FINISHES	1.71	29.13	51.08	12.03	92.24
PART K CEILING FINISHES	4.57	77.68	22.60	15.04	115.32
PART L ELECTRICAL WORK	9.00	171.00	82.96	38.09	292.05
PART M HEATING WORK	6.61	118.98	164.33	42.50	325.81
PART N ALTERATION WORK	20.00	340.00	32.10	55.82	427.92
Final total	178.80	2,990.86	3,427.36	962.73	8,512.95

	Ref	Unit	Qty	Hours	Hours £	Mat'ls £	O & P £	Total £
PART A								
PRELIMINARIES								
Concrete mixer	A1	wks	5.00					250.00
Small tools	A2	wks	7.00					280.00
Scaffolding (m²/weeks)	A3		80					480.00
Skip	A4	wks	3.00					330.00
Clean up	A5	hrs	6.00					78.00
Carried to summary								1,418.00
PART B								
SUBSTRUCTURE TO								
DPC LEVEL								
Excavate topsoil 150mm thick by hand	B1	m²	9.25	2.78	36.08	0.00	5.41	41.49
Excavate to reduce levels	B2	m³	2.77	6.93	90.03	0.00	13.50	103.53
Excavate for trench foundations by hand	B3	m³	1.22	3.17	41.24	0.00	6.19	47.42
Earthwork support to sides of trenches	B4	m²	10.80	4.32	56.16	19.98	11.42	87.56
Backfilling with excavated material	B5	m³	0.47	0.28	3.67	0.00	0.55	4.22
Hardcore 225mm thick	B6	m²	5.78	1.16	15.03	36.36	7.71	59.09
Hardcore filling to trench	B7	m³	0.09	0.09	1.17	6.29	1.12	8.58
Concrete grade (1:3:6) in foundations	B8	m³	1.16	0.86	11.18	99.32	16.57	127.07
Concrete grade (1:2:4) in bed 150mm thick	B9	m²	5.78	1.73	22.54	88.55	16.66	127.76
Concrete (1:2:4) in cavity wall filling	B10	m²	3.70	0.74	11.97	8.81	3.12	23.89
Carried forward				22.05	289.05	259.30	82.25	630.61

	Ref	Unit	Qty	Hours	Hours £	Mat'ls £	O & P £	Total £
Brought forward				22.05	289.05	259.30	82.25	630.61
Damp-proof membrane	B11	m²	6.13	0.25	3.19	14.22	2.61	20.02
Reinforcement ref A193 in foundation	B12	m²	3.70	0.44	5.77	7.47	1.99	15.23
Steel fabric reinforcement ref A193 in slab	B13	m²	5.78	0.87	11.27	11.56	3.42	26.26
Solid blockwork 140mm thick in cavity wall	B14	m²	4.81	6.25	106.30	80.23	27.98	214.51
Common bricks 112.5mm thick in cavity wall	B15	m²	3.70	6.29	106.93	60.87	25.17	192.96
Facing bricks in 112.5mm thick in skin of cavity wall	B16	m²	1.11	2.00	33.97	28.55	9.38	71.89
Form cavity 50mm wide in cavity wall	B17	m²	4.81	0.14	2.45	7.31	1.46	11.23
DPC 112mm wide	B18	m	7.40	0.37	6.29	7.55	2.08	15.91
DPC 140mm wide	B19	m	7.40	0.44	7.55	10.21	2.66	20.42
Bond in block wall	B20	m	1.30	0.57	9.72	3.78	2.03	15.53
Bond in half brick wall	B21	m	0.30	0.11	1.79	0.80	0.39	2.98
50mm thick insulation board	B22	m²	5.78	1.73	29.48	41.62	10.66	81.76
Carried to summary				41.52	613.76	533.48	172.09	1,319.32
PART C **EXTERNAL WALLS**								
Solid blockwork 140mm thick in cavity wall	C1	m²	13.00	16.90	287.30	216.84	75.62	579.76
Facing brickwork 112.5mm thick in cavity wall	C2	m²	13.00	23.40	397.80	371.15	115.34	884.29
75mm thick insulation in cavity wall	C3	m²	13.00	2.86	48.62	69.42	17.71	135.75
Carried forward				43.16	733.72	657.41	208.67	1,599.80

	Ref	Unit	Qty	Hours	Hours £	Mat'ls £	O & P £	Total £
Brought forward				43.16	733.72	657.41	208.67	1,599.80
Steel lintel 2400mm long	C4	nr	2.00	0.50	8.50	296.58	45.76	350.84
Steel lintel 1500mm long	C5	nr	2.00	0.40	6.80	242.72	37.43	286.95
Close cavity wall at jambs	C7	m	8.96	0.45	7.62	25.18	4.92	37.71
Close cavity wall at cills	C8	m	3.67	0.18	3.12	10.31	2.01	15.45
Close cavity wall at top	C9	m	7.40	0.37	6.29	20.79	4.06	31.15
DPC 112mm wide at jambs	C10	m	8.96	0.45	7.62	9.14	2.51	19.27
DPC 112mm wide at cills	C11	m	3.67	0.18	3.12	3.74	1.03	7.89
Carried to summary				45.69	776.78	1,265.88	306.40	2,349.06

PART D
FLAT ROOF

	Ref	Unit	Qty	Hours	Hours £	Mat'ls £	O & P £	Total £
200 x 50mm sawn softwood joists	D1	m	19.35	4.84	82.24	69.27	22.73	174.24
200 x 50mm sawn softwood sprocket pieces	D2	nr	8.00	1.12	19.04	14.32	5.00	38.36
18mm thick WPB grade decking	D3	m²	9.25	8.33	141.53	106.10	37.14	284.77
50 x 50mm (avg) wide sawn softwood firrings	D4	m	19.25	3.47	58.91	36.77	14.35	110.02
High density polyethylene vapour barrier 150mm thick	D5	m²	8.00	1.60	27.20	89.52	17.51	134.23
100 x 75mm sawn softwood wall plate	D6	m	8.00	2.40	40.80	21.92	9.41	72.13
100 x 75mm sawn softwood tilt fillet	D7	m	4.30	1.08	18.28	11.22	4.42	33.92
Build in ends of 200 x 50mm joists	D8	nr	9.00	3.15	53.55	1.98	8.33	63.86
Carried forward				25.97	441.53	351.10	118.90	911.53

	Ref	Unit	Qty	Hours	Hours £	Mat'ls £	O & P £	Total £
Brought forward				25.97	441.53	351.10	118.90	911.53
Rake out joint for flashing	D9	m	4.30	1.51	25.59	1.03	3.99	30.61
6mm thick soffit 150mm wide	D10	m	8.30	3.32	56.44	17.51	11.09	85.05
19mm wrought softwood fascia 200mm high	D11	m	8.30	4.15	70.55	22.49	13.96	107.00
Three layer fibre-based roofing felt	D12	m²	9.25	5.09	86.49	120.62	31.07	238.17
Felt turn-down 100mm girth	D13	m	8.30	0.83	14.11	11.62	3.86	29.59
Felt flashing 150mm girth	D14	m	4.30	0.43	7.31	7.40	2.21	16.91
112mm dia. PVC-U gutter	D15	m	4.30	1.12	19.01	13.93	4.94	37.88
Stop end	D16	nr	1.00	0.14	2.38	2.12	0.68	5.18
Stop end outlet	D17	nr	1.00	0.25	4.25	4.19	1.27	9.71
68mm diameter PVC-U down pipe	D18	m	2.50	0.63	10.63	14.25	3.73	28.61
Shoe	D19	nr	1.00	0.30	5.10	3.15	1.24	9.49
Paint fascia and soffit	D20	m²	2.49	0.50	8.47	3.98	1.87	14.32
Carried to summary				44.23	751.84	573.40	198.79	1,524.03
PART E **PITCHED ROOF**				N/A	N/A	N/A	N/A	N/A
PART F **WINDOWS AND** **EXTERNAL DOORS**								
PVC-U door size 840 x 1980mm complete (B)	F1	0.00	0.00	0.00	0.00	0.00	0.00	0.00
PVC-U sliding patio door size 1700 x 2075mm (C)	F2	nr	2.00	14.00	238.00	616.94	128.24	983.18
Carried forward				14.00	238.00	616.94	128.24	983.18

	Ref	Unit	Qty	Hours	Hours £	Mat'ls £	O & P £	Total £
Brought forward				14.00	238.00	616.94	128.24	983.18
PVC-U window size 1200 x 1200mm complete (A)	F3	nr	2.00	4.00	68.00	364.28	64.84	497.12
25 x 225mm wrought softwood window board	F4	m	2.40	0.72	12.24	9.65	3.28	25.17
Paint window board	F5	m	2.40	0.14	2.38	2.21	0.69	5.28
Carried to summary				18.86	320.62	993.08	197.05	1,510.75
PART G **INTERNAL** **PARTITIONS AND** **DOORS**				N/A	N/A	N/A	N/A	N/A
PART H **WALL FINISHES**								
19 x 100mm wrought softwood skirting	H1	m	5.54	0.94	16.01	13.91	4.49	34.40
12mm plasterboard fixed to walls with dabs	H2	m²	11.50	4.49	76.25	32.78	16.35	125.37
12mm plasterboard fixed to walls less than 300mm wide	H3	m	12.63	2.27	38.65	11.62	7.54	57.81
Two coats emulsion paint to walls	H4	m²	13.40	3.48	59.23	12.86	10.81	82.91
Paint skirting	H5	m	5.54	1.11	18.84	5.10	3.59	27.52
Carried to summary				12.29	208.97	76.26	42.78	328.01
PART J **FLOOR FINISHES**								
Cement and sand floor screed 40mm thick	J1	m²	5.78	1.45	24.57	27.51	7.81	59.89
Vinyl floor tiles, size 300 x 300mm	J2	m²	5.78	0.98	16.70	44.85	9.23	70.79
Carried to summary				2.43	41.27	72.37	17.05	130.68

	Ref	Unit	Qty	Hours	Hours £	Mat'ls £	O & P £	Total £
PART K **CEILING FINISHES**								
Plasterboard with taped butt joints fixed to joists	K1	m²	5.78	2.08	35.37	16.47	7.78	59.62
5mm skim coat to plasterboard ceilings	K2	m²	5.78	2.89	49.13	10.00	8.87	68.00
Two coats emulsion paint to ceilings	K3	m²	5.78	1.50	25.55	5.55	4.66	35.76
Carried to summary				6.47	110.05	32.02	21.31	163.38
PART L **ELECTRICAL WORK**								
13 amp double switched socket outlet with neon	L1	nr	2.00	1.60	30.40	17.62	7.20	55.22
Lighting point	L2	nr	2.00	1.40	26.60	14.68	6.19	47.47
Lighting switch	L3	nr	2.00	1.40	26.60	8.84	5.32	40.76
Lighting wiring	L4	m	6.00	1.20	22.80	11.16	5.09	39.05
Power cable	L5	m	14.00	4.20	79.80	37.94	17.66	135.40
Carried to summary				9.80	186.20	90.24	41.47	317.91
PART M **HEATING WORK**								
15mm copper pipe	M1	m	5.00	2.20	39.60	10.75	7.55	57.90
Elbow	M2	nr	4.00	2.24	40.32	9.76	7.51	57.59
Tee	M3	nr	1.00	0.68	12.24	2.44	2.20	16.88
Radiator, double convector size 1400 x 520mm	M4	nr	2.00	2.60	46.80	278.22	48.75	373.77
Break into existing pipe and insert tee	M5	nr	1.00	0.75	13.50	4.42	2.69	20.61
Carried to summary				8.47	152.46	305.59	68.71	526.76

	Ref	Unit	Qty	Hours	Hours £	Mat'ls £	O & P £	Total £

PART N
ALTERATION WORK

	Ref	Unit	Qty	Hours	Hours £	Mat'ls £	O & P £	Total £
Take out existing window size 1500 x 1000mm and lintel over, adapt opening to receive 1770 x 2000mm patio door and insert new lintel over (both measured separately) and make good	N1	nr	1.00	20.00	340.00	32.10	55.82	427.92
Carried to summary				20.00	340.00	32.10	55.82	427.92

SUMMARY

	Hours	Hours £	Mat'ls £	O & P £	Total £
PART A **PRELIMINARIES**	0.00	0.00	0.00	0.00	1,418.00
PART B SUBSTRUCTURE TO **DPC LEVEL**	41.52	613.76	533.48	172.09	1,319.33
PART C **EXTERNAL WALLS**	45.69	776.78	1,265.88	306.40	2,349.06
PART D **FLAT ROOF**	44.23	751.84	573.40	198.79	1,524.03
PART E **PITCHED ROOF**	0.00	0.00	0.00	0.00	0.00
PART F WINDOWS AND **EXTERNAL DOORS**	18.86	320.62	993.08	197.05	1,510.75
PART G INTERNAL **PARTITIONS AND DOORS**	0.00	0.00	0.00	0.00	0.00
PART H **WALL FINISHES**	12.29	208.97	76.26	42.78	328.01
PART J **FLOOR FINISHES**	2.43	41.27	72.37	17.05	130.68
PART K **CEILING FINISHES**	6.47	110.05	32.02	21.31	163.38
PART L **ELECTRICAL WORK**	9.80	186.20	90.24	41.47	317.91
PART M **HEATING WORK**	8.47	152.46	305.59	68.71	526.76
PART N **ALTERATION WORK**	20.00	340.00	32.10	55.82	427.92
Final total	209.76	3,501.95	3,974.42	1,121.46	10,015.82

	Ref	Unit	Qty	Hours	Hours £	Mat'ls £	O & P £	Total £
PART A **PRELIMINARIES**								
Concrete mixer	A1	wks	6.00					300.00
Small tools	A2	wks	8.00					320.00
Scaffolding (m²/weeks)	A3		90					540.00
Skip	A4	wks	4.00					440.00
Clean up	A5	hrs	8.00					104.00
Carried to summary								1,704.00
PART B **SUBSTRUCTURE TO** **DPC LEVEL**								
Excavate topsoil 150mm thick by hand	B1	m²	11.40	3.42	44.46	0.00	6.67	51.13
Excavate to reduce levels	B2	m³	3.42	8.55	111.15	0.00	16.67	127.82
Excavate for trench foundations by hand	B3	m³	1.39	3.61	46.98	0.00	7.05	54.03
Earthwork support to sides of trenches	B4	m²	12.26	4.90	63.75	22.68	12.96	99.40
Backfilling with excavated material	B5	m³	0.55	0.33	4.29	0.00	0.64	4.93
Hardcore 225mm thick	B6	m²	7.48	1.50	19.45	47.05	9.97	76.47
Hardcore filling to trench	B7	m³	0.12	0.09	1.17	6.29	1.12	8.58
Concrete grade (1:3:6) in foundations	B8	m³	1.13	0.86	11.18	96.75	16.19	124.12
Concrete grade (1:2:4) in bed 150mm thick	B9	m²	7.48	2.24	29.17	114.59	21.56	165.33
Concrete (1:2:4) in cavity wall filling	B10	m²	4.20	0.84	11.97	10.00	3.29	25.26
Carried forward				26.35	343.57	297.36	96.14	737.07

	Ref	Unit	Qty	Hours	Hours £	Mat'ls £	O & P £	Total £
Brought forward				26.35	343.57	297.36	96.14	737.07
Damp-proof membrane	B11	m²	7.88	0.32	4.10	18.28	3.36	25.74
Reinforcement ref A193 in foundation	B12	m²	4.20	0.50	6.55	8.48	2.26	17.29
Steel fabric reinforcement ref A193 in slab	B13	m²	7.48	1.12	14.59	14.96	4.43	33.98
Solid blockwork 140mm thick in cavity wall	B14	m²	5.46	7.10	120.67	91.07	31.76	243.50
Common bricks 112.5mm thick in cavity wall	B15	m²	4.20	7.14	121.38	69.09	28.57	219.04
Facing bricks in 112.5mm thick in skin of cavity wall	B16	m²	1.26	2.27	38.56	28.55	10.07	77.17
Form cavity 50mm wide in cavity wall	B17	m²	5.46	0.16	2.78	8.30	1.66	12.75
DPC 112mm wide	B18	m	8.40	0.42	7.14	8.57	2.36	18.06
DPC 140mm wide	B19	m	8.40	0.50	8.57	11.59	3.02	23.18
Bond in block wall	B20	m	1.30	0.57	9.72	3.78	2.03	15.53
Bond in half brick wall	B21	m	0.30	0.11	1.79	0.80	0.39	2.98
50mm thick insulation board	B22	m²	7.48	2.24	38.15	53.86	13.80	105.80
Carried to summary				48.80	717.56	614.70	199.84	1,532.10

**PART C
EXTERNAL WALLS**

	Ref	Unit	Qty	Hours	Hours £	Mat'ls £	O & P £	Total £
Solid blockwork 140mm thick in cavity wall	C1	m²	15.50	20.15	342.55	258.54	90.16	691.25
Facing brickwork 112.5mm thick in cavity wall	C2	m²	15.50	27.90	474.30	442.53	137.52	1,054.35
75mm thick insulation in cavity wall	C3	m²	15.50	3.41	57.97	82.77	21.11	161.85
Carried forward				51.46	874.82	783.84	248.80	1,907.45

	Ref	Unit	Qty	Hours	Hours £	Mat'ls £	O & P £	Total £
Brought forward				51.46	874.82	783.84	248.80	1,907.45
Steel lintel 2400mm long	C4	nr	2.00	0.50	8.50	296.58	45.76	350.84
Steel lintel 1500mm long	C5	nr	2.00	0.40	6.80	242.72	37.43	286.95
Close cavity wall at jambs	C7	m	8.96	0.45	7.62	25.18	4.92	37.71
Close cavity wall at cills	C8	m	3.67	0.18	3.12	10.31	2.01	15.45
Close cavity wall at top	C9	m	8.40	0.42	7.14	23.60	4.61	35.36
DPC 112mm wide at jambs	C10	m	8.96	0.45	7.62	9.14	2.51	19.27
DPC 112mm wide at cills	C11	m	3.67	0.18	3.12	3.74	1.03	7.89
Carried to summary				54.04	918.73	1,395.11	347.08	2,660.92

PART D
FLAT ROOF

	Ref	Unit	Qty	Hours	Hours £	Mat'ls £	O & P £	Total £
200 x 50mm sawn softwood joists	D1	m	23.65	5.91	100.51	84.67	27.78	212.96
200 x 50mm sawn softwood sprocket pieces	D2	nr	8.00	1.12	19.04	14.32	5.00	38.36
18mm thick WPB grade decking	D3	m²	14.40	12.96	220.32	165.17	57.82	443.31
50 x 50mm (avg) wide sawn softwood firrings	D4	m	23.65	4.26	72.37	45.17	17.63	135.17
High density polyethylene vapour barrier 150mm thick	D5	m²	10.00	2.00	34.00	111.90	21.89	167.79
100 x 75mm sawn softwood wall plate	D6	m	9.00	2.70	45.90	24.66	10.58	81.14
100 x 75mm sawn softwood tilt fillet	D7	m	5.30	1.33	22.53	13.83	5.45	41.81
Build in ends of 200 x 50mm joists	D8	nr	11.00	3.85	65.45	2.42	10.18	78.05
Carried forward				34.12	580.12	462.14	156.34	1,198.59

	Ref	Unit	Qty	Hours	Hours £	Mat'ls £	O & P £	Total £
Brought forward				34.12	580.12	462.14	156.34	1,198.59
Rake out joint for flashing	D9	m	5.30	1.86	31.54	1.27	4.92	37.73
6mm thick soffit 150mm wide	D10	m	9.30	3.72	63.24	19.62	12.43	95.29
19mm wrought softwood fascia 200mm high	D11	m	9.30	4.65	79.05	25.20	15.64	119.89
Three layer fibre-based roofing felt	D12	m²	11.40	6.27	106.59	148.66	38.29	293.53
Felt turn-down 100mm girth	D13	m	9.30	0.93	15.81	13.02	4.32	33.15
Felt flashing 150mm girth	D14	m	5.30	0.53	9.01	9.12	2.72	20.84
112mm diameter PVC-U gutter	D15	m	4.30	1.12	19.01	13.93	4.94	37.88
Stop end	D16	nr	1.00	0.14	2.38	2.12	0.68	5.18
Stop end outlet	D17	nr	1.00	0.25	4.25	4.19	1.27	9.71
68mm diameter PVC-U down pipe	D18	m	2.50	0.63	10.63	14.25	3.73	28.61
Shoe	D19	nr	1.00	0.30	5.10	3.15	1.24	9.49
Paint fascia and soffit	D20	m²	2.79	0.56	9.49	4.46	2.09	16.04
Carried to summary				55.07	936.20	721.14	248.60	1,905.93
PART E PITCHED ROOF				N/A	N/A	N/A	N/A	N/A
PART F WINDOWS AND EXTERNAL DOORS								
PVC-U door size 840 x 1980mm complete (B)	F1	0.00	0.00	0.00	0.00	0.00	0.00	0.00
PVC-U sliding patio door size 1700 x 2075mm (C)	F2	nr	2.00	14.00	238.00	616.94	128.24	983.18
Carried forward				14.00	238.00	616.94	128.24	983.18

	Ref	Unit	Qty	Hours	Hours £	Mat'ls £	O & P £	Total £
Brought forward				14.00	238.00	616.94	128.24	983.18
PVC-U window size 1200 x 1200mm complete (A)	F3	nr	2.00	4.00	68.00	364.28	64.84	497.12
25 x 225mm wrought softwood window board	F4	m	2.40	0.72	12.24	9.65	3.28	25.17
Paint window board	F5	m	2.40	0.14	2.38	2.21	0.69	5.28
Carried to summary				18.86	320.62	993.08	197.05	1,510.75
PART G INTERNAL PARTITIONS AND DOORS				N/A	N/A	N/A	N/A	N/A
PART H WALL FINISHES								
19 x 100mm wrought softwood skirting	H1	m	6.54	1.11	18.90	16.42	5.30	40.61
12mm plasterboard fixed to walls with dabs	H2	m²	14.00	5.46	92.82	39.90	19.91	152.63
12mm plasterboard fixed to walls less than 300mm wide	H3	m	12.63	2.27	38.65	11.62	7.54	57.81
Two coats emulsion paint to walls	H4	m²	15.90	4.13	70.28	15.26	12.83	98.37
Paint skirting	H5	m	6.54	1.31	22.24	6.02	4.24	32.49
Carried to summary				14.29	242.88	89.22	49.81	381.91
PART J FLOOR FINISHES								
Cement and sand floor screed 40mm thick	J1	m²	7.48	1.87	31.79	35.60	10.11	77.50
Vinyl floor tiles, size 300 x 300mm	J2	m²	7.48	1.27	21.62	58.04	11.95	91.61
Carried to summary				3.14	53.41	93.65	22.06	169.12

	Ref	Unit	Qty	Hours	Hours £	Mat'ls £	O & P £	Total £
PART K								
CEILING FINISHES								
Plasterboard with taped butt joints fixed to joists	K1	m²	7.48	2.69	45.78	21.32	10.06	77.16
5mm skim coat to plasterboard ceilings	K2	m²	7.48	3.74	63.58	12.94	11.48	88.00
Two coats emulsion paint to ceilings	K3	m²	7.48	1.94	33.06	7.18	6.04	46.28
Carried to summary				8.38	142.42	41.44	27.58	211.44
PART L								
ELECTRICAL WORK								
13 amp double switched socket outlet with neon	L1	nr	3.00	2.40	45.60	26.43	10.80	82.83
Lighting point	L2	nr	3.00	2.10	39.90	22.02	9.29	71.21
Lighting switch	L3	nr	2.00	1.40	26.60	8.84	5.32	40.76
Lighting wiring	L4	m	7.00	1.40	26.60	13.02	5.94	45.56
Power cable	L5	m	16.00	4.80	91.20	43.36	20.18	154.74
Carried to summary				12.10	229.90	113.67	51.54	395.11
PART M								
HEATING WORK								
15mm copper pipe	M1	m	6.00	2.64	47.52	12.90	9.06	69.48
Elbow	M2	nr	4.00	2.24	40.32	9.76	7.51	57.59
Tee	M3	nr	1.00	0.68	12.24	2.44	2.20	16.88
Radiator, double convector size 1400 x 520mm	M4	nr	2.00	2.60	46.80	278.22	48.75	373.77
Break into existing pipe and insert tee	M5	nr	1.00	0.75	13.50	4.42	2.69	20.61
Carried to summary				8.91	160.38	307.74	70.22	538.34

	Ref	Unit	Qty	Hours	Hours £	Mat'ls £	O & P £	Total £

PART N
ALTERATION WORK

	Ref	Unit	Qty	Hours	Hours £	Mat'ls £	O & P £	Total £
Take out existing window size 1500 x 1000mm and lintel over, adapt opening to receive 1770 x 2000mm patio door and insert new lintel over (both measured separately) and make good	N1	nr	1.00	20.00	340.00	32.10	55.82	427.92
Carried to summary				20.00	340.00	32.10	55.82	427.92

SUMMARY

	Hours	Hours £	Mat'ls £	O & P £	Total £
PART A **PRELIMINARIES**	0.00	0.00	0.00	0.00	1,704.00
PART B SUBSTRUCTURE TO **DPC LEVEL**	48.80	717.56	614.70	199.84	1,532.10
PART C **EXTERNAL WALLS**	54.04	918.73	1,395.11	347.08	2,660.92
PART D **FLAT ROOF**	55.07	936.20	721.14	248.60	1,905.93
PART E **PITCHED ROOF**	0.00	0.00	0.00	0.00	0.00
PART F WINDOWS AND **EXTERNAL DOORS**	18.86	320.62	993.08	197.05	1,510.75
PART G INTERNAL **PARTITIONS AND DOORS**	0.00	0.00	0.00	0.00	0.00
PART H **WALL FINISHES**	14.29	242.88	89.22	49.81	381.91
PART J **FLOOR FINISHES**	3.14	53.41	93.65	22.06	169.12
PART K **CEILING FINISHES**	8.38	142.42	41.44	27.58	211.44
PART L **ELECTRICAL WORK**	12.10	229..9	113.67	51.54	395.11
PART M **HEATING WORK**	8.91	160.38	307.74	70.22	538.34
PART N **ALTERATION WORK**	20.00	340.00	32.10	55.82	427.92
Final total	243.59	4,024.46	4,401.85	1,269.59	11,437.53

	Ref	Unit	Qty	Hours	Hours £	Mat'ls £	O & P £	Total £
PART A **PRELIMINARIES**								
Concrete mixer	A1	wks	6.00					300.00
Small tools	A2	wks	7.00					280.00
Scaffolding (m²/weeks)	A3		90					540.00
Skip	A4	wks	3.00					330.00
Clean up	A5	hrs	8.00					104.00
Carried to summary								1,554.00
PART B **SUBSTRUCTURE TO** **DPC LEVEL**								
Excavate topsoil 150mm thick by hand	B1	m²	10.40	3.12	40.56	0.00	6.08	46.64
Excavate to reduce levels	B2	m³	3.12	7.80	101.40	0.00	15.21	116.61
Excavate for trench foundations by hand	B3	m³	1.39	3.61	46.98	0.00	7.05	54.03
Earthwork support to sides of trenches	B4	m²	12.26	4.90	63.75	22.68	12.96	99.40
Backfilling with excavated material	B5	m³	0.47	0.28	3.67	0.00	0.55	4.22
Hardcore 225mm thick	B6	m²	6.48	1.30	16.85	40.76	8.64	66.25
Hardcore filling to trench	B7	m³	0.09	0.09	1.17	6.29	1.12	8.58
Concrete grade (1:3:6) in foundations	B8	m³	1.13	0.86	11.18	96.75	16.19	124.12
Concrete grade (1:2:4) in bed 150mm thick	B9	m²	7.48	2.24	29.17	114.59	21.56	165.33
Concrete (1:2:4) in cavity wall filling	B10	m²	4.20	0.84	11.97	10.00	3.29	25.26
Carried forward				25.05	326.70	291.07	92.67	710.44

	Ref	Unit	Qty	Hours	Hours £	Mat'ls £	O & P £	Total £
Brought forward				25.05	326.70	291.07	92.67	710.44
Damp-proof membrane	B11	m²	6.88	0.28	3.58	15.96	2.93	22.47
Reinforcement ref A193 in foundation	B12	m²	4.20	0.50	6.55	8.48	2.26	17.29
Steel fabric reinforcement ref A193 in slab	B13	m²	6.48	0.97	12.64	12.96	3.84	29.44
Solid blockwork 140mm thick in cavity wall	B14	m²	5.46	7.10	120.67	91.07	31.76	243.50
Common bricks 112.5mm thick in cavity wall	B15	m²	4.20	7.14	121.38	69.09	28.57	219.04
Facing bricks in 112.5mm thick in skin of cavity wall	B16	m²	1.26	2.27	38.56	28.55	10.07	77.17
Form cavity 50mm wide in cavity wall	B17	m²	5.46	0.16	2.78	8.30	1.66	12.75
DPC 112mm wide	B18	m	8.40	0.42	7.14	8.57	2.36	18.06
DPC 140mm wide	B19	m	8.40	0.50	8.57	11.59	3.02	23.18
Bond in block wall	B20	m	1.30	0.57	9.72	3.78	2.03	15.53
Bond in half brick wall	B21	m	0.30	0.11	1.79	0.80	0.39	2.98
50mm thick insulation board	B22	m²	6.48	1.94	33.05	46.66	11.96	91.66
Carried to summary				47.02	693.12	596.89	193.50	1,483.51

**PART C
EXTERNAL WALLS**

	Ref	Unit	Qty	Hours	Hours £	Mat'ls £	O & P £	Total £
Solid blockwork 140mm thick in cavity wall	C1	m²	15.50	20.15	342.55	258.54	90.16	691.25
Facing brickwork 112.5mm thick in cavity wall	C2	m²	15.50	27.90	474.30	442.53	137.52	1,054.35
75mm thick insulation in cavity wall	C3	m²	15.50	3.41	57.97	82.77	21.11	161.85
Carried forward				51.46	874.82	783.84	248.80	1,907.45

	Ref	Unit	Qty	Hours	Hours £	Mat'ls £	O & P £	Total £
Brought forward				51.46	874.82	783.84	248.80	1,907.45
Steel lintel 2400mm long	C4	nr	2.00	0.50	8.50	296.58	45.76	350.84
Steel lintel 1500mm long	C5	nr	2.00	0.40	6.80	242.72	37.43	286.95
Close cavity wall at jambs	C7	m	8.96	0.45	7.62	25.18	4.92	37.71
Close cavity wall at cills	C8	m	3.67	0.18	3.12	10.31	2.01	15.45
Close cavity wall at top	C9	m	8.40	0.42	7.14	23.60	4.61	35.36
DPC 112mm wide at jambs	C10	m	8.96	0.45	7.62	9.14	2.51	19.27
DPC 112mm wide at cills	C11	m	3.67	0.18	3.12	3.74	1.03	7.89
Carried to summary				54.04	918.73	1,395.11	347.08	2,660.92

PART D
FLAT ROOF

	Ref	Unit	Qty	Hours	Hours £	Mat'ls £	O & P £	Total £
200 x 50mm sawn softwood joists	D1	m	22.05	5.51	93.71	78.94	25.90	198.55
200 x 50mm sawn softwood sprocket pieces	D2	nr	12.00	1.68	28.56	21.48	7.51	57.55
18mm thick WPB grade decking	D3	m²	10.40	9.36	159.12	119.29	41.76	320.17
50 x 50mm (avg) wide sawn softwood firrings	D4	m	11.05	1.99	33.81	21.11	8.24	63.16
High density polyethylene vapour barrier 150mm thick	D5	m²	9.00	1.80	30.60	100.71	19.70	151.01
100 x 75mm sawn softwood wall plate	D6	m	9.00	2.70	45.90	24.66	10.58	81.14
100 x 75mm sawn softwood tilt fillet	D7	m	3.30	0.83	14.03	8.61	3.40	26.03
Build in ends of 200 x 50mm joists	D8	nr	7.00	2.45	41.65	1.54	6.48	49.67
Carried forward				26.32	447.38	376.34	123.56	947.27

	Ref	Unit	Qty	Hours	Hours £	Mat'ls £	O & P £	Total £
Brought forward				26.32	447.38	376.34	123.56	947.27
Rake out joint for flashing	D9	m	3.30	1.16	19.64	0.79	3.06	23.49
6mm thick soffit 150mm wide	D10	m	9.30	3.72	63.24	19.62	12.43	95.29
19mm wrought softwood fascia 200mm high	D11	m	9.30	4.65	79.05	25.20	15.64	119.89
Three layer fibre-based roofing felt	D12	m²	10.40	5.72	97.24	135.62	34.93	267.78
Felt turn-down 100mm girth	D13	m	9.30	0.93	15.81	13.02	4.32	33.15
Felt flashing 150mm girth	D14	m	3.30	0.33	5.61	5.68	1.69	12.98
112mm dia. PVC-U gutter	D15	m	3.30	0.86	14.59	10.69	3.79	29.07
Stop end	D16	nr	1.00	0.14	2.38	2.12	0.68	5.18
Stop end outlet	D17	nr	1.00	0.25	4.25	4.19	1.27	9.71
68mm diameter PVC-U down pipe	D18	m	2.50	0.63	10.63	14.25	3.73	28.61
Shoe	D19	nr	1.00	0.30	5.10	3.15	1.24	9.49
Paint fascia and soffit	D20	m²	2.79	0.56	9.49	4.46	2.09	16.04
Carried to summary				45.55	774.39	615.13	208.43	1,597.95
PART E **PITCHED ROOF**				N/A	N/A	N/A	N/A	N/A
PART F **WINDOWS AND** **EXTERNAL DOORS**								
PVC-U door size 840 x 1980mm complete (B)	F1	0.00	0.00	0.00	0.00	0.00	0.00	0.00
PVC-U sliding patio door size 1700 x 2075mm (C)	F2	nr	2.00	14.00	238.00	616.94	128.24	983.18
Carried forward				14.00	238.00	616.94	128.24	983.18

	Ref	Unit	Qty	Hours	Hours £	Mat'ls £	O & P £	Total £
Brought forward				14.00	238.00	616.94	128.24	983.18
PVC-U window size 1200 x 1200mm complete (A)	F3	nr	2.00	4.00	68.00	364.28	64.84	497.12
25 x 225mm wrought softwood window board	F4	m	2.40	0.72	12.24	9.65	3.28	25.17
Paint window board	F5	m	2.40	0.14	2.38	2.21	0.69	5.28
Carried to summary				18.86	320.62	993.08	197.05	1,510.75
PART G INTERNAL PARTITIONS AND DOORS				N/A	N/A	N/A	N/A	N/A
PART H WALL FINISHES								
19 x 100mm wrought softwood skirting	H1	m	6.54	1.11	18.90	16.42	5.30	40.61
12mm plasterboard fixed to walls with dabs	H2	m²	14.00	5.46	92.82	39.90	19.91	152.63
12mm plasterboard fixed to walls less than 300mm wide	H3	m	12.63	2.27	38.65	11.62	7.54	57.81
Two coats emulsion paint to walls	H4	m²	15.90	4.13	70.28	15.26	12.83	98.37
Paint skirting	H5	m	6.54	1.31	22.24	6.02	4.24	32.49
Carried to summary				14.29	242.88	89.22	49.81	381.91
PART J FLOOR FINISHES								
Cement and sand floor screed 40mm thick	J1	m²	6.48	1.62	27.54	30.84	8.76	67.14
Vinyl floor tiles, size 300 x 300mm	J2	m²	6.48	1.10	18.73	50.28	10.35	79.36
Carried to summary				2.72	46.27	81.13	19.11	146.51

	Ref	Unit	Qty	Hours	Hours £	Mat'ls £	O & P £	Total £
PART K **CEILING FINISHES**								
Plasterboard with taped butt joints fixed to joists	K1	m²	6.48	2.33	39.66	18.47	8.72	66.84
5mm skim coat to plasterboard ceilings	K2	m²	6.48	3.24	55.08	11.21	9.94	76.23
Two coats emulsion paint to ceilings	K3	m²	6.48	1.68	28.64	6.22	5.23	40.09
Carried to summary				7.26	123.38	35.90	23.89	183.17
PART L **ELECTRICAL WORK**								
13 amp double switched socket outlet with neon	L1	nr	2.00	1.60	30.40	17.62	7.20	55.22
Lighting point	L2	nr	1.00	0.70	13.30	7.34	3.10	23.74
Lighting switch	L3	nr	2.00	1.40	26.60	8.84	5.32	40.76
Lighting wiring	L4	m	6.00	1.20	22.80	11.16	5.09	39.05
Power cable	L5	m	14.00	4.20	79.80	37.94	17.66	135.40
Carried to summary				9.10	172.90	82.90	38.37	294.17
PART M **HEATING WORK**								
15mm copper pipe	M1	m	5.00	2.20	39.60	10.75	7.55	57.90
Elbow	M2	nr	4.00	2.24	40.32	9.76	7.51	57.59
Tee	M3	nr	1.00	0.68	12.24	2.44	2.20	16.88
Radiator, double convector size 1400 x 520mm	M4	nr	2.00	2.60	46.80	278.22	48.75	373.77
Break into existing pipe and insert tee	M5	nr	1.00	0.75	13.50	4.42	2.69	20.61
Carried to summary				8.47	152.46	305.59	68.71	526.76

	Ref	Unit	Qty	Hours	Hours £	Mat'ls £	O & P £	Total £

PART N
ALTERATION WORK

	Ref	Unit	Qty	Hours	Hours £	Mat'ls £	O & P £	Total £
Take out existing window size 1500 x 1000mm and lintel over, adapt opening to receive 1770 x 2000mm patio door and insert new lintel over (both measured separately) and make good	N1	nr	1.00	20.00	340.00	32.10	55.82	427.92
Carried to summary				20.00	340.00	32.10	55.82	427.92

SUMMARY

	Hours	Hours £	Mat'ls £	O & P £	Total £
PART A **PRELIMINARIES**	0.00	0.00	0.00	0.00	1,554.00
PART B SUBSTRUCTURE TO **DPC LEVEL**	47.02	693.12	596.89	193.50	1,483.51
PART C **EXTERNAL WALLS**	54.04	918.73	1,395.11	347.08	2,660.92
PART D **FLAT ROOF**	45.55	774.39	615.13	208.43	1,597.95
PART E **PITCHED ROOF**	0.00	0.00	0.00	0.00	0.00
PART F WINDOWS AND **EXTERNAL DOORS**	18.86	320.62	993.08	197.05	1,510.75
PART G INTERNAL **PARTITIONS AND DOORS**	0.00	0.00	0.00	0.00	0.00
PART H **WALL FINISHES**	14.29	242.88	89.22	49.81	381.91
PART J **FLOOR FINISHES**	2.72	46.27	81.13	19.11	146.21
PART K **CEILING FINISHES**	7.26	123.38	35.90	23.89	183.87
PART L **ELECTRICAL WORK**	9.10	172.90	82.90	38.37	294.17
PART M **HEATING WORK**	8.47	152.46	305.59	68.71	526.76
PART N **ALTERATION WORK**	20.00	340.00	32.10	55.82	427.92
Final total	227.31	3,784.75	4,227.05	1,201.76	10,767.96

	Ref	Unit	Qty	Hours	Hours £	Mat'ls £	O & P £	Total £
PART A **PRELIMINARIES**								
Concrete mixer	A1	wks	7.00					350.00
Small tools	A2	wks	8.00					320.00
Scaffolding (m²/weeks)	A3		100					600.00
Skip	A4	wks	4.00					440.00
Clean up	A5	hrs	6.00					78.00
Carried to summary								1,788.00
PART B **SUBSTRUCTURE TO** **DPC LEVEL**								
Excavate topsoil 150mm thick by hand	B1	m²	14.19	4.26	55.34	0.00	8.30	63.64
Excavate to reduce levels	B2	m³	4.06	10.15	131.95	0.00	19.79	151.74
Excavate for trench foundations by hand	B3	m³	1.39	3.61	46.98	0.00	7.05	54.03
Earthwork support to sides of trenches	B4	m²	13.72	5.49	71.34	25.38	14.51	111.23
Backfilling with excavated material	B5	m³	0.55	0.33	4.29	0.00	0.64	4.93
Hardcore 225mm thick	B6	m²	9.18	1.84	23.87	57.74	12.24	93.85
Hardcore filling to trench	B7	m³	0.12	0.09	1.17	6.29	1.12	8.58
Concrete grade (1:3:6) in foundations	B8	m³	1.27	0.86	11.18	108.74	17.99	137.91
Concrete grade (1:2:4) in bed 150mm thick	B9	m²	9.18	2.75	35.80	140.64	26.47	202.91
Concrete (1:2:4) in cavity wall filling	B10	m²	4.70	0.94	11.97	11.19	3.47	26.63
Carried forward				30.32	393.90	349.98	111.58	855.45

	Ref	Unit	Qty	Hours	Hours £	Mat'ls £	O & P £	Total £
Brought forward				30.32	393.90	349.98	111.58	855.45
Damp-proof membrane	B11	m²	9.68	0.39	5.03	22.46	4.12	31.61
Reinforcement ref A193 in foundation	B12	m²	4.70	0.56	7.33	9.49	2.52	19.35
Steel fabric reinforcement ref A193 in slab	B13	m²	9.18	1.38	17.90	18.36	5.44	41.70
Solid blockwork 140mm thick in cavity wall	B14	m²	6.11	7.94	135.03	101.91	35.54	272.49
Common bricks 112.5mm thick in cavity wall	B15	m²	4.70	7.99	135.83	77.32	31.97	245.12
Facing bricks in 112.5mm thick in skin of cavity wall	B16	m²	1.41	2.54	43.15	28.55	10.75	82.45
Form cavity 50mm wide in cavity wall	B17	m²	5.46	0.16	2.78	8.30	1.66	12.75
DPC 112mm wide	B18	m	9.40	0.47	7.99	9.59	2.64	20.21
DPC 140mm wide	B19	m	9.40	0.56	9.59	12.97	3.38	25.94
Bond in block wall	B20	m	1.30	0.57	9.72	3.78	2.03	15.53
Bond in half brick wall	B21	m	0.30	0.11	1.79	0.80	0.39	2.98
50mm thick insulation board	B22	m²	9.18	2.75	46.82	66.10	16.94	129.85
Carried to summary				55.75	816.86	709.61	228.97	1,755.44

PART C
EXTERNAL WALLS

	Ref	Unit	Qty	Hours	Hours £	Mat'ls £	O & P £	Total £
Solid blockwork 140mm thick in cavity wall	C1	m²	18.00	23.40	397.80	300.24	104.71	802.75
Facing brickwork 112.5mm thick in cavity wall	C2	m²	18.00	32.40	550.80	513.90	159.71	1,224.41
75mm thick insulation in cavity wall	C3	m²	18.00	3.96	67.32	96.12	24.52	187.96
Carried forward				59.76	1,015.92	910.26	288.93	2,215.11

	Ref	Unit	Qty	Hours	Hours £	Mat'ls £	O & P £	Total £
Brought forward				59.76	1,015.92	910.26	288.93	2,215.11
Steel lintel 2400mm long	C4	nr	2.00	0.50	8.50	296.58	45.76	350.84
Steel lintel 1500mm long	C5	nr	2.00	0.40	6.80	242.72	37.43	286.95
Close cavity wall at jambs	C7	m	8.96	0.45	7.62	25.18	4.92	37.71
Close cavity wall at cills	C8	m	3.67	0.18	3.12	10.31	2.01	15.45
Close cavity wall at top	C9	m	9.40	0.47	7.99	26.41	5.16	39.56
DPC 112mm wide at jambs	C10	m	8.96	0.45	7.62	9.14	2.51	19.27
DPC 112mm wide at cills	C11	m	3.67	0.18	3.12	3.74	1.03	7.89
Carried to summary				62.39	1,060.68	1,524.35	387.75	2,972.78

PART D
FLAT ROOF

	Ref	Unit	Qty	Hours	Hours £	Mat'ls £	O & P £	Total £
200 x 50mm sawn softwood joists	D1	m	28.35	7.09	120.49	101.49	33.30	255.28
200 x 50mm sawn softwood sprocket pieces	D2	nr	12.00	1.68	28.56	21.48	7.51	57.55
18mm thick WPB grade decking	D3	m²	13.55	12.20	207.32	155.42	54.41	417.14
50 x 50mm (avg) wide sawn softwood firrings	D4	m	28.35	5.10	86.75	54.15	21.13	162.03
High density polyethylene vapour barrier 150mm thick	D5	m²	12.00	2.40	40.80	134.28	26.26	201.34
100 x 75mm sawn softwood wall plate	D6	m	10.00	3.00	51.00	27.40	11.76	90.16
100 x 75mm sawn softwood tilt fillet	D7	m	4.30	1.08	18.28	11.22	4.42	33.92
Build in ends of 200 x 50mm joists	D8	nr	9.00	3.15	53.55	1.98	8.33	63.86
Carried forward				35.69	606.74	507.42	167.12	1,281.29

	Ref	Unit	Qty	Hours	Hours £	Mat'ls £	O & P £	Total £
Brought forward				35.69	606.74	507.42	167.12	1,281.29
Rake out joint for flashing	D9	m	4.30	1.51	25.59	1.03	3.99	30.61
6mm thick soffit 150mm wide	D10	m	10.30	4.12	70.04	21.73	13.77	105.54
19mm wrought softwood fascia 200mm high	D11	m	10.30	5.15	87.55	27.91	17.32	132.78
Three layer fibre-based roofing felt	D12	m²	13.55	7.45	126.69	176.69	45.51	348.89
Felt turn-down 100mm girth	D13	m	10.30	1.03	17.51	14.42	4.79	36.72
Felt flashing 150mm girth	D14	m	4.30	0.43	7.31	7.40	2.21	16.91
112mm dia. PVC-U gutter	D15	m	4.30	1.12	19.01	13.93	4.94	37.88
Stop end	D16	nr	1.00	0.14	2.38	2.12	0.68	5.18
Stop end outlet	D17	nr	1.00	0.25	4.25	4.19	1.27	9.71
68mm diameter PVC-U down pipe	D18	m	2.50	0.63	10.63	14.25	3.73	28.61
Shoe	D19	nr	1.00	0.30	5.10	3.15	1.24	9.49
Paint fascia and soffit	D20	m²	3.09	0.62	10.51	4.94	2.32	17.77
Carried to summary				58.43	993.29	799.20	268.87	2,061.36
PART E PITCHED ROOF				N/A	N/A	N/A	N/A.	N/A
PART F WINDOWS AND EXTERNAL DOORS								
PVC-U door size 840 x 1980mm complete (B)	F1	0.00	0.00	0.00	0.00	0.00	0.00	0.00
PVC-U sliding patio door size 1700 x 2075mm (C)	F2	nr	2.00	14.00	238.00	616.94	128.24	983.18
Carried forward				14.00	238.00	616.94	128.24	983.18

	Ref	Unit	Qty	Hours	Hours £	Mat'ls £	O & P £	Total £
Brought forward				14.00	238.00	616.94	128.24	983.18
PVC-U window size 1200 x 1200mm complete (A)	F3	nr	2.00	4.00	68.00	364.28	64.84	497.12
25 x 225mm wrought softwood window board	F4	m	2.40	0.72	12.24	9.65	3.28	25.17
Paint window board	F5	m	2.40	0.14	2.38	2.21	0.69	5.28
Carried to summary				18.86	320.62	993.08	197.05	1,510.75
PART G INTERNAL PARTITIONS AND DOORS				N/A	N/A	N/A	N/A	N/A
PART H WALL FINISHES								
19 x 100mm wrought softwood skirting	H1	m	7.54	1.28	21.79	18.93	6.11	46.82
12mm plasterboard fixed to walls with dabs	H2	m²	16.50	6.44	109.40	47.03	23.46	179.88
12mm plasterboard fixed to walls less than 300mm wide	H3	m	12.63	2.27	38.65	11.62	7.54	57.81
Two coats emulsion paint to walls	H4	m²	18.40	4.78	81.33	17.66	14.85	113.84
Paint skirting	H5	m	7.54	1.51	25.64	6.94	4.89	37.46
Carried to summary				16.28	276.80	102.17	56.85	435.81
PART J FLOOR FINISHES								
Cement and sand floor screed 40mm thick	J1	m²	9.18	2.30	39.02	43.70	12.41	95.12
Vinyl floor tiles, size 300 x 300mm	J2	m²	9.18	1.56	26.53	71.24	14.67	112.43
Carried to summary				3.86	65.55	114.93	27.07	207.55

	Ref	Unit	Qty	Hours	Hours £	Mat'ls £	O & P £	Total £
PART K **CEILING FINISHES**								
Plasterboard with taped butt joints fixed to joists	K1	m²	9.18	3.30	56.18	26.16	12.35	94.70
5mm skim coat to plasterboard ceilings	K2	m²	9.18	4.59	78.03	15.88	14.09	108.00
Two coats emulsion paint to ceilings	K3	m²	9.18	2.39	40.58	8.81	7.41	56.80
Carried to summary				10.28	174.79	50.86	33.85	259.49
PART L **ELECTRICAL WORK**								
13 amp double switched socket outlet with neon	L1	nr	3.00	2.40	45.60	26.43	10.80	82.83
Lighting point	L2	nr	2.00	1.40	26.60	14.68	6.19	47.47
Lighting switch	L3	nr	2.00	1.40	26.60	8.84	5.32	40.76
Lighting wiring	L4	m	6.00	1.20	22.80	11.16	5.09	39.05
Power cable	L5	m	16.00	4.80	91.20	43.36	20.18	154.74
Carried to summary				11.20	212.80	104.47	47.59	364.86
PART M **HEATING WORK**								
15mm copper pipe	M1	m	6.00	2.64	47.52	12.90	9.06	69.48
Elbow	M2	nr	4.00	2.24	40.32	9.76	7.51	57.59
Tee	M3	nr	1.00	0.68	12.24	2.44	2.20	16.88
Radiator, double convector size 1400 x 520mm	M4	nr	2.00	2.60	46.80	278.22	48.75	373.77
Break into existing pipe and insert tee	M5	nr	1.00	0.75	13.50	4.42	2.69	20.61
Carried to summary				8.91	160.38	307.74	70.22	538.34

	Ref	Unit	Qty	Hours	Hours £	Mat'ls £	O & P £	Total £
PART N **ALTERATION WORK**								
Take out existing window size 1500 x 1000mm and lintel over, adapt opening to receive 1770 x 2000mm patio door and insert new lintel over (both measured separately) and make good	N1	nr	1.00	20.00	340.00	32.10	55.82	427.92
Carried to summary				20.00	340.00	32.10	55.82	427.92

SUMMARY

	Hours	Hours £	Mat'ls £	O & P £	Total £
PART A **PRELIMINARIES**	0.00	0.00	0.00	0.00	1,788.00
PART B SUBSTRUCTURE TO **DPC LEVEL**	55.75	816.86	709.61	228.97	1,755.44
PART C **EXTERNAL WALLS**	62.39	1,060.68	1,524.35	387.75	2,972.78
PART D **FLAT ROOF**	58.43	993.29	799.20	268.87	2,061.36
PART E **PITCHED ROOF**	0.00	0.00	0.00	0.00	0.00
PART F WINDOWS AND **EXTERNAL DOORS**	18.86	320.62	993.08	197.05	1,510.75
PART G INTERNAL **PARTITIONS AND DOORS**	0.00	0.00	0.00	0.00	0.00
PART H **WALL FINISHES**	16.28	276.80	102.17	56.85	435.81
PART J **FLOOR FINISHES**	3.86	65.55	114.93	27.07	207.55
PART K **CEILING FINISHES**	10.28	174.79	50.86	33.85	259.49
PART L **ELECTRICAL WORK**	11.20	212.80	104.47	47.59	364.86
PART M **HEATING WORK**	8.91	160.38	307.74	70.22	538.34
PART N **ALTERATION WORK**	20.00	340.00	32.10	55.82	427.92
Final total	265.96	4,421.77	4,738.51	1,374.04	12,322.30

	Ref	Unit	Qty	Hours	Hours £	Mat'ls £	O & P £	Total £
PART A **PRELIMINARIES**								
Concrete mixer	A1	wks	8.00					400.00
Small tools	A2	wks	9.00					360.00
Scaffolding (m²/weeks)	A3		110					660.00
Skip	A4	wks	5.00					550.00
Clean up	A5	hrs	8.00					104.00
Carried to summary								2,074.00
PART B **SUBSTRUCTURE TO** **DPC LEVEL**								
Excavate topsoil 150mm thick by hand	B1	m²	17.50	5.25	68.25	0.00	10.24	78.49
Excavate to reduce levels	B2	m³	5.00	12.50	162.50	0.00	24.38	186.88
Excavate for trench foundations by hand	B3	m³	1.72	4.47	58.14	0.00	8.72	66.86
Earthwork support to sides of trenches	B4	m²	15.18	6.07	78.94	28.08	16.05	123.07
Backfilling with excavated material	B5	m³	0.63	0.38	4.91	0.00	0.74	5.65
Hardcore 225mm thick	B6	m²	11.88	2.38	30.89	74.73	15.84	121.46
Hardcore filling to trench	B7	m³	0.14	0.09	1.17	6.29	1.12	8.58
Concrete grade (1:3:6) in foundations	B8	m³	1.40	0.86	11.18	119.87	19.66	150.71
Concrete grade (1:2:4) in bed 150mm thick	B9	m²	11.88	3.56	46.33	182.00	34.25	262.58
Concrete (1:2:4) in cavity wall filling	B10	m²	5.20	1.04	11.97	12.38	3.65	28.00
Carried forward				36.60	474.28	423.34	134.64	1,032.26

	Ref	Unit	Qty	Hours	Hours £	Mat'ls £	O & P £	Total £
Brought forward				36.60	474.28	423.34	134.64	1,032.26
Damp-proof membrane	B11	m²	12.38	0.50	6.44	28.72	5.27	40.43
Reinforcement ref A193 in foundation	B12	m²	5.20	0.62	8.11	10.50	2.79	21.41
Steel fabric reinforcement ref A193 in slab	B13	m²	11.88	1.78	23.17	23.76	7.04	53.96
Solid blockwork 140mm thick in cavity wall	B14	m²	6.76	8.79	149.40	112.76	39.32	301.48
Common bricks 112.5mm thick in cavity wall	B15	m²	5.20	8.84	150.28	85.54	35.37	271.19
Facing bricks in 112.5mm thick in skin of cavity wall	B16	m²	1.56	2.81	47.74	28.55	11.44	87.73
Form cavity 50mm wide in cavity wall	B17	m²	6.76	0.20	3.45	10.28	2.06	15.78
DPC 112mm wide	B18	m	10.40	0.52	8.84	10.61	2.92	22.37
DPC 140mm wide	B19	m	10.40	0.62	10.61	14.35	3.74	28.70
Bond in block wall	B20	m	1.30	0.57	9.72	3.78	2.03	15.53
Bond in half brick wall	B21	m	0.30	0.11	1.79	0.80	0.39	2.98
50mm thick insulation board	B22	m²	11.88	3.56	60.59	85.54	21.92	168.04
Carried to summary				65.53	954.40	838.53	268.94	2,061.87

PART C
EXTERNAL WALLS

	Ref	Unit	Qty	Hours	Hours £	Mat'ls £	O & P £	Total £
Solid blockwork 140mm thick in cavity wall	C1	m²	20.50	26.65	453.05	341.94	119.25	914.24
Facing brickwork 112.5mm thick in cavity wall	C2	m²	20.50	36.90	627.30	585.28	181.89	1,394.46
75mm thick insulation in cavity wall	C3	m²	20.50	4.51	76.67	109.47	27.92	214.06
Carried forward				68.06	1,157.02	1,036.69	329.06	2,522.76

	Ref	Unit	Qty	Hours	Hours £	Mat'ls £	O & P £	Total £
Brought forward				68.06	1,157.02	1,036.69	329.06	2,522.76
Steel lintel 2400mm long	C4	nr	2.00	0.50	8.50	296.58	45.76	350.84
Steel lintel 1500mm long	C5	nr	2.00	0.40	6.80	242.72	37.43	286.95
Close cavity wall at jambs	C7	m	8.96	0.45	7.62	25.18	4.92	37.71
Close cavity wall at cills	C8	m	3.67	0.18	3.12	10.31	2.01	15.45
Close cavity wall at top	C9	m	10.40	0.52	8.84	29.22	5.71	43.77
DPC 112mm wide at jambs	C10	m	8.96	0.45	7.62	9.14	2.51	19.27
DPC 112mm wide at cills	C11	m	3.67	0.18	3.12	3.74	1.03	7.89
Carried to summary				70.74	1,202.63	1,653.58	428.43	3,284.64

PART D
FLAT ROOF

	Ref	Unit	Qty	Hours	Hours £	Mat'ls £	O & P £	Total £
200 x 50mm sawn softwood joists	D1	m	34.65	8.66	147.26	124.05	40.70	312.01
200 x 50mm sawn softwood sprocket pieces	D2	nr	12.00	1.68	28.56	21.48	7.51	57.55
18mm thick WPB grade decking	D3	m²	16.70	15.03	255.51	191.55	67.06	514.12
50 x 50mm (avg) wide sawn softwood firrings	D4	m	34.65	6.24	106.03	66.18	25.83	198.04
High density polyethylene vapour barrier 150mm thick	D5	m²	15.00	3.00	51.00	167.85	32.83	251.68
100 x 75mm sawn softwood wall plate	D6	m	11.00	3.30	56.10	30.14	12.94	99.18
100 x 75mm sawn softwood tilt fillet	D7	m	5.30	1.33	22.53	13.83	5.45	41.81
Build in ends of 200 x 50mm joists	D8	nr	11.00	3.85	65.45	2.42	10.18	78.05
Carried forward				43.08	732.44	617.50	202.49	1,552.43

	Ref	Unit	Qty	Hours	Hours £	Mat'ls £	O & P £	Total £
Brought forward				43.08	732.44	617.50	202.49	1,552.43
Rake out joint for flashing	D9	m	5.30	1.86	31.54	1.27	4.92	37.73
6mm thick soffit 150mm wide	D10	m	11.30	4.52	76.84	23.84	15.10	115.79
19mm wrought softwood fascia 200mm high	D11	m	11.30	5.65	96.05	30.62	19.00	145.67
Three layer fibre-based roofing felt	D12	m²	16.70	9.19	156.15	217.77	56.09	430.00
Felt turn-down 100mm girth	D13	m	11.30	1.13	19.21	15.82	5.25	40.28
Felt flashing 150mm girth	D14	m	5.30	0.53	9.01	9.12	2.72	20.84
112mm dia. PVC-U gutter	D15	m	5.30	1.38	23.43	17.17	6.09	46.69
Stop end	D16	nr	1.00	0.14	2.38	2.12	0.68	5.18
Stop end outlet	D17	nr	1.00	0.25	4.25	4.19	1.27	9.71
68mm diameter PVC-U down pipe	D18	m	2.50	0.63	10.63	14.25	3.73	28.61
Shoe	D19	nr	1.00	0.30	5.10	3.15	1.24	9.49
Paint fascia and soffit	D20	m²	3.39	0.68	11.53	5.42	2.54	19.49
Carried to summary				69.33	1,178.53	962.25	321.12	2,461.90
PART E PITCHED ROOF				N/A	N/A	N/A	N/A	N/A
PART F WINDOWS AND EXTERNAL DOORS								
PVC-U door size 840 x 1980mm complete (B)	F1	0.00	0.00	0.00	0.00	0.00	0.00	0.00
PVC-U sliding patio door size 1700 x 2075mm (C)	F2	nr	2.00	14.00	238.00	616.94	128.24	983.18
Carried forward				14.00	238.00	616.94	128.24	983.18

	Ref	Unit	Qty	Hours	Hours £	Mat'ls £	O & P £	Total £
Brought forward				14.00	238.00	616.94	128.24	983.18
PVC-U window size 1200 x 1200mm complete (A)	F3	nr	2.00	4.00	68.00	364.28	64.84	497.12
25 x 225mm wrought softwood window board	F4	m	2.40	0.72	12.24	9.65	3.28	25.17
Paint window board	F5	m	2.40	0.14	2.38	2.21	0.69	5.28
Carried to summary				18.86	320.62	993.08	197.05	1,510.75
PART G INTERNAL PARTITIONS AND DOORS				N/A	N/A	N/A	N/A	N/A
PART H WALL FINISHES								
19 x 100mm wrought softwood skirting	H1	m	8.54	1.45	24.68	21.44	6.92	53.03
12mm plasterboard fixed to walls with dabs	H2	m²	19.00	7.41	125.97	54.15	27.02	207.14
12mm plasterboard fixed to walls less than 300mm wide	H3	m	13.74	2.47	42.04	12.64	8.20	62.89
Two coats emulsion paint to walls	H4	m²	21.02	5.47	92.91	20.18	16.96	130.05
Paint skirting	H5	m	8.54	1.71	29.04	7.86	5.53	42.43
Carried to summary				18.51	314.64	116.26	64.64	495.54
PART J FLOOR FINISHES								
Cement and sand floor screed 40mm thick	J1	m²	11.88	2.97	50.49	56.55	16.06	123.09
Vinyl floor tiles, size 300 x 300mm	J2	m²	11.88	2.02	34.33	92.19	18.98	145.50
Carried to summary				4.99	84.82	148.74	35.03	268.59

	Ref	Unit	Qty	Hours	Hours £	Mat'ls £	O & P £	Total £
PART K								
CEILING FINISHES								
Plasterboard with taped butt joints fixed to joists	K1	m²	11.88	4.28	72.71	33.86	15.98	122.55
5mm skim coat to plasterboard ceilings	K2	m²	11.88	5.94	100.98	20.55	18.23	139.76
Two coats emulsion paint to ceilings	K3	m²	11.88	3.09	52.51	11.40	9.59	73.50
Carried to summary				13.31	226.20	65.82	43.80	335.81
PART L								
ELECTRICAL WORK								
13 amp double switched socket outlet with neon	L1	nr	3.00	2.40	45.60	26.43	10.80	82.83
Lighting point	L2	nr	2.00	1.40	26.60	14.68	6.19	47.47
Lighting switch	L3	nr	2.00	1.40	26.60	8.84	5.32	40.76
Lighting wiring	L4	m	7.00	1.40	26.60	13.02	5.94	45.56
Power cable	L5	m	18.00	5.40	102.60	48.78	22.71	174.09
Carried to summary				12.00	228.00	111.75	50.96	390.71
PART M								
HEATING WORK								
15mm copper pipe	M1	m	7.00	3.08	55.44	15.05	10.57	81.06
Elbow	M2	nr	4.00	2.24	40.32	9.76	7.51	57.59
Tee	M3	nr	1.00	0.68	12.24	2.44	2.20	16.88
Radiator, double convector size 1400 x 520mm	M4	nr	2.00	2.60	46.80	278.22	48.75	373.77
Break into existing pipe and insert tee	M5	nr	1.00	0.75	13.50	4.42	2.69	20.61
Carried to summary				9.35	168.30	309.89	71.73	549.92

	Ref	Unit	Qty	Hours	Hours £	Mat'ls £	O & P £	Total £
PART N								
ALTERATION WORK								
Take out existing window size 1500 x 1000mm and lintel over, adapt opening to receive 1770 x 2000mm patio door and insert new lintel over (both measured separately) and make good	N1	nr	1.00	20.00	340.00	32.10	55.82	427.92
Carried to summary				20.00	340.00	32.10	55.82	427.92

SUMMARY

	Hours	Hours £	Mat'ls £	O & P £	Total £
PART A **PRELIMINARIES**	0.00	0.00	0.00	0.00	2,074.00
PART B SUBSTRUCTURE TO **DPC LEVEL**	65.53	954.40	838.53	268.94	2,061.87
PART C **EXTERNAL WALLS**	70.74	1,202.63	1,653.58	428.43	3,284.64
PART D **FLAT ROOF**	69.33	1,178.53	962.25	321.12	2,461.90
PART E **PITCHED ROOF**	0.00	0.00	0.00	0.00	0.00
PART F WINDOWS AND **EXTERNAL DOORS**	18.86	320.62	993.08	197.05	1,510.75
PART G INTERNAL **PARTITIONS AND DOORS**	0.00	0.00	0.00	0.00	0.00
PART H **WALL FINISHES**	18.51	314.64	116.26	64.64	495.54
PART J **FLOOR FINISHES**	4.99	84.82	148.74	35.03	268.59
PART K **CEILING FINISHES**	13.31	226.20	65.82	43.80	335.81
PART L **ELECTRICAL WORK**	12.00	228.00	111.75	50.96	390.71
PART M **HEATING WORK**	9.35	168.30	309.89	71.73	549.92
PART N **ALTERATION WORK**	20.00	340.00	32.10	55.82	427.92
Final total	302.62	5,018.14	5,232.00	1,537.52	13,861.65

	Ref	Unit	Qty	Hours	Hours £	Mat'ls £	O & P £	Total £
PART A **PRELIMINARIES**								
Concrete mixer	A1	wks	9.00					450.00
Small tools	A2	wks	10.00					400.00
Scaffolding (m²/weeks)	A3		130					780.00
Skip	A4	wks	6.00					660.00
Clean up	A5	hrs	10.00					130.00
Carried to summary								2,420.00
PART B **SUBSTRUCTURE TO** **DPC LEVEL**								
Excavate topsoil 150mm thick by hand	B1	m²	19.85	5.96	77.42	0.00	11.61	89.03
Excavate to reduce levels	B2	m³	5.95	14.88	193.38	0.00	29.01	222.38
Excavate for trench foundations by hand	B3	m³	1.88	4.89	63.54	0.00	9.53	73.08
Earthwork support to sides of trenches	B4	m²	16.64	6.66	86.53	30.78	17.60	134.91
Backfilling with excavated material	B5	m³	0.70	0.42	5.46	0.00	0.82	6.28
Hardcore 225mm thick	B6	m²	14.58	2.92	37.91	91.71	19.44	149.06
Hardcore filling to trench	B7	m³	0.15	0.09	1.17	6.29	1.12	8.58
Concrete grade (1:3:6) in foundations	B8	m³	1.54	0.86	11.18	131.85	21.46	164.49
Concrete grade (1:2:4) in bed 150mm thick	B9	m²	14.58	4.37	56.86	223.37	42.03	322.26
Concrete (1:2:4) in cavity wall filling	B10	m²	5.70	1.14	11.97	13.57	3.83	29.37
Carried forward				42.17	545.41	497.57	156.45	1,199.43

	Ref	Unit	Qty	Hours	Hours £	Mat'ls £	O & P £	Total £
Brought forward				42.17	545.41	497.57	156.45	1,199.43
Damp-proof membrane	B11	m²	15.13	0.61	7.87	35.10	6.45	49.41
Reinforcement ref A193 in foundation	B12	m²	5.70	0.68	8.89	11.51	3.06	23.47
Steel fabric reinforcement ref A193 in slab	B13	m²	14.58	2.19	28.43	29.16	8.64	66.23
Solid blockwork 140mm thick in cavity wall	B14	m²	7.41	9.63	163.76	123.60	43.10	330.46
Common bricks 112.5mm thick in cavity wall	B15	m²	5.70	9.69	164.73	93.77	38.77	297.27
Facing bricks in 112.5mm thick in skin of cavity wall	B16	m²	1.71	3.08	52.33	28.55	12.13	93.01
Form cavity 50mm wide in cavity wall	B17	m²	7.41	0.22	3.78	11.26	2.26	17.30
DPC 112mm wide	B18	m	11.40	0.57	9.69	11.63	3.20	24.52
DPC 140mm wide	B19	m	11.40	0.68	11.63	15.73	4.10	31.46
Bond in block wall	B20	m	1.30	0.57	9.72	3.78	2.03	15.53
Bond in half brick wall	B21	m	0.30	0.11	1.79	0.80	0.39	2.98
50mm thick insulation board	B22	m²	14.58	4.37	74.36	104.98	26.90	206.23
Carried to summary				74.58	1,082.38	967.44	307.47	2,357.30

**PART C
EXTERNAL WALLS**

	Ref	Unit	Qty	Hours	Hours £	Mat'ls £	O & P £	Total £
Solid blockwork 140mm thick in cavity wall	C1	m²	19.90	25.87	439.79	331.93	115.76	887.48
Facing brickwork 112.5mm thick in cavity wall	C2	m²	19.90	35.82	608.94	568.15	176.56	1,353.65
75mm thick insulation in cavity wall	C3	m²	19.90	4.38	74.43	106.27	27.10	207.80
Carried forward				66.07	1,123.16	1,006.34	319.42	2,448.92

	Ref	Unit	Qty	Hours	Hours £	Mat'ls £	O & P £	Total £
Brought forward				66.07	1,123.16	1,006.34	319.42	2,448.92
Steel lintel 2400mm long	C4	nr	2.00	0.50	8.50	296.58	45.76	350.84
Steel lintel 1500mm long	C5	nr	3.00	0.60	10.20	364.08	56.14	430.42
Close cavity wall at jambs	C7	m	17.32	0.87	14.72	48.67	9.51	72.90
Close cavity wall at cills	C8	m	5.71	0.29	4.85	16.05	3.13	24.03
Close cavity wall at top	C9	m	11.40	0.57	9.69	32.03	6.26	47.98
DPC 112mm wide at jambs	C10	m	17.32	0.87	14.72	17.67	4.86	37.25
DPC 112mm wide at cills	C11	m	5.71	0.29	4.85	5.82	1.60	12.28
Carried to summary				70.04	1,190.70	1,787.24	446.69	3,424.63

PART D
FLAT ROOF

	Ref	Unit	Qty	Hours	Hours £	Mat'ls £	O & P £	Total £
200 x 50mm sawn softwood joists	D1	m	40.95	10.24	174.04	146.60	48.10	368.73
200 x 50mm sawn softwood sprocket pieces	D2	nr	12.00	1.68	28.56	21.48	7.51	57.55
18mm thick WPB grade decking	D3	m²	19.85	17.87	303.71	227.68	79.71	611.09
50 x 50mm (avg) wide sawn softwood firrings	D4	m	40.95	7.37	125.31	78.21	30.53	234.05
High density polyethylene vapour barrier 150mm thick	D5	m²	18.00	3.60	61.20	201.42	39.39	302.01
100 x 75mm sawn softwood wall plate	D6	m	12.00	3.60	61.20	32.88	14.11	108.19
100 x 75mm sawn softwood tilt fillet	D7	m	6.30	1.58	26.78	16.44	6.48	49.70
Build in ends of 200 x 50mm joists	D8	nr	13.00	4.55	77.35	2.86	12.03	92.24
Carried forward				50.48	858.13	727.58	237.86	1,823.57

	Ref	Unit	Qty	Hours	Hours £	Mat'ls £	O & P £	Total £
Brought forward				50.48	858.13	727.58	237.86	1,823.57
Rake out joint for flashing	D9	m	6.30	2.21	37.49	1.51	5.85	44.85
6mm thick soffit 150mm wide	D10	m	12.30	4.92	83.64	25.95	16.44	126.03
19mm wrought softwood fascia 200mm high	D11	m	12.30	6.15	104.55	33.33	20.68	158.57
Three layer fibre-based roofing felt	D12	m²	19.85	10.92	185.60	258.84	66.67	511.11
Felt turn-down 100mm girth	D13	m	12.30	1.23	20.91	17.22	5.72	43.85
Felt flashing 150mm girth	D14	m	6.30	0.63	10.71	10.84	3.23	24.78
112mm dia. PVC-U gutter	D15	m	6.30	1.64	27.85	20.41	7.24	55.50
Stop end	D16	nr	1.00	0.14	2.38	2.12	0.68	5.18
Stop end outlet	D17	nr	1.00	0.25	4.25	4.19	1.27	9.71
68mm diameter PVC-U down pipe	D18	m	2.50	0.63	10.63	14.25	3.73	28.61
Shoe	D19	nr	1.00	0.30	5.10	3.15	1.24	9.49
Paint fascia and soffit	D20	m²	3.69	0.74	12.55	5.90	2.77	21.22
Carried to summary				80.22	1,363.77	1,125.30	373.36	2,862.44
PART E PITCHED ROOF				N/A	N/A	N/A	N/A	N/A
PART F WINDOWS AND EXTERNAL DOORS								
PVC-U door size 840 x 1980mm complete (B)	F1	0.00	1.00	2.00	34.00	278.67	46.90	359.57
PVC-U sliding patio door size 1700 x 2075mm (C)	F2	nr	2.00	14.00	238.00	616.94	128.24	983.18
Carried forward				16.00	272.00	895.61	175.14	1,342.75

	Ref	Unit	Qty	Hours	Hours £	Mat'ls £	O & P £	Total £
Brought forward				16.00	272.00	895.61	175.14	1,342.75
PVC-U window size 1200 x 1200mm complete (A)	F3	nr	3.00	6.00	102.00	546.42	97.26	745.68
25 x 225mm wrought softwood window board	F4	m	3.60	1.08	18.36	14.47	4.92	37.76
Paint window board	F5	m	3.60	0.14	2.38	3.31	0.85	6.55
Carried to summary				23.22	394.74	1,459.81	278.18	2,132.74
PART G INTERNAL PARTITIONS AND DOORS				N/A	N/A	N/A	N/A	N/A
PART H WALL FINISHES								
19 x 100mm wrought softwood skirting	H1	m	8.70	1.48	25.14	21.84	7.05	54.03
12mm plasterboard fixed to walls with dabs	H2	m²	18.40	7.18	121.99	52.44	26.16	200.60
12mm plasterboard fixed to walls less than 300mm wide	H3	m	23.03	4.15	70.47	21.19	13.75	105.41
Two coats emulsion paint to walls	H4	m²	21.85	5.68	96.58	20.98	17.63	135.19
Paint skirting	H5	m	8.70	1.74	29.58	8.00	5.64	43.22
Carried to summary				20.22	343.76	124.44	70.23	538.44
PART J FLOOR FINISHES								
Cement and sand floor screed 40mm thick	J1	m²	14.58	3.65	61.97	69.40	19.70	151.07
Vinyl floor tiles, size 300 x 300mm	J2	m²	14.58	2.48	42.14	113.14	23.29	178.57
Carried to summary				6.12	104.10	182.54	43.00	329.64

	Ref	Unit	Qty	Hours	Hours £	Mat'ls £	O & P £	Total £
PART K **CEILING FINISHES**								
Plasterboard with taped butt joints fixed to joists	K1	m²	14.58	5.25	89.23	41.55	19.62	150.40
5mm skim coat to plasterboard ceilings	K2	m²	14.58	7.29	123.93	25.22	22.37	171.53
Two coats emulsion paint to ceilings	K3	m²	14.58	3.79	64.44	14.00	11.77	90.21
Carried to summary				16.33	277.60	80.77	53.76	412.13
PART L **ELECTRICAL WORK**								
13 amp double switched socket outlet with neon	L1	nr	4.00	3.20	60.80	35.24	14.41	110.45
Lighting point	L2	nr	2.00	1.40	26.60	14.68	6.19	47.47
Lighting switch	L3	nr	3.00	2.10	39.90	13.26	7.97	61.13
Lighting wiring	L4	m	7.00	1.40	26.60	13.02	5.94	45.56
Power cable	L5	m	20.00	6.00	114.00	54.20	25.23	193.43
Carried to summary				14.10	267.90	130.40	59.75	458.05
PART M **HEATING WORK**								
15mm copper pipe	M1	m	8.00	3.52	63.36	17.20	12.08	92.64
Elbow	M2	nr	4.00	2.24	40.32	9.76	7.51	57.59
Tee	M3	nr	1.00	0.68	12.24	2.44	2.20	16.88
Radiator, double convector size 1400 x 520mm	M4	nr	2.00	2.60	46.80	278.22	48.75	373.77
Break into existing pipe and insert tee	M5	nr	1.00	0.75	13.50	4.42	2.69	20.61
Carried to summary				9.79	176.22	312.04	73.24	561.50

	Ref	Unit	Qty	Hours	Hours £	Mat'ls £	O & P £	Total £
PART N								
ALTERATION WORK								
Take out existing window size 1500 x 1000mm and lintel over, adapt opening to receive 1770 x 2000mm patio door and insert new lintel over (both measured separately) and make good	N1	nr	1.00	20.00	340.00	32.10	55.82	427.92
Carried to summary				20.00	340.00	32.10	55.82	427.92

SUMMARY

	Hours	Hours £	Mat'ls £	O & P £	Total £
PART A **PRELIMINARIES**	0.00	0.00	0.00	0.00	2,420.00
PART B SUBSTRUCTURE TO **DPC LEVEL**	74.58	1,082.38	967.44	307.47	2,357.29
PART C **EXTERNAL WALLS**	70.04	1,190.70	1,787.24	446.69	3,424.63
PART D **FLAT ROOF**	80.22	1,363.77	1,125.30	373.36	2,862.44
PART E **PITCHED ROOF**	0.00	0.00	0.00	0.00	0.00
PART F WINDOWS AND **EXTERNAL DOORS**	23.22	394.74	1,459.81	278.18	2,132.74
PART G INTERNAL **PARTITIONS AND DOORS**	0.00	0.00	0.00	0.00	0.00
PART H **WALL FINISHES**	20.22	343.76	124.44	70.23	538.44
PART J **FLOOR FINISHES**	6.12	104.10	182.54	43.00	329.64
PART K **CEILING FINISHES**	16.33	277.60	80.77	53.76	412.13
PART L **ELECTRICAL WORK**	14.10	267.90	130.40	59.75	458.05
PART M **HEATING WORK**	9.79	176.22	312.04	73.24	561.50
PART N **ALTERATION WORK**	20.00	340.00	32.10	55.82	427.92
Final total	334.62	4,024.46	6,202.08	1,761.50	15,924.78

	Ref	Unit	Qty	Hours	Hours £	Mat'ls £	O & P £	Total £
PART A **PRELIMINARIES**								
Concrete mixer	A1	wks	8.00					400.00
Small tools	A2	wks	9.00					360.00
Scaffolding (m²/weeks)	A3		130					780.00
Skip	A4	wks	5.00					550.00
Clean up	A5	hrs	8.00					104.00
Carried to summary								2,194.00
PART B **SUBSTRUCTURE TO** **DPC LEVEL**								
Excavate topsoil 150mm thick by hand	B1	m²	17.85	5.36	69.62	0.00	10.44	80.06
Excavate to reduce levels	B2	m³	5.35	13.38	173.88	0.00	26.08	199.96
Excavate for trench foundations by hand	B3	m³	1.88	4.89	63.54	0.00	9.53	73.08
Earthwork support to sides of trenches	B4	m²	16.64	6.66	86.53	30.78	17.60	134.91
Backfilling with excavated material	B5	m³	0.63	0.38	4.91	0.00	0.74	5.65
Hardcore 225mm thick	B6	m²	13.69	2.74	35.59	86.11	18.26	139.96
Hardcore filling to trench	B7	m³	0.14	0.09	1.17	6.29	1.12	8.58
Concrete grade (1:3:6) in foundations	B8	m³	1.54	0.86	11.18	131.85	21.46	164.49
Concrete grade (1:2:4) in bed 150mm thick	B9	m²	13.69	4.11	53.39	209.73	39.47	302.59
Concrete (1:2:4) in cavity wall filling	B10	m²	5.70	1.14	11.97	13.57	3.83	29.37
Carried forward				39.59	511.78	478.34	148.52	1,138.63

	Ref	Unit	Qty	Hours	Hours £	Mat'ls £	O & P £	Total £
Brought forward				39.59	511.78	478.34	148.52	1,138.63
Damp-proof membrane	B11	m²	13.13	0.53	6.83	30.46	5.59	42.88
Reinforcement ref A193 in foundation	B12	m²	5.70	0.68	8.89	11.51	3.06	23.47
Steel fabric reinforcement ref A193 in slab	B13	m²	13.69	2.05	26.70	27.38	8.11	62.19
Solid blockwork 140mm thick in cavity wall	B14	m²	7.41	9.63	163.76	123.60	43.10	330.46
Common bricks 112.5mm thick in cavity wall	B15	m²	5.70	9.69	164.73	93.77	38.77	297.27
Facing bricks in 112.5mm thick in skin of cavity wall	B16	m²	1.71	3.08	52.33	28.55	12.13	93.01
Form cavity 50mm wide in cavity wall	B17	m²	7.41	0.22	3.78	11.26	2.26	17.30
DPC 112mm wide	B18	m	11.40	0.57	9.69	11.63	3.20	24.52
DPC 140mm wide	B19	m	11.40	0.68	11.63	15.73	4.10	31.46
Bond in block wall	B20	m	1.30	0.57	9.72	3.78	2.03	15.53
Bond in half brick wall	B21	m	0.30	0.11	1.79	0.80	0.39	2.98
50mm thick insulation board	B22	m²	13.69	4.11	69.82	98.57	25.26	193.65
Carried to summary				71.51	1,041.44	935.38	296.52	2,273.34

**PART C
EXTERNAL WALLS**

	Ref	Unit	Qty	Hours	Hours £	Mat'ls £	O & P £	Total £
Solid blockwork 140mm thick in cavity wall	C1	m²	19.90	25.87	439.79	331.93	115.76	887.48
Facing brickwork 112.5mm thick in cavity wall	C2	m²	19.90	35.82	608.94	568.15	176.56	1,353.65
75mm thick insulation in cavity wall	C3	m²	19.90	4.38	74.43	106.27	27.10	207.80
Carried forward				66.07	1,123.16	1,006.34	319.42	2,448.92

	Ref	Unit	Qty	Hours	Hours £	Mat'ls £	O & P £	Total £
Brought forward				66.07	1,123.16	1,006.34	319.42	2,448.92
Steel lintel 2400mm long	C4	nr	2.00	0.50	8.50	296.58	45.76	350.84
Steel lintel 1500mm long	C5	nr	3.00	0.60	10.20	364.08	56.14	430.42
Steel lintel 1150mm long	C6	nr	1.00	0.20	3.40	98.72	15.32	117.44
Close cavity wall at jambs	C7	m	17.32	0.87	14.72	48.67	9.51	72.90
Close cavity wall at cills	C8	m	5.71	0.29	4.85	16.05	3.13	24.03
Close cavity wall at top	C9	m	11.40	0.57	9.69	32.03	6.26	47.98
DPC 112mm wide at jambs	C10	m	17.32	0.87	14.72	17.67	4.86	37.25
DPC 112mm wide at cills	C11	m	5.71	0.29	4.85	5.82	1.60	12.28
Carried to summary				70.04	1,190.70	1,787.24	446.69	3,424.63

PART D
FLAT ROOF

	Ref	Unit	Qty	Hours	Hours £	Mat'ls £	O & P £	Total £
200 x 50mm sawn softwood joists	D1	m	37.35	9.34	158.74	133.71	43.87	336.32
200 x 50mm sawn softwood sprocket pieces	D2	nr	16.00	2.24	38.08	28.64	10.01	76.73
18mm thick WPB grade decking	D3	m²	17.85	16.07	273.11	204.74	71.68	549.52
50 x 50mm (avg) wide sawn softwood firrings	D4	m	37.35	6.72	114.29	71.34	27.84	213.47
High density polyethylene vapour barrier 150mm thick	D5	m²	16.00	3.20	54.40	179.04	35.02	268.46
100 x 75mm sawn softwood wall plate	D6	m	12.00	3.60	61.20	32.88	14.11	108.19
100 x 75mm sawn softwood tilt fillet	D7	m	4.30	1.08	18.28	11.22	4.42	33.92
Build in ends of 200 x 50mm joists	D8	nr	9.00	3.15	53.55	1.98	8.33	63.86
Carried forward				45.39	771.64	663.55	215.28	1,650.47

	Ref	Unit	Qty	Hours	Hours £	Mat'ls £	O & P £	Total £
Brought forward				45.39	771.64	663.55	215.28	1,650.47
Rake out joint for flashing	D9	m	4.30	1.51	25.59	1.03	3.99	30.61
6mm thick soffit 150mm wide	D10	m	12.30	4.92	83.64	25.95	16.44	126.03
19mm wrought softwood fascia 200mm high	D11	m	12.30	6.15	104.55	33.33	20.68	158.57
Three layer fibre-based roofing felt	D12	m²	17.85	9.82	166.90	232.76	59.95	459.61
Felt turn-down 100mm girth	D13	m	12.30	1.23	20.91	17.22	5.72	43.85
Felt flashing 150mm girth	D14	m	4.30	0.43	7.31	7.40	2.21	16.91
112mm dia. PVC-U gutter	D15	m	4.30	1.12	19.01	13.93	4.94	37.88
Stop end	D16	nr	1.00	0.14	2.38	2.12	0.68	5.18
Stop end outlet	D17	nr	1.00	0.25	4.25	4.19	1.27	9.71
68mm diameter PVC-U down pipe	D18	m	2.50	0.63	10.63	14.25	3.73	28.61
Shoe	D19	nr	1.00	0.30	5.10	3.15	1.24	9.49
Paint fascia and soffit	D20	m²	3.69	0.74	12.55	5.90	2.77	21.22
Carried to summary				72.61	1,234.44	1,024.80	338.89	2,598.12
PART E PITCHED ROOF				N/A	N/A	N/A	N/A	N/A
PART F WINDOWS AND EXTERNAL DOORS								
PVC-U door size 840 x 1980mm complete (B)	F1	0.00	1.00	2.00	34.00	278.67	46.90	359.57
PVC-U sliding patio door size 1700 x 2075mm (C)	F2	nr	2.00	14.00	238.00	616.94	128.24	983.18
Carried forward				16.00	272.00	895.61	175.14	1,342.75

	Ref	Unit	Qty	Hours	Hours £	Mat'ls £	O & P £	Total £
Brought forward				16.00	272.00	895.61	175.14	1,342.75
PVC-U window size 1200 x 1200mm complete (A)	F3	nr	3.00	6.00	102.00	546.42	97.26	745.68
25 x 225mm wrought softwood window board	F4	m	3.60	1.08	18.36	14.47	4.92	37.76
Paint window board	F5	m	3.60	0.14	2.38	3.31	0.85	6.55
Carried to summary				23.22	394.74	1,459.81	278.18	2,132.74
PART G INTERNAL PARTITIONS AND DOORS				N/A	N/A	N/A	N/A	N/A
PART H WALL FINISHES								
19 x 100mm wrought softwood skirting	H1	m	8.70	1.48	25.14	21.84	7.05	54.03
12mm plasterboard fixed to walls with dabs	H2	m²	18.00	7.02	119.34	51.30	25.60	196.24
12mm plasterboard fixed to walls less than 300mm wide	H3	m	23.03	4.15	70.47	21.19	13.75	105.41
Two coats emulsion paint to walls	H4	m²	21.85	5.68	96.58	20.98	17.63	135.19
Paint skirting	H5	m	8.70	1.74	29.58	8.00	5.64	43.22
Carried to summary				20.07	341.11	123.30	69.66	534.08
PART J FLOOR FINISHES								
Cement and sand floor screed 40mm thick	J1	m²	12.58	3.15	53.47	59.88	17.00	130.35
Vinyl floor tiles, size 300 x 300mm	J2	m²	12.58	2.14	36.36	97.62	20.10	154.07
Carried to summary				5.28	89.82	157.50	37.10	284.42

	Ref	Unit	Qty	Hours	Hours £	Mat'ls £	O & P £	Total £
PART K								
CEILING FINISHES								
Plasterboard with taped butt joints fixed to joists	K1	m²	12.58	4.53	76.99	35.85	16.93	129.77
5mm skim coat to plasterboard ceilings	K2	m²	12.58	6.29	106.93	21.76	19.30	148.00
Two coats emulsion paint to ceilings	K3	m²	12.58	3.27	55.60	12.08	10.15	77.83
Carried to summary				14.09	239.52	69.69	46.38	355.60
PART L								
ELECTRICAL WORK								
13 amp double switched socket outlet with neon	L1	nr	3.00	2.40	45.60	26.43	10.80	82.83
Lighting point	L2	nr	2.00	1.40	26.60	14.68	6.19	47.47
Lighting switch	L3	nr	3.00	2.10	39.90	13.26	7.97	61.13
Lighting wiring	L4	m	8.00	1.60	30.40	14.88	6.79	52.07
Power cable	L5	m	18.00	5.40	102.60	48.78	22.71	174.09
Carried to summary				12.90	245.10	118.03	54.46	417.60
PART M								
HEATING WORK								
15mm copper pipe	M1	m	9.00	3.96	71.28	19.35	13.59	104.22
Elbow	M2	nr	4.00	2.24	40.32	9.76	7.51	57.59
Tee	M3	nr	1.00	0.68	12.24	2.44	2.20	16.88
Radiator, double convector size 1400 x 520mm	M4	nr	2.00	2.60	46.80	278.22	48.75	373.77
Break into existing pipe and insert tee	M5	nr	1.00	0.75	13.50	4.42	2.69	20.61
Carried to summary				10.23	184.14	314.19	74.75	573.08

	Ref	Unit	Qty	Hours	Hours £	Mat'ls £	O & P £	Total £
PART N								
ALTERATION WORK								
Take out existing window size 1500 x 1000mm and lintel over, adapt opening to receive 1770 x 2000mm patio door and insert new lintel over (both measured separately) and make good	N1	nr	1.00	20.00	340.00	32.10	55.82	427.92
Carried to summary				20.00	340.00	32.10	55.82	427.92

SUMMARY

	Hours	Hours £	Mat'ls £	O & P £	Total £
PART A PRELIMINARIES	0.00	0.00	0.00	0.00	2,194.00
PART B SUBSTRUCTURE TO DPC LEVEL	71.51	1,041.44	935.38	296.52	2,273.34
PART C EXTERNAL WALLS	70.04	1,190.70	1,787.24	446.69	3,424.63
PART D FLAT ROOF	72.61	1,234.44	1,024.80	338.89	2,598.12
PART E PITCHED ROOF	0.00	0.00	0.00	0.00	0.00
PART F WINDOWS AND EXTERNAL DOORS	23.22	394.74	1,459.81	278.18	2,132.74
PART G INTERNAL PARTITIONS AND DOORS	0.00	0.00	0.00	0.00	0.00
PART H WALL FINISHES	20.07	341.11	123.30	69.66	534.08
PART J FLOOR FINISHES	5.28	89.82	157.50	37.10	284.42
PART K CEILING FINISHES	14.09	239.52	69.69	46.38	355.60
PART L ELECTRICAL WORK	12.90	245.10	118.03	54.46	417.60
PART M HEATING WORK	10.23	184.14	314.19	74.75	573.08
PART N ALTERATION WORK	20.00	340.00	32.10	55.82	427.92
Final total	319.95	4,024.46	6,022.04	1,698.45	15,215.53

	Ref	Unit	Qty	Hours	Hours £	Mat'ls £	O & P £	Total £
PART A **PRELIMINARIES**								
Concrete mixer	A1	wks	9.00					450.00
Small tools	A2	wks	10.00					400.00
Scaffolding (m²/weeks)	A3		140					840.00
Skip	A4	wks	6.00					660.00
Clean up	A5	hrs	10.00					130.00
Carried to summary								2,480.00
PART B **SUBSTRUCTURE TO** **DPC LEVEL**								
Excavate topsoil 150mm thick by hand	B1	m²	22.01	6.60	85.84	0.00	12.88	98.71
Excavate to reduce levels	B2	m³	6.60	16.50	214.50	0.00	32.18	246.68
Excavate for trench foundations by hand	B3	m³	2.05	5.33	69.29	0.00	10.39	79.68
Earthwork support to sides of trenches	B4	m²	18.10	7.24	94.12	33.49	19.14	146.75
Backfilling with excavated material	B5	m³	0.70	0.42	5.46	0.00	0.82	6.28
Hardcore 225mm thick	B6	m²	16.28	3.26	42.33	102.40	21.71	166.44
Hardcore filling to trench	B7	m³	0.15	0.09	1.17	6.29	1.12	8.58
Concrete grade (1:3:6) in foundations	B8	m³	1.67	0.86	11.18	142.99	23.12	177.29
Concrete grade (1:2:4) in bed 150mm thick	B9	m²	16.28	4.88	63.49	249.41	46.94	359.84
Concrete (1:2:4) in cavity wall filling	B10	m²	6.20	1.24	11.97	14.76	4.01	30.73
Carried forward				46.42	599.35	549.33	172.30	1,320.98

	Ref	Unit	Qty	Hours	Hours £	Mat'ls £	O & P £	Total £
Brought forward				46.42	599.35	549.33	172.30	1,320.98
Damp-proof membrane	B11	m²	16.88	0.68	8.78	39.16	7.19	55.13
Reinforcement ref A193 in foundation	B12	m²	6.20	0.74	9.67	12.52	3.33	25.53
Steel fabric reinforcement ref A193 in slab	B13	m²	16.28	2.44	31.75	32.56	9.65	73.95
Solid blockwork 140mm thick in cavity wall	B14	m²	8.06	10.48	178.13	134.44	46.89	359.45
Common bricks 112.5mm thick in skin of cavity wall	B15	m²	6.20	10.54	179.18	101.99	42.18	323.35
Facing bricks in 112.5mm thick in skin of cavity wall	B16	m²	1.86	3.35	56.92	28.55	12.82	98.29
Form cavity 50mm wide in cavity wall	B17	m²	8.06	0.24	4.11	12.25	2.45	18.82
DPC 112mm wide	B18	m	12.40	0.62	10.54	12.65	3.48	26.67
DPC 140mm wide	B19	m	12.40	0.74	12.65	17.11	4.46	34.22
Bond in block wall	B20	m	1.30	0.57	9.72	3.78	2.03	15.53
Bond in half brick wall	B21	m	0.30	0.11	1.79	0.80	0.39	2.98
50mm thick insulation board	B22	m²	16.28	4.88	83.03	117.22	30.04	230.28
Carried to summary				81.82	1,185.60	1,062.37	337.20	2,585.17

PART C
EXTERNAL WALLS

	Ref	Unit	Qty	Hours	Hours £	Mat'ls £	O & P £	Total £
Solid blockwork 140mm thick in cavity wall	C1	m²	20.96	27.25	463.22	349.61	121.92	934.75
Facing brickwork 112.5mm thick in cavity wall	C2	m²	20.96	37.73	641.38	598.41	185.97	1,425.75
75mm thick insulation in cavity wall	C3	m²	20.96	4.61	78.39	111.93	28.55	218.86
Carried forward				69.59	1,182.98	1,059.95	336.44	2,579.37

	Ref	Unit	Qty	Hours	Hours £	Mat'ls £	O & P £	Total £
Brought forward				69.59	1,182.98	1,059.95	336.44	2,579.37
Steel lintel 2400mm long	C4	nr	2.00	0.50	8.50	296.58	45.76	350.84
Steel lintel 1500mm long	C5	nr	3.00	0.60	10.20	364.08	56.14	430.42
Steel lintel 1150mm long	C6	nr	1.00	0.20	3.40	98.72	15.32	117.44
Close cavity wall at jambs	C7	m	19.72	0.99	16.76	55.41	10.83	83.00
Close cavity wall at cills	C8	m	6.91	0.35	5.87	19.42	3.79	29.08
Close cavity wall at top	C9	m	12.40	0.62	10.54	34.84	6.81	52.19
DPC 112mm wide at jambs	C10	m	19.72	0.99	16.76	20.11	5.53	42.41
DPC 112mm wide at cills	C11	m	6.91	0.35	5.87	7.05	1.94	14.86
Carried to summary				73.97	1,257.49	1,857.44	467.24	3,582.18

PART D
FLAT ROOF

	Ref	Unit	Qty	Hours	Hours £	Mat'ls £	O & P £	Total £
200 x 50mm sawn softwood joists	D1	m	45.65	11.41	194.01	163.43	53.62	411.06
200 x 50mm sawn softwood sprocket pieces	D2	nr	16.00	2.24	38.08	28.64	10.01	76.73
18mm thick WPB grade decking	D3	m²	22.00	19.80	336.60	252.34	88.34	677.28
50 x 50mm (avg) wide sawn softwood firrings	D4	m	45.65	8.22	139.69	87.19	34.03	260.91
High density polyethylene vapour barrier 150mm thick	D5	m²	20.00	4.00	68.00	223.80	43.77	335.57
100 x 75mm sawn softwood wall plate	D6	m	13.00	3.90	66.30	35.62	15.29	117.21
100 x 75mm sawn softwood tilt fillet	D7	m	5.30	1.33	22.53	13.83	5.45	41.81
Build in ends of 200 x 50mm joists	D8	nr	11.00	3.85	65.45	2.42	10.18	78.05
Carried forward				54.74	930.66	807.27	260.69	1,998.62

	Ref	Unit	Qty	Hours	Hours £	Mat'ls £	O & P £	Total £
Brought forward				54.74	930.66	807.27	260.69	1,998.62
Rake out joint for flashing	D9	m	5.30	1.86	31.54	1.27	4.92	37.73
6mm thick soffit 150mm wide	D10	m	13.30	5.32	90.44	28.06	17.78	136.28
19mm wrought softwood fascia 200mm high	D11	m	13.30	6.65	113.05	36.04	22.36	171.46
Three layer fibre-based roofing felt	D12	m²	22.00	12.10	205.70	286.88	73.89	566.47
Felt turn-down 100mm girth	D13	m	13.30	1.33	22.61	18.62	6.18	47.41
Felt flashing 150mm girth	D14	m	5.30	0.53	9.01	9.12	2.72	20.84
112mm dia. PVC-U gutter	D15	m	5.30	1.38	23.43	17.17	6.09	46.69
Stop end	D16	nr	1.00	0.14	2.38	2.12	0.68	5.18
Stop end outlet	D17	nr	1.00	0.25	4.25	4.19	1.27	9.71
68mm diameter PVC-U down pipe	D18	m	2.50	0.63	10.63	14.25	3.73	28.61
Shoe	D19	nr	1.00	0.30	5.10	3.15	1.24	9.49
Paint fascia and soffit	D20	m²	3.99	0.80	13.57	6.38	2.99	22.94
Carried to summary				86.02	1,462.35	1,234.53	404.53	3,101.41
PART E **PITCHED ROOF**				N/A	N/A	N/A	N/A	N/A
PART F **WINDOWS AND** **EXTERNAL DOORS**								
PVC-U door size 840 x 1980mm complete (B)	F1	0.00	1.00	2.00	34.00	278.67	46.90	359.57
PVC-U sliding patio door size 1700 x 2075mm (C)	F2	nr	2.00	14.00	238.00	616.94	128.24	983.18
Carried forward				16.00	272.00	895.61	175.14	1,342.75

	Ref	Unit	Qty	Hours	Hours £	Mat'ls £	O & P £	Total £
Brought forward				16.00	272.00	895.61	175.14	1,342.75
PVC-U window size 1200 x 1200mm complete (A)	F3	nr	4.00	8.00	136.00	728.56	129.68	994.24
25 x 225mm wrought softwood window board	F4	m	4.80	1.44	24.48	19.30	6.57	50.34
Paint window board	F5	m	4.80	0.14	2.38	4.42	1.02	7.82
Carried to summary				25.58	434.86	1,647.88	312.41	2,395.15
PART G INTERNAL PARTITIONS AND DOORS				N/A	N/A	N/A	N/A	N/A
PART H WALL FINISHES								
19 x 100mm wrought softwood skirting	H1	m	9.70	1.65	28.03	24.35	7.86	60.24
12mm plasterboard fixed to walls with dabs	H2	m²	19.60	7.64	129.95	55.86	27.87	213.68
12mm plasterboard fixed to walls less than 300mm wide	H3	m	26.63	4.79	81.49	24.50	15.90	121.89
Two coats emulsion paint to walls	H4	m²	23.46	6.10	103.69	22.52	18.93	145.15
Paint skirting	H5	m	9.70	1.94	32.98	8.92	6.29	48.19
Carried to summary				22.13	376.14	136.15	76.84	589.14
PART J FLOOR FINISHES								
Cement and sand floor screed 40mm thick	J1	m²	16.28	4.07	69.19	77.49	22.00	168.69
Vinyl floor tiles, size 300 x 300mm	J2	m²	16.28	2.77	47.05	126.33	26.01	199.39
Carried to summary				6.84	116.24	203.83	48.01	368.07

	Ref	Unit	Qty	Hours	Hours £	Mat'ls £	O & P £	Total £
PART K								
CEILING FINISHES								
Plasterboard with taped butt joints fixed to joists	K1	m²	16.28	5.86	99.63	46.40	21.90	167.94
5mm skim coat to plasterboard ceilings	K2	m²	16.28	8.14	138.38	28.16	24.98	191.53
Two coats emulsion paint to ceilings	K3	m²	16.28	4.23	71.96	15.63	13.14	100.72
Carried to summary				18.23	309.97	90.19	60.02	460.19
PART L								
ELECTRICAL WORK								
13 amp double switched socket outlet with neon	L1	nr	4.00	3.20	60.80	35.24	14.41	110.45
Lighting point	L2	nr	3.00	2.10	39.90	22.02	9.29	71.21
Lighting switch	L3	nr	3.00	2.10	39.90	13.26	7.97	61.13
Lighting wiring	L4	m	9.00	1.80	34.20	16.74	7.64	58.58
Power cable	L5	m	20.00	6.00	114.00	54.20	25.23	193.43
Carried to summary				15.20	288.80	141.46	64.54	494.80
PART M								
HEATING WORK								
15mm copper pipe	M1	m	10.00	4.40	79.20	21.50	15.11	115.81
Elbow	M2	nr	4.00	2.24	40.32	9.76	7.51	57.59
Tee	M3	nr	1.00	0.68	12.24	2.44	2.20	16.88
Radiator, double convector size 1400 x 520mm	M4	nr	3.00	3.90	70.20	417.33	73.13	560.66
Break into existing pipe and insert tee	M5	nr	1.00	0.75	13.50	4.42	2.69	20.61
Carried to summary				11.97	215.46	455.45	100.64	771.55

	Ref	Unit	Qty	Hours	Hours £	Mat'ls £	O & P £	Total £
PART N **ALTERATION WORK**								
Take out existing window size 1500 x 1000mm and lintel over, adapt opening to receive 1770 x 2000mm patio door and insert new lintel over (both measured separately) and make good	N1	nr	1.00	20.00	340.00	32.10	55.82	427.92
Carried to summary				20.00	340.00	32.10	55.82	427.92

SUMMARY

	Hours	Hours £	Mat'ls £	O & P £	Total £
PART A **PRELIMINARIES**	0.00	0.00	0.00	0.00	2,480.00
PART B SUBSTRUCTURE TO **DPC LEVEL**	81.82	1,185.60	1,062.37	337.20	2,585.17
PART C **EXTERNAL WALLS**	73.97	1,257.49	1,857.44	467.24	3,582.17
PART D **FLAT ROOF**	86.02	1,462.35	1,234.53	404.53	3,101.41
PART E **PITCHED ROOF**	0.00	0.00	0.00	0.00	0.00
PART F WINDOWS AND **EXTERNAL DOORS**	25.58	434.86	1,647.88	312.41	2,395.15
PART G INTERNAL **PARTITIONS AND DOORS**	0.00	0.00	0.00	0.00	0.00
PART H **WALL FINISHES**	22.13	376.14	136.15	76.84	589.14
PART J **FLOOR FINISHES**	6.84	116.24	203.83	48.01	368.07
PART K **CEILING FINISHES**	18.23	309.97	90.19	60.02	460.19
PART L **ELECTRICAL WORK**	15.20	288.80	141.46	64.54	494.80
PART M **HEATING WORK**	11.97	215.46	455.45	100.64	771.55
PART N **ALTERATION WORK**	20.00	340.00	32.10	55.82	427.92
Final total	361.76	5,986.91	6,861.40	1,927.24	17,255.56

	Ref	Unit	Qty	Hours	Hours £	Mat'ls £	O & P £	Total £

PART A
PRELIMINARIES

	Ref	Unit	Qty	Hours	Hours £	Mat'ls £	O & P £	Total £
Concrete mixer	A1	wks	10.00					500.00
Small tools	A2	wks	11.00					440.00
Scaffolding (m²/weeks)	A3		150					900.00
Skip	A4	wks	7.00					770.00
Clean up	A5	hrs	10.00					130.00
Carried to summary								2,740.00

PART B
SUBSTRUCTURE TO
DPC LEVEL

	Ref	Unit	Qty	Hours	Hours £	Mat'ls £	O & P £	Total £
Excavate topsoil 150mm thick by hand	B1	m²	26.15	7.85	101.99	0.00	15.30	117.28
Excavate to reduce levels	B2	m³	7.85	19.63	255.13	0.00	38.27	293.39
Excavate for trench foundations by hand	B3	m³	2.21	5.75	74.70	0.00	11.20	85.90
Earthwork support to sides of trenches	B4	m²	19.56	7.82	101.71	36.19	20.68	158.58
Backfilling with excavated material	B5	m³	0.77	0.46	6.01	0.00	0.90	6.91
Hardcore 225mm thick	B6	m²	19.98	4.00	51.95	125.67	26.64	204.27
Hardcore filling to trench	B7	m³	0.17	0.09	1.17	6.29	1.12	8.58
Concrete grade (1:3:6) in foundations	B8	m³	1.81	0.86	11.18	154.97	24.92	191.08
Concrete grade (1:2:4) in bed 150mm thick	B9	m²	19.98	5.99	77.92	306.09	57.60	441.62
Concrete (1:2:4) in cavity wall filling	B10	m²	6.70	1.34	11.97	15.95	4.19	32.10
Carried forward				53.78	693.72	645.16	200.83	1,539.71

	Ref	Unit	Qty	Hours	Hours £	Mat'ls £	O & P £	Total £
Brought forward				53.78	693.72	645.16	200.83	1,539.71
Damp-proof membrane	B11	m²	20.63	0.83	10.73	47.86	8.79	67.38
Reinforcement ref A193 in foundation	B12	m²	6.70	0.80	10.45	13.53	3.60	27.58
Steel fabric reinforcement ref A193 in slab	B13	m²	19.98	3.00	38.96	39.96	11.84	90.76
Solid blockwork 140mm thick in cavity wall	B14	m²	8.71	11.32	192.49	145.28	50.67	388.44
Common bricks 112.5mm thick in cavity wall	B15	m²	6.70	11.39	193.63	110.22	45.58	349.42
Facing bricks in 112.5mm thick in skin of cavity wall	B16	m²	2.01	3.62	61.51	28.55	13.51	103.56
Form cavity 50mm wide in cavity wall	B17	m²	8.71	0.26	4.44	13.24	2.65	20.33
DPC 112mm wide	B18	m	13.40	0.67	11.39	13.67	3.76	28.82
DPC 140mm wide	B19	m	13.40	0.80	13.67	18.49	4.82	36.98
Bond in block wall	B20	m	1.30	0.57	9.72	3.78	2.03	15.53
Bond in half brick wall	B21	m	0.30	0.11	1.79	0.80	0.39	2.98
50mm thick insulation board	B22	m²	19.98	5.99	101.90	143.86	36.86	282.62
Carried to summary				93.15	1,344.39	1,224.41	385.32	2,954.12

**PART C
EXTERNAL WALLS**

	Ref	Unit	Qty	Hours	Hours £	Mat'ls £	O & P £	Total £
Solid blockwork 140mm thick in cavity wall	C1	m²	22.46	29.20	496.37	374.63	130.65	1,001.65
Facing brickwork 112.5mm thick in cavity wall	C2	m²	22.46	40.43	687.28	641.23	199.28	1,527.79
75mm thick insulation in cavity wall	C3	m²	22.46	4.94	84.00	119.94	30.59	234.53
Carried forward				74.57	1,267.64	1,135.80	360.52	2,763.96

	Ref	Unit	Qty	Hours	Hours £	Mat'ls £	O & P £	Total £
Brought forward				74.57	1,267.64	1,135.80	360.52	2,763.96
Steel lintel 2400mm long	C4	nr	2.00	0.50	8.50	296.58	45.76	350.84
Steel lintel 1500mm long	C5	nr	4.00	0.80	13.60	485.44	74.86	573.90
Steel lintel 1150mm long	C6	nr	1.00	0.20	3.40	98.72	15.32	117.44
Close cavity wall at jambs	C7	m	19.72	0.99	16.76	55.41	10.83	83.00
Close cavity wall at cills	C8	m	6.91	0.35	5.87	19.42	3.79	29.08
Close cavity wall at top	C9	m	13.40	0.67	11.39	37.65	7.36	56.40
DPC 112mm wide at jambs	C10	m	19.72	0.99	16.76	20.11	5.53	42.41
DPC 112mm wide at cills	C11	m	6.91	0.35	5.87	7.05	1.94	14.86
Carried to summary				79.20	1,346.40	2,057.47	510.58	3,914.45

PART D
FLAT ROOF

	Ref	Unit	Qty	Hours	Hours £	Mat'ls £	O & P £	Total £
200 x 50mm sawn softwood joists	D1	m	53.95	13.49	229.29	193.14	63.36	485.79
200 x 50mm sawn softwood sprocket pieces	D2	nr	16.00	2.24	38.08	28.64	10.01	76.73
18mm thick WPB grade decking	D3	m²	26.15	23.54	400.10	299.94	105.01	805.04
50 x 50mm (avg) wide sawn softwood firrings	D4	m	53.95	9.71	165.09	103.04	40.22	308.35
High density polyethylene vapour barrier 150mm thick	D5	m²	24.00	4.80	81.60	268.56	52.52	402.68
100 x 75mm sawn softwood wall plate	D6	m	14.00	4.20	71.40	38.36	16.46	126.22
100 x 75mm sawn softwood tilt fillet	D7	m	6.30	1.58	26.78	16.44	6.48	49.70
Build in ends of 200 x 50mm joists	D8	nr	13.00	4.55	77.35	2.86	12.03	92.24
Carried forward				64.10	1,089.67	950.99	306.10	2,346.76

104 One storey extension, size 4 x 6m, flat roof

	Ref	Unit	Qty	Hours	Hours £	Mat'ls £	O & P £	Total £
Brought forward				64.10	1,089.67	950.99	306.10	2,346.76
Rake out joint for flashing	D9	m	6.30	2.21	37.49	1.51	5.85	44.85
6mm thick soffit 150mm wide	D10	m	14.30	5.72	97.24	30.17	19.11	146.52
19mm wrought softwood fascia 200mm high	D11	m	14.30	7.15	121.55	38.75	24.05	184.35
Three layer fibre-based roofing felt	D12	m²	26.15	14.38	244.50	341.00	87.82	673.32
Felt turn-down 100mm girth	D13	m	14.30	1.43	24.31	20.02	6.65	50.98
Felt flashing 150mm girth	D14	m	6.30	0.63	10.71	10.84	3.23	24.78
112mm dia. PVC-U gutter	D15	m	6.30	1.64	27.85	20.41	7.24	55.50
Stop end	D16	nr	1.00	0.14	2.38	2.12	0.68	5.18
Stop end outlet	D17	nr	1.00	0.25	4.25	4.19	1.27	9.71
68mm diameter PVC-U down pipe	D18	m	2.50	0.63	10.63	14.25	3.73	28.61
Shoe	D19	nr	1.00	0.30	5.10	3.15	1.24	9.49
Paint fascia and soffit	D20	m²	4.29	0.86	14.59	6.86	3.22	24.67
Carried to summary				99.43	1,690.26	1,444.27	470.18	3,604.70
PART E PITCHED ROOF				N/A	N/A	N/A	N/A	N/A
PART F WINDOWS AND EXTERNAL DOORS								
PVC-U door size 840 x 1980mm complete (B)	F1	0.00	1.00	2.00	34.00	278.67	46.90	359.57
PVC-U sliding patio door size 1700 x 2075mm (C)	F2	nr	2.00	14.00	238.00	616.94	128.24	983.18
Carried forward				16.00	272.00	895.61	175.14	1,342.75

	Ref	Unit	Qty	Hours	Hours £	Mat'ls £	O & P £	Total £
Brought forward				16.00	272.00	895.61	175.14	1,342.75
PVC-U window size 1200 x 1200mm complete (A)	F3	nr	4.00	8.00	136.00	728.56	129.68	994.24
25 x 225mm wrought softwood window board	F4	m	4.80	1.44	24.48	19.30	6.57	50.34
Paint window board	F5	m	4.80	0.14	2.38	4.42	1.02	7.82
Carried to summary				25.58	434.86	1,647.88	312.41	2,395.15
PART G INTERNAL PARTITIONS AND DOORS				N/A	N/A	N/A	N/A	N/A
PART H WALL FINISHES								
19 x 100mm wrought softwood skirting	H1	m	10.07	1.71	29.10	25.28	8.16	62.53
12mm plasterboard fixed to walls with dabs	H2	m²	21.96	8.56	145.59	62.59	31.23	239.41
12mm plasterboard fixed to walls less than 300mm wide	H3	m	26.63	4.79	81.49	24.50	15.90	121.89
Two coats emulsion paint to walls	H4	m²	25.96	6.75	114.74	24.92	20.95	160.61
Paint skirting	H5	m	10.70	2.14	36.38	9.84	6.93	53.16
Carried to summary				23.96	407.31	147.13	83.17	637.60
PART J FLOOR FINISHES								
Cement and sand floor screed 40mm thick	J1	m²	19.88	4.97	84.49	94.63	26.87	205.99
Vinyl floor tiles, size 300 x 300mm	J2	m²	19.88	3.38	57.45	154.27	31.76	243.48
Carried to summary				8.35	141.94	248.90	58.63	449.47

	Ref	Unit	Qty	Hours	Hours £	Mat'ls £	O & P £	Total £
PART K **CEILING FINISHES**								
Plasterboard with taped butt joints fixed to joists	K1	m²	19.88	7.16	121.67	56.66	26.75	205.07
5mm skim coat to plasterboard ceilings	K2	m²	19.88	9.94	168.98	34.39	30.51	233.88
Two coats emulsion paint to ceilings	K3	m²	19.88	5.17	87.87	19.08	16.04	123.00
Carried to summary				22.27	378.52	110.14	73.30	561.95
PART L **ELECTRICAL WORK**								
13 amp double switched socket outlet with neon	L1	nr	5.00	4.00	76.00	44.05	18.01	138.06
Lighting point	L2	nr	3.00	2.10	39.90	22.02	9.29	71.21
Lighting switch	L3	nr	3.00	2.10	39.90	13.26	7.97	61.13
Lighting wiring	L4	m	11.00	2.20	41.80	20.46	9.34	71.60
Power cable	L5	m	22.00	6.60	125.40	59.62	27.75	212.77
Carried to summary				17.00	323.00	159.41	72.36	554.77
PART M **HEATING WORK**								
15mm copper pipe	M1	m	11.00	4.84	87.12	23.65	16.62	127.39
Elbow	M2	nr	4.00	2.24	40.32	9.76	7.51	57.59
Tee	M3	nr	1.00	0.68	12.24	2.44	2.20	16.88
Radiator, double convector size 1400 x 520mm	M4	nr	3.00	3.90	70.20	417.33	73.13	560.66
Break into existing pipe and insert tee	M5	nr	1.00	0.75	13.50	4.42	2.69	20.61
Carried to summary				12.41	223.38	457.60	102.15	783.13

	Ref	Unit	Qty	Hours	Hours £	Mat'ls £	O & P £	Total £
PART N								
ALTERATION WORK								
Take out existing window size 1500 x 1000mm and lintel over, adapt opening to receive 1770 x 2000mm patio door and insert new lintel over (both measured separately) and make good	N1	nr	1.00	20.00	340.00	32.10	55.82	427.92
Carried to summary				20.00	340.00	32.10	55.82	427.92

SUMMARY

	Hours	Hours £	Mat'ls £	O & P £	Total £
PART A PRELIMINARIES	0.00	0.00	0.00	0.00	2,740.00
PART B SUBSTRUCTURE TO DPC LEVEL	93.15	1,344.39	1,224.41	385.32	2,954.12
PART C EXTERNAL WALLS	79.20	1,346.40	2,057.47	510.58	3,914.45
PART D FLAT ROOF	99.43	1,690.26	1,444.27	470.18	3,604.70
PART E PITCHED ROOF	0.00	0.00	0.00	0.00	0.00
PART F WINDOWS AND EXTERNAL DOORS	25.58	434.86	1,647.88	312.41	2,395.15
PART G INTERNAL PARTITIONS AND DOORS	0.00	0.00	0.00	0.00	0.00
PART H WALL FINISHES	23.96	407.31	147.13	83.17	637.60
PART J FLOOR FINISHES	8.35	141.94	248.90	58.63	449.47
PART K CEILING FINISHES	22.27	378.52	110.14	73.30	561.95
PART L ELECTRICAL WORK	17.00	323.00	159.41	72.36	554.77
PART M HEATING WORK	12.41	223.38	457.60	102.15	783.13
PART N ALTERATION WORK	20.00	340.00	32.10	55.82	427.92
Final total	293.57	4,024.46	7,529.31	2,123.92	19,023.26

	Ref	Unit	Qty	Hours	Hours £	Mat'ls £	O & P £	Total £
PART A								
PRELIMINARIES								
Concrete mixer	A1	wks	4.00					200.00
Small tools	A2	wks	6.00					240.00
Scaffolding (m²/weeks)	A3		70					420.00
Skip	A4	wks	2.00					220.00
Clean up	A5	hrs	4.00					52.00
Carried to summary								1,132.00
PART B								
SUBSTRUCTURE TO								
DPC LEVEL								
Excavate topsoil 150mm thick by hand	B1	m²	7.10	2.13	27.69	0.00	4.15	31.84
Excavate to reduce levels	B2	m³	2.13	5.33	69.23	0.00	10.38	79.61
Excavate for trench foundations by hand	B3	m³	1.06	2.76	35.83	0.00	5.37	41.20
Earthwork support to sides of trenches	B4	m²	9.34	3.74	48.57	17.28	9.88	75.72
Backfilling with excavated material	B5	m³	0.40	0.24	3.12	0.00	0.47	3.59
Hardcore 225mm thick	B6	m²	4.08	0.82	10.61	25.66	5.44	41.71
Hardcore filling to trench	B7	m³	0.08	0.04	0.52	6.29	1.02	7.83
Concrete grade (1:3:6) in foundations	B8	m³	0.86	1.16	15.09	73.63	13.31	102.04
Concrete grade (1:2:4) in bed 150mm thick	B9	m²	4.08	1.22	15.91	62.51	11.76	90.18
Concrete (1:2:4) in cavity wall filling	B10	m²	3.20	0.64	11.97	7.62	2.94	22.52
Carried forward				18.07	238.53	192.99	64.73	496.25

	Ref	Unit	Qty	Hours	Hours £	Mat'ls £	O & P £	Total £
Brought forward				18.07	238.53	192.99	64.73	496.25
Damp-proof membrane	B11	m²	4.38	0.18	2.28	10.16	1.87	14.31
Reinforcement ref A193 in foundation	B12	m²	3.22	0.39	5.02	6.50	1.73	13.26
Steel fabric reinforcement ref A193 in slab	B13	m²	4.08	0.61	7.96	8.16	2.42	18.53
Solid blockwork 140mm thick in cavity wall	B14	m²	4.16	5.41	91.94	69.39	24.20	185.52
Common bricks 112.5mm thick in cavity wall	B15	m²	3.20	5.44	92.48	52.64	21.77	166.89
Facing bricks in 112.5mm thick in skin of cavity wall	B16	m²	0.96	1.73	29.38	28.55	8.69	66.61
Form cavity 50mm wide in cavity wall	B17	m²	4.16	0.12	2.12	6.32	1.27	9.71
DPC 112mm wide	B18	m	6.40	0.32	5.44	6.53	1.80	13.76
DPC 140mm wide	B19	m	6.40	0.38	6.53	8.83	2.30	17.66
Bond in block wall	B20	m	1.30	0.57	9.72	3.78	2.03	15.53
Bond in half brick wall	B21	m	0.30	0.11	1.79	0.80	0.39	2.98
50mm thick insulation board	B22	m²	4.08	1.22	20.81	29.38	7.53	57.71
Carried to summary				34.55	513.99	424.04	140.70	1,078.73

**PART C
EXTERNAL WALLS**

	Ref	Unit	Qty	Hours	Hours £	Mat'ls £	O & P £	Total £
Solid blockwork 140mm thick in cavity wall	C1	m²	13.50	17.55	298.35	225.18	78.53	602.06
Facing brickwork 112.5mm thick in cavity wall	C2	m²	13.50	24.30	413.10	385.43	119.78	918.30
75mm thick insulation in cavity wall	C3	m²	13.50	2.97	50.49	72.09	18.39	140.97
Carried forward				44.82	761.94	682.70	216.70	1,661.33

	Ref	Unit	Qty	Hours	Hours £	Mat'ls £	O & P £	Total £
Brought forward				44.82	761.94	682.70	216.70	1,661.33
Steel lintel 2400mm long	C4	nr	2.00	0.50	8.50	296.58	45.76	350.84
Steel lintel 1500mm long	C5	nr	2.00	0.40	6.80	242.72	37.43	286.95
Close cavity wall at jambs	C7	m	8.96	0.45	7.62	25.18	4.92	37.71
Close cavity wall at cills	C8	m	3.67	0.18	3.12	10.31	2.01	15.45
Close cavity wall at top	C9	m	6.40	0.32	5.44	17.98	3.51	26.94
DPC 112mm wide at jambs	C10	m	8.96	0.45	7.62	9.14	2.51	19.27
DPC 112mm wide at cills	C11	m	3.67	0.18	3.12	3.74	1.03	7.89
Carried to summary				47.30	804.15	1,288.35	313.88	2,406.38
PART D **FLAT ROOF**				N/A	N/A	N/A	N/A	N/A
PART E **PITCHED ROOF**								
100 x 75mm sawn softwood wall plate	E1	m	3.00	0.90	15.30	7.83	3.47	26.60
200 x 50mm sawn softwood pole plate	E2	nr	3.30	0.99	16.83	10.96	4.17	31.95
100 x 50mm sawn softwood rafters	E3	m	17.50	3.50	59.50	43.40	15.44	118.34
100 x 50mm sawn softwood purlin	E4	m	3.30	0.66	11.22	9.57	3.12	23.91
150 x 50mm softwood joists	E5	m	12.50	2.50	42.50	39.00	12.23	93.73
150 x 50mm sawn softwood sprockets	E6	nr	14.00	1.68	28.56	21.84	7.56	57.96
100mm layer of insulation quilt laid over and between joists	E7	m²	6.00	2.88	48.96	69.60	17.78	136.34
Carried forward				13.11	222.87	202.20	63.76	488.83

	Ref	Unit	Qty	Hours	Hours £	Mat'ls £	O & P £	Total £
Brought forward				13.11	222.87	202.20	63.76	488.83
6mm softwood soffit 150mm wide	E8	m	8.30	3.32	56.44	17.51	11.09	85.05
19mm wrought softwood fascia/ barge board 200mm high	E9	m	3.30	1.65	28.05	8.94	5.55	42.54
Marley Plain roof tiles on felt and battens	E10	m²	11.55	21.95	373.07	418.57	118.75	910.38
Double eaves course	E11	m	3.30	1.16	19.64	13.89	5.03	38.56
Verge with tile undercloak	E12	m	7.00	1.75	29.75	46.27	11.40	87.42
Lead flashing code 5, 200mm girth	E13	m	3.30	1.98	33.66	28.45	9.32	71.42
Rake out joint for flashing	E14	m	3.30	1.16	19.64	0.79	3.06	23.49
112mm diameter PVC-U gutter	E15	m	3.30	0.86	14.59	10.69	3.79	29.07
Stop end	E16	nr	1.00	0.14	2.38	2.12	0.68	5.18
Stop end outlet	E17	nr	1.00	0.25	4.25	4.19	1.27	9.71
68mm diameter PVC-U down pipe	E18	m	2.50	0.63	10.63	14.25	3.73	28.61
Shoe	E19	nr	1.00	0.30	5.10	3.15	1.24	9.49
Paint fascia and soffit	E20	m²	2.59	0.52	8.81	4.14	1.94	14.89
Carried to summary				48.76	828.85	775.17	240.60	1,844.63

**PART F
WINDOWS AND
EXTERNAL DOORS**

	Ref	Unit	Qty	Hours	Hours £	Mat'ls £	O & P £	Total £
PVC-U door size 840 x 1980mm complete (B)	F1	0.00	0.00	0.00	0.00	0.00	0.00	0.00
PVC-U sliding patio door size 1700 x 2075mm (C)	F2	nr	2.00	14.00	238.00	616.94	128.24	983.18
Carried forward				14.00	238.00	616.94	128.24	983.18

	Ref	Unit	Qty	Hours	Hours £	Mat'ls £	O & P £	Total £
Brought forward				14.00	238.00	616.94	128.24	983.18
PVC-U window size 1200 x 1200mm complete (A)	F3	nr	2.00	4.00	68.00	364.28	64.84	497.12
25 x 225mm wrought softwood window board	F4	m	2.40	0.72	12.24	9.65	3.28	25.17
Paint window board	F5	m	2.40	0.14	2.38	2.21	0.69	5.28
Carried to summary				18.86	320.62	993.08	197.05	1,510.75
PART G INTERNAL PARTITIONS AND DOORS			N/A	N/A	N/A	N/A	N/A	N/A
PART H WALL FINISHES								
19 x 100mm wrought softwood skirting	H1	m	4.54	0.77	13.12	11.40	3.68	28.19
12mm plasterboard fixed to walls with dabs	H2	m²	9.00	3.51	59.67	25.65	12.80	98.12
12mm plasterboard fixed to walls ess than 300mm wide	H3	m	12.63	2.27	38.65	11.62	7.54	57.81
Two coats emulsion paint to walls	H4	m²	10.90	2.83	48.18	10.46	8.80	67.44
Paint skirting	H5	m	4.54	0.91	15.44	4.18	2.94	22.55
Carried to summary				10.30	175.05	63.31	35.75	274.11
PART J FLOOR FINISHES								
Cement and sand floor screed 40mm thick	J1	m²	4.08	1.02	17.34	19.42	5.51	42.27
Vinyl floor tiles, size 300 x 300mm	J2	m²	4.08	0.69	11.79	31.66	6.52	49.97
Carried to summary				1.71	29.13	51.08	12.03	92.24

	Ref	Unit	Qty	Hours	Hours £	Mat'ls £	O & P £	Total £
PART K **CEILING FINISHES**								
Plasterboard with taped butt joints fixed to joists	K1	m²	4.08	1.47	24.97	11.63	5.49	42.09
5mm skim coat to plasterboard ceilings	K2	m²	4.08	2.04	34.68	7.06	6.26	48.00
Two coats emulsion paint to ceilings	K3	m²	4.08	1.06	18.03	3.92	3.29	25.24
Carried to summary				4.57	77.68	22.60	15.04	115.33
PART L **ELECTRICAL WORK**								
13 amp double switched socket outlet with neon	L1	nr	2.00	1.60	30.40	17.62	7.20	55.22
Lighting point	L2	nr	2.00	1.40	26.60	14.68	6.19	47.47
Lighting switch	L3	nr	2.00	1.40	26.60	8.84	5.32	40.76
Lighting wiring	L4	m	5.00	1.00	19.00	9.30	4.25	32.55
Power cable	L5	m	12.00	3.60	68.40	32.52	15.14	116.06
Carried to summary				9.00	171.00	82.96	38.09	292.05
PART M **HEATING WORK**								
15mm copper pipe	M1	m	4.00	1.76	31.68	8.60	6.04	46.32
Elbow	M2	nr	4.00	2.24	40.32	9.76	7.51	57.59
Tee	M3	nr	1.00	0.56	10.08	2.44	1.88	14.40
Radiator, double convector size 1400 x 520mm	M4	nr	1.00	1.30	23.40	139.11	24.38	186.89
Break into existing pipe and insert tee	M5	nr	1.00	0.75	13.50	4.42	2.69	20.61
Carried to summary				6.61	118.98	164.33	42.50	325.81

	Ref	Unit	Qty	Hours	Hours £	Mat'ls £	O & P £	Total £
PART N								
ALTERATION WORK								
Take out existing window size 1500 x 1000mm and lintel over, adapt opening to receive 1770 x 2000mm patio door and insert new lintel over (both measured separately) and make good	N1	nr	1.00	20.00	340.00	32.10	55.82	427.92
Carried to summary				20.00	340.00	32.10	55.82	427.92

SUMMARY

	Hours	Hours £	Mat'ls £	O & P £	Total £
PART A **PRELIMINARIES**	0.00	0.00	0.00	0.00	1,132.00
PART B SUBSTRUCTURE TO **DPC LEVEL**	34.55	513.99	424.04	140.70	1,078.73
PART C **EXTERNAL WALLS**	47.30	804.15	1,288.35	313.88	2,406.38
PART D **FLAT ROOF**	0.00	0.00	0.00	0.00	0.00
PART E **PITCHED ROOF**	48.76	828.85	775.17	240.60	1,844.63
PART F WINDOWS AND **EXTERNAL DOORS**	18.86	320.62	993.08	197.05	1,510.75
PART G INTERNAL **PARTITIONS AND DOORS**	0.00	0.00	0.00	0.00	0.00
PART H **WALL FINISHES**	10.30	175.05	63.31	35.75	274.11
PART J **FLOOR FINISHES**	1.71	29.13	51.08	12.03	92.24
PART K **CEILING FINISHES**	4.57	77.68	22.60	15.04	115.32
PART L **ELECTRICAL WORK**	9.00	171.00	82.96	38.09	292.05
PART M **HEATING WORK**	6.61	118.98	164.33	42.50	325.81
PART N **ALTERATION WORK**	20.00	340.00	32.10	55.82	427.92
Final total	201.66	3,379.45	3,897.02	1,091.47	9,499.95

	Ref	Unit	Qty	Hours	Hours £	Mat'ls £	O & P £	Total £
PART A **PRELIMINARIES**								
Concrete mixer	A1	wks	5.00					250.00
Small tools	A2	wks	7.00					280.00
Scaffolding (m²/weeks)	A3		80					480.00
Skip	A4	wks	3.00					330.00
Clean up	A5	hrs	6.00					78.00
Carried to summary								1,418.00
PART B **SUBSTRUCTURE TO** **DPC LEVEL**								
Excavate topsoil 150mm thick by hand	B1	m²	9.25	2.78	36.08	0.00	5.41	41.49
Excavate to reduce levels	B2	m³	2.77	6.93	90.03	0.00	13.50	103.53
Excavate for trench foundations by hand	B3	m³	1.22	3.17	41.24	0.00	6.19	47.42
Earthwork support to sides of trenches	B4	m²	10.80	4.32	56.16	19.98	11.42	87.56
Backfilling with excavated material	B5	m³	0.47	0.28	3.67	0.00	0.55	4.22
Hardcore 225mm thick	B6	m²	5.78	1.16	15.03	36.36	7.71	59.09
Hardcore filling to trench	B7	m³	0.09	0.05	0.59	6.29	1.03	7.91
Concrete grade (1:3:6) in foundations	B8	m³	1.00	1.35	17.55	85.62	15.48	118.65
Concrete grade (1:2:4) in bed 150mm thick	B9	m²	5.78	1.73	22.54	88.55	16.66	127.76
Concrete (1:2:4) in cavity wall filling	B10	m²	3.70	0.74	11.97	8.81	3.12	23.89
Carried forward				22.50	294.84	245.60	81.07	621.50

	Ref	Unit	Qty	Hours	Hours £	Mat'ls £	O & P £	Total £
Brought forward				22.50	294.84	245.60	81.07	621.50
Damp-proof membrane	B11	m²	6.13	0.25	3.19	14.22	2.61	20.02
Reinforcement ref A193 in foundation	B12	m²	3.70	0.44	5.77	7.47	1.99	15.23
Steel fabric reinforcement ref A193 in slab	B13	m²	5.78	0.87	11.27	11.56	3.42	26.26
Solid blockwork 140mm thick in cavity wall	B14	m²	4.81	6.25	106.30	80.23	27.98	214.51
Common bricks 112.5mm thick in cavity wall	B15	m²	3.70	6.29	106.93	60.87	25.17	192.96
Facing bricks in 112.5mm thick in skin of cavity wall	B16	m²	1.11	2.00	33.97	28.55	9.38	71.89
Form cavity 50mm wide in cavity wall	B17	m²	4.81	0.14	2.45	7.31	1.46	11.23
DPC 112mm wide	B18	m	7.40	0.37	6.29	7.55	2.08	15.91
DPC 140mm wide	B19	m	7.40	0.44	7.55	10.21	2.66	20.42
Bond in block wall	B20	m	1.30	0.57	9.72	3.78	2.03	15.53
Bond in half brick wall	B21	m	0.30	0.11	1.79	0.80	0.39	2.98
50mm thick insulation board	B22	m²	5.78	1.73	29.48	41.62	10.66	81.76
Carried to summary				41.97	619.54	519.78	170.90	1,310.22

PART C
EXTERNAL WALLS

	Ref	Unit	Qty	Hours	Hours £	Mat'ls £	O & P £	Total £
Solid blockwork 140mm thick in cavity wall	C1	m²	16.00	20.80	353.60	266.88	93.07	713.55
Facing brickwork 112.5mm thick in cavity wall	C2	m²	16.00	28.80	489.60	456.80	141.96	1,088.36
75mm thick insulation in cavity wall	C3	m²	16.00	3.52	59.84	85.44	21.79	167.07
Carried forward				53.12	903.04	809.12	256.82	1,968.98

	Ref	Unit	Qty	Hours	Hours £	Mat'ls £	O & P £	Total £
Brought forward				53.12	903.04	809.12	256.82	1,968.98
Steel lintel 2400mm long	C4	nr	2.00	0.50	8.50	296.58	45.76	350.84
Steel lintel 1500mm long	C5	nr	2.00	0.40	6.80	242.72	37.43	286.95
Close cavity wall at jambs	C7	m	8.96	0.45	7.62	25.18	4.92	37.71
Close cavity wall at cills	C8	m	3.67	0.18	3.12	10.31	2.01	15.45
Close cavity wall at top	C9	m	7.40	0.37	6.29	20.79	4.06	31.15
DPC 112mm wide at jambs	C10	m	8.96	0.45	7.62	9.14	2.51	19.27
DPC 112mm wide at cills	C11	m	3.67	0.18	3.12	3.74	1.03	7.89
Carried to summary				55.65	946.10	1,417.59	354.55	2,718.24
PART D **FLAT ROOF**				N/A	N/A	N/A	N/A	N/A
PART E **PITCHED ROOF**								
100 x 75mm sawn softwood wall plate	E1	m	4.00	1.20	20.40	10.44	4.63	35.47
200 x 50mm sawn softwood pole plate	E2	nr	4.30	1.29	21.93	14.28	5.43	41.64
100 x 50mm sawn softwood rafters	E3	m	17.50	3.50	59.50	43.40	15.44	118.34
100 x 50mm sawn softwood purlin	E4	m	4.30	0.86	14.62	12.47	4.06	31.15
150 x 50mm softwood joists	E5	m	12.50	2.50	42.50	39.00	12.23	93.73
150 x 50mm sawn softwood sprockets	E6	nr	14.00	1.68	28.56	21.84	7.56	57.96
100mm layer of insulation quilt laid over and between joists	E7	m²	8.00	3.84	65.28	92.80	23.71	181.79
Carried forward				14.87	252.79	234.23	73.05	560.07

	Ref	Unit	Qty	Hours	Hours £	Mat'ls £	O & P £	Total £
Brought forward				14.87	252.79	234.23	73.05	560.07
6mm softwood soffit 150mm wide	E8	m	9.30	3.72	63.24	19.62	12.43	95.29
19mm wrought softwood fascia/ barge board 200mm high	E9	m	4.30	2.15	36.55	11.65	7.23	55.43
Marley Plain roof tiles on felt and battens	E10	m²	15.05	28.60	486.12	545.41	154.73	1,186.26
Double eaves course	E11	m	4.30	1.51	25.59	18.10	6.55	50.24
Verge with tile undercloak	E12	m	7.00	1.75	29.75	46.27	11.40	87.42
Lead flashing code 5, 200mm girth	E13	m	4.30	2.58	43.86	37.07	12.14	93.06
Rake out joint for flashing	E14	m	4.30	1.51	25.59	1.03	3.99	30.61
112mm diameter PVC-U gutter	E15	m	4.30	1.12	19.01	13.93	4.94	37.88
Stop end	E6	nr	1.00	0.14	2.38	2.12	0.68	5.18
Stop end outlet	E17	nr	1.00	0.25	4.25	4.19	1.27	9.71
68mm diameter PVC-U down pipe	E18	m	2.50	0.63	10.63	14.25	3.73	28.61
Shoe	E19	nr	1.00	0.30	5.10	3.15	1.24	9.49
Paint fascia and soffit	E20	m²	2.79	0.56	9.49	4.46	2.09	16.04
Carried to summary				59.67	1,014.32	955.49	295.47	2,265.28

PART F
WINDOWS AND
EXTERNAL DOORS

	Ref	Unit	Qty	Hours	Hours £	Mat'ls £	O & P £	Total £
PVC-U door size 840 x 1980mm complete (B)	F1	0.00	0.00	0.00	0.00	0.00	0.00	0.00
PVC-U sliding patio door size 1700 x 2075mm (C)	F2	nr	2.00	14.00	238.00	616.94	128.24	983.18
Carried forward				14.00	238.00	616.94	128.24	983.18

	Ref	Unit	Qty	Hours	Hours £	Mat'ls £	O & P £	Total £
Brought forward				14.00	238.00	616.94	128.24	983.18
PVC-U window size 1200 x 1200mm complete (A)	F3	nr	2.00	4.00	68.00	364.28	64.84	497.12
25 x 225mm wrought softwood window board	F4	m	2.40	0.72	12.24	9.65	3.28	25.17
Paint window board	F5	m	2.40	0.14	2.38	2.21	0.69	5.28
Carried to summary				18.86	320.62	993.08	197.05	1,510.75
PART G INTERNAL PARTITIONS AND DOORS			N/A	N/A	N/A	N/A	N/A	N/A
PART H WALL FINISHES								
19 x 100mm wrought softwood skirting	H1	m	5.54	0.94	16.01	13.91	4.49	34.40
12mm plasterboard fixed to walls with dabs	H2	m²	11.50	4.49	76.25	32.78	16.35	125.37
12mm plasterboard fixed to walls less than 300mm wide	H3	m	12.63	2.27	38.65	11.62	7.54	57.81
Two coats emulsion paint to walls	H4	m²	13.40	3.48	59.23	12.86	10.81	82.91
Paint skirting	H5	m	5.54	1.11	18.84	5.10	3.59	27.52
Carried to summary				12.29	208.97	76.26	42.78	328.01
PART J FLOOR FINISHES								
Cement and sand floor screed 40mm thick	J1	m²	5.78	1.45	24.57	27.51	7.81	59.89
Vinyl floor tiles, size 300 x 300mm	J2	m²	5.78	0.98	16.70	44.85	9.23	70.79
Carried to summary				2.43	41.27	72.37	17.05	130.68

	Ref	Unit	Qty	Hours	Hours £	Mat'ls £	O & P £	Total £
PART K **CEILING FINISHES**								
Plasterboard with taped butt joints fixed to joists	K1	m²	5.78	2.08	35.37	16.47	7.78	59.62
5mm skim coat to plasterboard ceilings	K2	m²	5.78	2.89	49.13	10.00	8.87	68.00
Two coats emulsion paint to ceilings	K3	m²	5.78	1.50	25.55	5.55	4.66	35.76
Carried to summary				6.47	110.05	32.02	21.31	163.38
PART L **ELECTRICAL WORK**								
13 amp double switched socket outlet with neon	L1	nr	2.00	1.60	30.40	17.62	7.20	55.22
Lighting point	L2	nr	2.00	1.40	26.60	14.68	6.19	47.47
Lighting switch	L3	nr	2.00	1.40	26.60	8.84	5.32	40.76
Lighting wiring	L4	m	6.00	1.20	22.80	11.16	5.09	39.05
Power cable	L5	m	14.00	4.20	79.80	37.94	17.66	135.40
Carried to summary				9.80	186.20	90.24	41.47	317.91
PART M **HEATING WORK**								
15mm copper pipe	M1	m	5.00	2.20	39.60	10.75	7.55	57.90
Elbow	M2	nr	4.00	2.24	40.32	9.76	7.51	57.59
Tee	M3	nr	1.00	0.56	10.08	2.44	1.88	14.40
Radiator, double convector size 1400 x 520mm	M4	nr	2.00	2.60	46.80	278.22	48.75	373.77
Break into existing pipe and insert tee	M5	nr	1.00	0.75	13.50	4.42	2.69	20.61
Carried to summary				8.35	150.30	305.59	68.38	524.27

	Ref	Unit	Qty	Hours	Hours £	Mat'ls £	O & P £	Total £

PART N
ALTERATION WORK

	Ref	Unit	Qty	Hours	Hours £	Mat'ls £	O & P £	Total £
Take out existing window size 1500 x 1000mm and lintel over, adapt opening to receive 1770 x 2000mm patio door and insert new lintel over (both measured separately) and make good	N1	nr	1.00	20.00	340.00	32.10	55.82	427.92
Carried to summary				20.00	340.00	32.10	55.82	427.92

SUMMARY

	Hours	Hours £	Mat'ls £	O & P £	Total £
PART A PRELIMINARIES	0.00	0.00	0.00	0.00	1,418.00
PART B SUBSTRUCTURE TO DPC LEVEL	41.97	619.54	519.78	170.90	1,310.22
PART C EXTERNAL WALLS	55.65	946.10	1,417.59	354.55	2,718.24
PART D FLAT ROOF	0.00	0.00	0.00	0.00	0.00
PART E PITCHED ROOF	59.67	1,014.32	955.49	295.47	2,265.28
PART F WINDOWS AND EXTERNAL DOORS	18.86	320.62	993.08	197.05	1,510.75
PART G INTERNAL PARTITIONS AND DOORS	0.00	0.00	0.00	0.00	0.00
PART H WALL FINISHES	12.29	208.97	76.26	42.78	328.01
PART J FLOOR FINISHES	2.43	41.27	72.37	17.05	130.68
PART K CEILING FINISHES	6.47	110.05	32.02	21.31	163.38
PART L ELECTRICAL WORK	9.80	186.20	90.24	41.47	317.91
PART M HEATING WORK	8.35	150.30	305.59	68.38	524.27
PART N ALTERATION WORK	20.00	340.00	32.10	55.82	427.92
Final total	235.49	3,937.37	4,494.52	1,264.77	11,114.66

	Ref	Unit	Qty	Hours	Hours £	Mat'ls £	O & P £	Total £
PART A **PRELIMINARIES**								
Concrete mixer	A1	wks	6.00					300.00
Small tools	A2	wks	8.00					320.00
Scaffolding (m²/weeks)	A3		90					540.00
Skip	A4	wks	4.00					440.00
Clean up	A5	hrs	8.00					104.00
Carried to summary								1,704.00
PART B **SUBSTRUCTURE TO** **DPC LEVEL**								
Excavate topsoil 150mm thick by hand	B1	m²	11.40	3.42	44.46	0.00	6.67	51.13
Excavate to reduce levels	B2	m³	3.42	8.55	111.15	0.00	16.67	127.82
Excavate for trench foundations by hand	B3	m³	1.39	3.61	46.98	0.00	7.05	54.03
Earthwork support to sides of trenches	B4	m²	12.26	4.90	63.75	22.68	12.96	99.40
Backfilling with excavated material	B5	m³	0.55	0.33	4.29	0.00	0.64	4.93
Hardcore 225mm thick	B6	m²	7.48	1.50	19.45	47.05	9.97	76.47
Hardcore filling to trench	B7	m³	0.12	0.06	0.78	6.29	1.06	8.13
Concrete grade (1:3:6) in foundations	B8	m³	1.13	1.53	19.83	96.75	17.49	134.07
Concrete grade (1:2:4) in bed 150mm thick	B9	m²	7.48	2.24	29.17	114.59	21.56	165.33
Concrete (1:2:4) in cavity wall filling	B10	m²	4.20	0.84	11.97	10.00	3.29	25.26
Carried forward				26.98	351.84	297.36	97.38	746.58

	Ref	Unit	Qty	Hours	Hours £	Mat'ls £	O & P £	Total £
Brought forward				26.98	351.84	297.36	97.38	746.58
Damp-proof membrane	B11	m²	7.88	0.32	4.10	18.28	3.36	25.74
Reinforcement ref A193 in foundation	B12	m²	4.20	0.50	6.55	8.48	2.26	17.29
Steel fabric reinforcement ref A193 in slab	B13	m²	7.48	1.12	14.59	14.96	4.43	33.98
Solid blockwork 140mm thick in cavity wall	B14	m²	5.46	7.10	120.67	91.07	31.76	243.50
Common bricks 112.5mm thick in cavity wall	B15	m²	4.20	7.14	121.38	69.09	28.57	219.04
Facing bricks in 112.5mm thick in skin of cavity wall	B16	m²	1.26	2.27	38.56	28.55	10.07	77.17
Form cavity 50mm wide in cavity wall	B17	m²	5.46	0.16	2.78	8.30	1.66	12.75
DPC 112mm wide	B18	m	8.40	0.42	7.14	8.57	2.36	18.06
DPC 140mm wide	B19	m	8.40	0.50	8.57	11.59	3.02	23.18
Bond in block wall	B20	m	1.30	0.57	9.72	3.78	2.03	15.53
Bond in half brick wall	B21	m	0.30	0.11	1.79	0.80	0.39	2.98
50mm thick insulation board	B22	m²	7.48	2.24	38.15	53.86	13.80	105.80
Carried to summary				49.44	725.82	614.70	201.08	1,541.60

PART C
EXTERNAL WALLS

	Ref	Unit	Qty	Hours	Hours £	Mat'ls £	O & P £	Total £
Solid blockwork 140mm thick in cavity wall	C1	m²	18.50	24.05	408.85	308.58	107.61	825.04
Facing brickwork 112.5mm thick in cavity wall	C2	m²	18.50	33.30	566.10	528.18	164.14	1,258.42
75mm thick insulation in cavity wall	C3	m²	18.50	4.07	69.19	98.79	25.20	193.18
Carried forward				61.42	1,044.14	935.55	296.95	2,276.64

	Ref	Unit	Qty	Hours	Hours £	Mat'ls £	O & P £	Total £
Brought forward				61.42	1,044.14	935.55	296.95	2,276.64
Steel lintel 2400mm long	C4	nr	2.00	0.50	8.50	296.58	45.76	350.84
Steel lintel 1500mm long	C5	nr	2.00	0.40	6.80	242.72	37.43	286.95
Close cavity wall at jambs	C7	m	8.96	0.45	7.62	25.18	4.92	37.71
Close cavity wall at cills	C8	m	3.67	0.18	3.12	10.31	2.01	15.45
Close cavity wall at top	C9	m	8.40	0.42	7.14	23.60	4.61	35.36
DPC 112mm wide at jambs	C10	m	8.96	0.45	7.62	9.14	2.51	19.27
DPC 112mm wide at cills	C11	m	3.67	0.18	3.12	3.74	1.03	7.89
Carried to summary				64.00	1,088.05	1,546.82	395.23	3,030.10
PART D **FLAT ROOF**				N/A	N/A	N/A	N/A	N/A
PART E **PITCHED ROOF**								
100 x 75mm sawn softwood wall plate	E1	m	5.00	1.50	25.50	13.05	5.78	44.33
200 x 50mm sawn softwood pole plate	E2	nr	5.30	1.59	27.03	17.60	6.69	51.32
100 x 50mm sawn softwood rafters	E3	m	17.50	3.50	59.50	43.40	15.44	118.34
100 x 50mm sawn softwood purlin	E4	m	5.30	1.06	18.02	15.37	5.01	38.40
150 x 50mm softwood joists	E5	m	12.50	2.50	42.50	39.00	12.23	93.73
150 x 50mm sawn softwood sprockets	E6	nr	14.00	1.68	28.56	21.84	7.56	57.96
100mm layer of insulation quilt laid over and between joists	E7	m²	10.00	4.80	81.60	116.00	29.64	227.24
Carried forward				16.63	282.71	266.26	82.34	631.31

	Ref	Unit	Qty	Hours	Hours £	Mat'ls £	O & P £	Total £
Brought forward				16.63	282.71	266.26	82.34	631.31
6mm softwood soffit 150mm wide	E8	m	10.30	4.12	70.04	21.73	13.77	105.54
19mm wrought softwood fascia/ barge board 200mm high	E9	m	5.30	2.65	45.05	14.36	8.91	68.32
Marley Plain roof tiles on felt and battens	E10	m²	18.55	35.25	599.17	672.25	190.71	1,462.13
Double eaves course	E11	m	5.30	1.86	31.54	22.31	8.08	61.93
Verge with tile undercloak	E12	m	7.00	1.75	29.75	46.27	11.40	87.42
Lead flashing code 5, 200mm girth	E13	m	5.30	3.18	54.06	45.69	14.96	114.71
Rake out joint for flashing	E14	m	5.30	1.86	31.54	1.27	4.92	37.73
112mm diameter PVC-U gutter	E15	m	5.30	1.38	23.43	17.17	6.09	46.69
Stop end	E16	nr	1.00	0.14	2.38	2.12	0.68	5.18
Stop end outlet	E17	nr	1.00	0.25	4.25	4.19	1.27	9.71
68mm diameter PVC-U down pipe	E18	m	2.50	0.63	10.63	14.25	3.73	28.61
Shoe	E19	nr	1.00	0.30	5.10	3.15	1.24	9.49
Paint fascia and soffit	E20	m²	3.09	0.62	10.51	4.94	2.32	17.77
Carried to summary				70.60	1,200.13	1,135.97	350.42	2,686.52

**PART F
WINDOWS AND
EXTERNAL DOORS**

	Ref	Unit	Qty	Hours	Hours £	Mat'ls £	O & P £	Total £
PVC-U door size 840 x 1980mm complete (B)	F1	0.00	0.00	0.00	0.00	0.00	0.00	0.00
PVC-U sliding patio door size 1700 x 2075mm (C)	F2	nr	2.00	14.00	238.00	616.94	128.24	983.18
Carried forward				14.00	238.00	616.94	128.24	983.18

	Ref	Unit	Qty	Hours	Hours £	Mat'ls £	O & P £	Total £
Brought forward				14.00	238.00	616.94	128.24	983.18
PVC-U window size 1200 x 1200mm complete (A)	F3	nr	2.00	4.00	68.00	364.28	64.84	497.12
25 x 225mm wrought softwood window board	F4	m	2.40	0.72	12.24	9.65	3.28	25.17
Paint window board	F5	m	2.40	0.14	2.38	2.21	0.69	5.28
Carried to summary				18.86	320.62	993.08	197.05	1,510.75
PART G INTERNAL PARTITIONS AND DOORS			N/A	N/A	N/A	N/A	N/A	N/A
PART H WALL FINISHES								
19 x 100mm wrought softwood skirting	H1	m	6.54	1.11	18.90	16.42	5.30	40.61
12mm plasterboard fixed to walls with dabs	H2	m²	14.00	5.46	92.82	39.90	19.91	152.63
12mm plasterboard fixed to walls less than 300mm wide	H3	m	12.63	2.27	38.65	11.62	7.54	57.81
Two coats emulsion paint to walls	H4	m²	15.90	4.13	70.28	15.26	12.83	98.37
Paint skirting	H5	m	6.54	1.31	22.24	6.02	4.24	32.49
Carried to summary				14.29	242.88	89.22	49.81	381.91
PART J FLOOR FINISHES								
Cement and sand floor screed 40mm thick	J1	m²	7.48	1.87	31.79	35.60	10.11	77.50
Vinyl floor tiles, size 300 x 300mm	J2	m²	7.48	1.27	21.62	58.04	11.95	91.61
Carried to summary				3.14	53.41	93.65	22.06	169.12

	Ref	Unit	Qty	Hours	Hours £	Mat'ls £	O & P £	Total £
PART K **CEILING FINISHES**								
Plasterboard with taped butt joints fixed to joists	K1	m²	7.48	2.69	45.78	21.32	10.06	77.16
5mm skim coat to plasterboard ceilings	K2	m²	7.48	3.74	63.58	12.94	11.48	88.00
Two coats emulsion paint to ceilings	K3	m²	7.48	1.94	33.06	7.18	6.04	46.28
Carried to summary				8.38	142.42	41.44	27.58	211.44
PART L **ELECTRICAL WORK**								
13 amp double switched socket outlet with neon	L1	nr	3.00	2.40	45.60	26.43	10.80	82.83
Lighting point	L2	nr	3.00	2.10	39.90	22.02	9.29	71.21
Lighting switch	L3	nr	2.00	1.40	26.60	8.84	5.32	40.76
Lighting wiring	L4	m	7.00	1.40	26.60	13.02	5.94	45.56
Power cable	L5	m	16.00	4.80	91.20	43.36	20.18	154.74
Carried to summary				12.10	229.90	113.67	51.54	395.11
PART M **HEATING WORK**								
15mm copper pipe	M1	m	6.00	2.64	47.52	12.90	9.06	69.48
Elbow	M2	nr	4.00	2.24	40.32	9.76	7.51	57.59
Tee	M3	nr	1.00	0.56	10.08	2.44	1.88	14.40
Radiator, double convector size 1400 x 520mm	M4	nr	2.00	2.60	46.80	278.22	48.75	373.77
Break into existing pipe and insert tee	M5	nr	1.00	0.75	13.50	4.42	2.69	20.61
Carried to summary				8.79	158.22	307.74	69.89	535.85

	Ref	Unit	Qty	Hours	Hours £	Mat'ls £	O & P £	Total £

PART N
ALTERATION WORK

	Ref	Unit	Qty	Hours	Hours £	Mat'ls £	O & P £	Total £
Take out existing window size 1500 x 1000mm and lintel over, adapt opening to receive 1770 x 2000mm patio door and insert new lintel over (both measured separately) and make good	N1	nr	1.00	20.00	340.00	32.10	55.82	427.92
Carried to summary				20.00	340.00	32.10	55.82	427.92

SUMMARY

	Hours	Hours £	Mat'ls £	O & P £	Total £
PART A **PRELIMINARIES**	0.00	0.00	0.00	0.00	1,704.00
PART B SUBSTRUCTURE TO **DPC LEVEL**	49.44	725.82	614.70	201.08	1,541.60
PART C **EXTERNAL WALLS**	64.00	1,088.05	1,546.82	395.23	3,030.10
PART D **FLAT ROOF**	0.00	0.00	0.00	0.00	0.00
PART E **PITCHED ROOF**	70.60	1,200.13	1,135.97	350.42	2,686.52
PART F WINDOWS AND **EXTERNAL DOORS**	18.86	320.62	993.08	197.05	1,510.75
PART G INTERNAL **PARTITIONS AND DOORS**	0.00	0.00	0.00	0.00	0.00
PART H **WALL FINISHES**	14.29	242.88	89.22	49.81	381.91
PART J **FLOOR FINISHES**	3.14	53.41	93.65	22.06	169.12
PART K **CEILING FINISHES**	8.38	142.42	41.44	27.58	211.44
PART L **ELECTRICAL WORK**	12.10	229.90	113.67	51.54	395.11
PART M **HEATING WORK**	8.79	158.22	307.74	69.89	535.85
PART N **ALTERATION WORK**	20.00	340.00	32.10	55.82	427.92
Final total	269.60	4,501.45	4,968.39	1,420.48	12,594.32

	Ref	Unit	Qty	Hours	Hours £	Mat'ls £	O & P £	Total £
PART A **PRELIMINARIES**								
Concrete mixer	A1	wks	6.00					300.00
Small tools	A2	wks	7.00					280.00
Scaffolding (m²/weeks)	A3		90					540.00
Skip	A4	wks	3.00					330.00
Clean up	A5	hrs	8.00					104.00
Carried to summary								1,554.00
PART B **SUBSTRUCTURE TO** **DPC LEVEL**								
Excavate topsoil 150mm thick by hand	B1	m²	10.40	3.12	40.56	0.00	6.08	46.64
Excavate to reduce levels	B2	m³	3.12	7.80	101.40	0.00	15.21	116.61
Excavate for trench foundations by hand	B3	m³	1.39	3.61	46.98	0.00	7.05	54.03
Earthwork support to sides of trenches	B4	m²	12.26	4.90	63.75	22.68	12.96	99.40
Backfilling with excavated material	B5	m³	0.47	0.28	3.67	0.00	0.55	4.22
Hardcore 225mm thick	B6	m²	6.48	1.30	16.85	40.76	8.64	66.25
Hardcore filling to trench	B7	m³	0.09	0.05	0.59	6.29	1.03	7.91
Concrete grade (1:3:6) in foundations	B8	m³	1.13	1.53	19.83	96.75	17.49	134.07
Concrete grade (1:2:4) in bed 150mm thick	B9	m²	6.48	1.94	25.27	99.27	18.68	143.23
Concrete (1:2:4) in cavity wall filling	B10	m²	4.20	0.84	11.97	10.00	3.29	25.26
Carried forward				25.37	330.87	275.75	90.99	697.61

	Ref	Unit	Qty	Hours	Hours £	Mat'ls £	O & P £	Total £
Brought forward				25.37	330.87	275.75	90.99	697.61
Damp-proof membrane	B11	m²	6.88	0.28	3.58	15.96	2.93	22.47
Reinforcement ref A193 in foundation	B12	m²	4.20	0.50	6.55	8.48	2.26	17.29
Steel fabric reinforcement ref A193 in slab	B13	m²	6.48	0.97	12.64	12.96	3.84	29.44
Solid blockwork 140mm thick in cavity wall	B14	m²	5.46	7.10	120.67	91.07	31.76	243.50
Common bricks 112.5mm thick in cavity wall	B15	m²	4.20	7.14	121.38	69.09	28.57	219.04
Facing bricks in 112.5mm thick in skin of cavity wall	B16	m²	1.26	2.27	38.56	28.55	10.07	77.17
Form cavity 50mm wide in cavity wall	B17	m²	5.46	0.16	2.78	8.30	1.66	12.75
DPC 112mm wide	B18	m	8.40	0.42	7.14	8.57	2.36	18.06
DPC 140mm wide	B19	m	8.40	0.50	8.57	11.59	3.02	23.18
Bond in block wall	B20	m	1.30	0.57	9.72	3.78	2.03	15.53
Bond in half brick wall	B21	m	0.30	0.11	1.79	0.80	0.39	2.98
50mm thick insulation board	B22	m²	6.48	1.94	33.05	46.66	11.96	91.66
Carried to summary				47.34	697.28	581.57	191.83	1,470.68
PART C **EXTERNAL WALLS**								
Solid blockwork 140mm thick in cavity wall	C1	m²	21.50	27.95	475.15	358.62	125.07	958.84
Facing brickwork 112.5mm thick in cavity wall	C2	m²	21.50	38.70	657.90	613.83	190.76	1,462.48
75mm thick insulation in cavity wall	C3	m²	21.50	4.73	80.41	114.81	29.28	224.50
Carried forward				71.38	1,213.46	1,087.26	345.11	2,645.82

	Ref	Unit	Qty	Hours	Hours £	Mat'ls £	O & P £	Total £
Brought forward				71.38	1,213.46	1,087.26	345.11	2,645.82
Steel lintel 2400mm long	C4	nr	2.00	0.50	8.50	296.58	45.76	350.84
Steel lintel 1500mm long	C5	nr	2.00	0.40	6.80	242.72	37.43	286.95
Close cavity wall at jambs	C7	m	8.96	0.45	7.62	25.18	4.92	37.71
Close cavity wall at cills	C8	m	3.67	0.18	3.12	10.31	2.01	15.45
Close cavity wall at top	C9	m	8.40	0.42	7.14	23.60	4.61	35.36
DPC 112mm wide at jambs	C10	m	8.96	0.45	7.62	9.14	2.51	19.27
DPC 112mm wide at cills	C11	m	3.67	0.18	3.12	3.74	1.03	7.89
Carried to summary				73.96	1,257.37	1,698.53	443.39	3,399.29
PART D **FLAT ROOF**				N/A	N/A	N/A	N/A	N/A
PART E **PITCHED ROOF**								
100 x 75mm sawn softwood wall plate	E1	m	3.00	0.90	15.30	7.83	3.47	26.60
200 x 50mm sawn softwood pole plate	E2	nr	3.30	0.99	16.83	10.96	4.17	31.95
100 x 50mm sawn softwood rafters	E3	m	31.50	6.30	107.10	78.12	27.78	213.00
100 x 50mm sawn softwood purlin	E4	m	3.30	0.66	11.22	9.57	3.12	23.91
150 x 50mm softwood joists	E5	m	24.50	4.90	83.30	76.44	23.96	183.70
150 x 50mm sawn softwood sprockets	E6	nr	18.00	2.16	36.72	28.08	9.72	74.52
100mm layer of insulation quilt laid over and between joists	E7	m²	9.00	4.32	73.44	104.40	26.68	204.52
Carried forward				20.23	343.91	315.40	98.90	758.20

	Ref	Unit	Qty	Hours	Hours £	Mat'ls £	O & P £	Total £
Brought forward				20.23	343.91	315.40	98.90	758.20
6mm softwood soffit 150mm wide	E8	m	10.30	4.12	70.04	21.73	13.77	105.54
19mm wrought softwood fascia/ barge board 200mm high	E9	m	3.30	1.65	28.05	8.94	5.55	42.54
Marley Plain roof tiles on felt and battens	E10	m²	14.85	28.22	479.66	538.16	152.67	1,170.49
Double eaves course	E11	m	3.30	1.16	19.64	13.89	5.03	38.56
Verge with tile undercloak	E12	m	9.00	2.25	38.25	59.49	14.66	112.40
Lead flashing code 5, 200mm girth	E13	m	3.30	1.98	33.66	28.45	9.32	71.42
Rake out joint for flashing	E14	m	3.30	1.16	19.64	0.79	3.06	23.49
112mm diameter PVC-U gutter	E15	m	3.30	0.86	14.59	10.69	3.79	29.07
Stop end	E16	nr	1.00	0.14	2.38	2.12	0.68	5.18
Stop end outlet	E17	nr	1.00	0.25	4.25	4.19	1.27	9.71
68mm diameter PVC-U down pipe	E18	m	2.50	0.63	10.63	14.25	3.73	28.61
Shoe	E19	nr	1.00	0.30	5.10	3.15	1.24	9.49
Paint fascia and soffit	E20	m²	3.09	0.62	10.51	4.94	2.32	17.77
Carried to summary				63.55	1,080.31	1,026.20	315.97	2,422.46

**PART F
WINDOWS AND
EXTERNAL DOORS**

	Ref	Unit	Qty	Hours	Hours £	Mat'ls £	O & P £	Total £
PVC-U door size 840 x 1980mm complete (B)	F1	0.00	0.00	0.00	0.00	0.00	0.00	0.00
PVC-U sliding patio door size 1700 x 2075mm (C)	F2	nr	2.00	14.00	238.00	616.94	128.24	983.18
Carried forward				14.00	238.00	616.94	128.24	983.18

	Ref	Unit	Qty	Hours	Hours £	Mat'ls £	O & P £	Total £
Brought forward				14.00	238.00	616.94	128.24	983.18
PVC-U window size 1200 x 1200mm complete (A)	F3	nr	2.00	4.00	68.00	364.28	64.84	497.12
25 x 225mm wrought softwood window board	F4	m	2.40	0.72	12.24	9.65	3.28	25.17
Paint window board	F5	m	2.40	0.14	2.38	2.21	0.69	5.28
Carried to summary				18.86	320.62	993.08	197.05	1,510.75
PART G INTERNAL PARTITIONS AND DOORS			N/A	N/A	N/A	N/A	N/A	N/A
PART H WALL FINISHES								
19 x 100mm wrought softwood skirting	H1	m	6.54	1.11	18.90	16.42	5.30	40.61
12mm plasterboard fixed to walls with dabs	H2	m²	14.00	5.46	92.82	39.90	19.91	152.63
12mm plasterboard fixed to walls less than 300mm wide	H3	m	12.63	2.27	38.65	11.62	7.54	57.81
Two coats emulsion paint to walls	H4	m²	15.90	4.13	70.28	15.26	12.83	98.37
Paint skirting	H5	m	6.54	1.31	22.24	6.02	4.24	32.49
Carried to summary				14.29	242.88	89.22	49.81	381.91
PART J FLOOR FINISHES								
Cement and sand floor screed 40mm thick	J1	m²	6.48	1.62	27.54	30.84	8.76	67.14
Vinyl floor tiles, size 300 x 300mm	J2	m²	6.48	1.10	18.73	50.28	10.35	79.36
Carried to summary				2.72	46.27	81.13	19.11	146.51

	Ref	Unit	Qty	Hours	Hours £	Mat'ls £	O & P £	Total £
PART K **CEILING FINISHES**								
Plasterboard with taped butt joints fixed to joists	K1	m²	6.48	2.33	39.66	18.47	8.72	66.84
5mm skim coat to plasterboard ceilings	K2	m²	6.48	3.24	55.08	11.21	9.94	76.23
Two coats emulsion paint to ceilings	K3	m²	6.48	1.68	28.64	6.22	5.23	40.09
Carried to summary				7.26	123.38	35.90	23.89	183.17
PART L **ELECTRICAL WORK**								
13 amp double switched socket outlet with neon	L1	nr	2.00	1.60	30.40	17.62	7.20	55.22
Lighting point	L2	nr	1.00	0.70	13.30	7.34	3.10	23.74
Lighting switch	L3	nr	2.00	1.40	26.60	8.84	5.32	40.76
Lighting wiring	L4	m	6.00	1.20	22.80	11.16	5.09	39.05
Power cable	L5	m	14.00	4.20	79.80	37.94	17.66	135.40
Carried to summary				9.10	172.90	82.90	38.37	294.17
PART M **HEATING WORK**								
15mm copper pipe	M1	m	5.00	2.20	39.60	10.75	7.55	57.90
Elbow	M2	nr	4.00	2.24	40.32	9.76	7.51	57.59
Tee	M3	nr	1.00	0.56	10.08	2.44	1.88	14.40
Radiator, double convector size 1400 x 520mm	M4	nr	2.00	2.60	46.80	278.22	48.75	373.77
Break into existing pipe and insert tee	M5	nr	1.00	0.75	13.50	4.42	2.69	20.61
Carried to summary				8.35	150.30	305.59	68.38	524.27

	Ref	Unit	Qty	Hours	Hours £	Mat'ls £	O & P £	Total £

PART N
ALTERATION WORK

	Ref	Unit	Qty	Hours	Hours £	Mat'ls £	O & P £	Total £
Take out existing window size 1500 x 1000mm and lintel over, adapt opening to receive 1770 x 2000mm patio door and insert new lintel over (both measured separately) and make good	N1	nr	1.00	20.00	340.00	32.10	55.82	427.92
Carried to summary				20.00	340.00	32.10	55.82	427.92

SUMMARY

	Hours	Hours £	Mat'ls £	O & P £	Total £
PART A **PRELIMINARIES**	0.00	0.00	0.00	0.00	1,554.00
PART B SUBSTRUCTURE TO **DPC LEVEL**	47.34	697.28	581.57	191.83	1,470.68
PART C **EXTERNAL WALLS**	73.96	1,257.37	1,698.53	443.39	3,399.29
PART D **FLAT ROOF**	0.00	0.00	0.00	0.00	0.00
PART E **PITCHED ROOF**	63.55	1,080.31	1,026.20	315.97	2,422.46
PART F WINDOWS AND **EXTERNAL DOORS**	18.86	320.62	993.08	197.05	1,510.75
PART G INTERNAL **PARTITIONS AND DOORS**	0.00	0.00	0.00	0.00	0.00
PART H **WALL FINISHES**	14.29	242.88	89.22	49.81	381.91
PART J **FLOOR FINISHES**	2.72	46.27	81.13	19.11	146.51
PART K **CEILING FINISHES**	7.26	123.38	35.90	23.89	183.17
PART L **ELECTRICAL WORK**	9.10	172.90	82.90	38.37	294.17
PART M **HEATING WORK**	8.35	150.30	305.59	68.38	524.27
PART N **ALTERATION WORK**	20.00	340.00	32.10	55.82	427.92
Final total	265.43	4,431.31	4,926.22	1,403.61	12,315.12

	Ref	Unit	Qty	Hours	Hours £	Mat'ls £	O & P £	Total £
PART A **PRELIMINARIES**								
Concrete mixer	A1	wks	7.00					350.00
Small tools	A2	wks	8.00					320.00
Scaffolding (m²/weeks)	A3		100					600.00
Skip	A4	wks	4.00					440.00
Clean up	A5	hrs	6.00					78.00
Carried to summary								1,788.00
PART B **SUBSTRUCTURE TO** **DPC LEVEL**								
Excavate topsoil 150mm thick by hand	B1	m²	14.19	4.26	55.34	0.00	8.30	63.64
Excavate to reduce levels	B2	m³	4.06	10.15	131.95	0.00	19.79	151.74
Excavate for trench foundations by hand	B3	m³	1.39	3.61	46.98	0.00	7.05	54.03
Earthwork support to sides of trenches	B4	m²	13.72	5.49	71.34	25.38	14.51	111.23
Backfilling with excavated material	B5	m³	1.55	0.93	12.09	0.00	1.81	13.90
Hardcore 225mm thick	B6	m²	9.18	1.84	23.87	57.74	12.24	93.85
Hardcore filling to trench	B7	m³	0.12	0.06	0.78	6.29	1.06	8.13
Concrete grade (1:3:6) in foundations	B8	m³	1.27	1.71	22.29	108.74	19.65	150.68
Concrete grade (1:2:4) in bed 150mm thick	B9	m²	9.18	2.75	35.80	140.64	26.47	202.91
Concrete (1:2:4) in cavity wall filling	B10	m²	4.70	0.94	11.97	11.19	3.47	26.63
Carried forward				31.74	412.42	349.98	114.36	876.75

	Ref	Unit	Qty	Hours	Hours £	Mat'ls £	O & P £	Total £
Brought forward				31.74	412.42	349.98	114.36	876.75
Damp-proof membrane	B11	m²	9.63	0.39	5.01	22.34	4.10	31.45
Reinforcement ref A193 in foundation	B12	m²	4.70	0.56	7.33	9.49	2.52	19.35
Steel fabric reinforcement ref A193 in slab	B13	m²	9.18	1.38	17.90	18.36	5.44	41.70
Solid blockwork 140mm thick in cavity wall	B14	m²	6.11	7.94	135.03	101.91	35.54	272.49
Common bricks 112.5mm thick in cavity wall	B15	m²	4.70	7.99	135.83	77.32	31.97	245.12
Facing bricks in 112.5mm thick in skin of cavity wall	B16	m²	1.41	2.54	43.15	28.55	10.75	82.45
Form cavity 50mm wide in cavity wall	B17	m²	5.46	0.16	2.78	8.30	1.66	12.75
DPC 112mm wide	B18	m	9.40	0.47	7.99	9.59	2.64	20.21
DPC 140mm wide	B19	m	9.40	0.56	9.59	12.97	3.38	25.94
Bond in block wall	B20	m	1.30	0.57	9.72	3.78	2.03	15.53
Bond in half brick wall	B21	m	0.30	0.11	1.79	0.80	0.39	2.98
50mm thick insulation board	B22	m²	9.18	2.75	46.82	66.10	16.94	129.85
Carried to summary				57.17	835.35	709.49	231.73	1,776.57

**PART C
EXTERNAL WALLS**

	Ref	Unit	Qty	Hours	Hours £	Mat'ls £	O & P £	Total £
Solid blockwork 140mm thick in cavity wall	C1	m²	24.00	31.20	530.40	400.32	139.61	1,070.33
Facing brickwork 112.5mm thick in cavity wall	C2	m²	24.00	43.20	734.40	685.20	212.94	1,632.54
75mm thick insulation in cavity wall	C3	m²	24.00	5.28	89.76	128.16	32.69	250.61
Carried forward				79.68	1,354.56	1,213.68	385.24	2,953.48

	Ref	Unit	Qty	Hours	Hours £	Mat'ls £	O & P £	Total £
Brought forward				79.68	1,354.56	1,213.68	385.24	2,953.48
Steel lintel 2400mm long	C4	nr	2.00	0.50	8.50	296.58	45.76	350.84
Steel lintel 1500mm long	C5	nr	2.00	0.40	6.80	242.72	37.43	286.95
Close cavity wall at jambs	C7	m	8.96	0.45	7.62	25.18	4.92	37.71
Close cavity wall at cills	C8	m	3.67	0.18	3.12	10.31	2.01	15.45
Close cavity wall at top	C9	m	9.40	0.47	7.99	26.41	5.16	39.56
DPC 112mm wide at jambs	C10	m	8.96	0.45	7.62	9.14	2.51	19.27
DPC 112mm wide at cills	C11	m	3.67	0.18	3.12	3.74	1.03	7.89
Carried to summary				82.31	1,399.32	1,827.77	484.06	3,711.15
PART D **FLAT ROOF**				N/A	N/A	N/A	N/A	N/A
PART E **PITCHED ROOF**								
100 x 75mm sawn softwood wall plate	E1	m	4.00	1.20	20.40	10.44	4.63	35.47
200 x 50mm sawn softwood pole plate	E2	nr	4.30	1.29	21.93	14.28	5.43	41.64
100 x 50mm sawn softwood rafters	E3	m	31.50	6.30	107.10	78.12	27.78	213.00
100 x 50mm sawn softwood purlin	E4	m	4.30	0.86	14.62	12.47	4.06	31.15
150 x 50mm softwood joists	E5	m	24.50	4.90	83.30	76.44	23.96	183.70
150 x 50mm sawn softwood sprockets	E6	nr	18.00	2.16	36.72	28.08	9.72	74.52
100mm layer of insulation quilt laid over and between oists	E7	m²	12.00	5.76	97.92	139.20	35.57	272.69
Carried forward				22.47	381.99	359.03	111.15	852.17

	Ref	Unit	Qty	Hours	Hours £	Mat'ls £	O & P £	Total £
Brought forward				22.47	381.99	359.03	111.15	852.17
6mm softwood soffit 150mm wide	E8	m	11.30	4.52	76.84	23.84	15.10	115.79
19mm wrought softwood fascia/ barge board 200mm high	E9	m	4.30	2.15	36.55	11.65	7.23	55.43
Marley Plain roof tiles on felt and battens	E10	m²	19.55	37.15	631.47	708.49	200.99	1,540.95
Double eaves course	E11	m	4.30	1.51	25.59	18.10	6.55	50.24
Verge with tile undercloak	E12	m	9.00	2.25	38.25	59.49	14.66	112.40
Lead flashing code 5, 200mm girth	E13	m	4.30	2.58	43.86	37.07	12.14	93.06
Rake out joint for flashing	E14	m	4.30	1.51	25.59	1.03	3.99	30.61
112mm diameter PVC-U gutter	E15	m	4.30	1.12	19.01	13.93	4.94	37.88
Stop end	E16	nr	1.00	0.14	2.38	2.12	0.68	5.18
Stop end outlet	E17	nr	1.00	0.25	4.25	4.19	1.27	9.71
68mm diameter PVC-U down pipe	E18	m	2.50	0.63	10.63	14.25	3.73	28.61
Shoe	E19	nr	1.00	0.30	5.10	3.15	1.24	9.49
Paint fascia and soffit	E20	m²	3.09	0.62	10.51	4.94	2.32	17.77
Carried to summary				77.18	1,311.99	1,261.29	385.99	2,959.28

**PART F
WINDOWS AND
EXTERNAL DOORS**

	Ref	Unit	Qty	Hours	Hours £	Mat'ls £	O & P £	Total £
PVC-U door size 840 x 1980mm complete (B)	F1	0.00	0.00	0.00	0.00	0.00	0.00	0.00
PVC-U sliding patio door size 1700 x 2075mm (C)	F2	nr	2.00	14.00	238.00	616.94	128.24	983.18
Carried forward				14.00	238.00	616.94	128.24	983.18

	Ref	Unit	Qty	Hours	Hours £	Mat'ls £	O & P £	Total £
Brought forward				14.00	238.00	616.94	128.24	983.18
PVC-U window size 1200 x 1200mm complete (A)	F3	nr	2.00	4.00	68.00	364.28	64.84	497.12
25 x 225mm wrought softwood window board	F4	m	2.40	0.72	12.24	9.65	3.28	25.17
Paint window board	F5	m	2.40	0.14	2.38	2.21	0.69	5.28
Carried to summary				18.86	320.62	993.08	197.05	1,510.75
PART G INTERNAL PARTITIONS AND DOORS			N/A	N/A	N/A	N/A	N/A	N/A
PART H WALL FINISHES								
19 x 100mm wrought softwood skirting	H1	m	7.54	1.28	21.79	18.93	6.11	46.82
12mm plasterboard fixed to walls with dabs	H2	m²	16.50	6.44	109.40	47.03	23.46	179.88
12mm plasterboard fixed to walls less than 300mm wide	H3	m	12.63	2.27	38.65	11.62	7.54	57.81
Two coats emulsion paint to walls	H4	m²	18.40	4.78	81.33	17.66	14.85	113.84
Paint skirting	H5	m	7.54	1.51	25.64	6.94	4.89	37.46
Carried to summary				16.28	276.80	102.17	56.85	435.81
PART J FLOOR FINISHES								
Cement and sand floor screed 40mm thick	J1	m²	9.18	2.30	39.02	43.70	12.41	95.12
Vinyl floor tiles, size 300 x 300mm	J2	m²	9.18	1.56	26.53	71.24	14.67	112.43
Carried to summary				3.86	65.55	114.93	27.07	207.55

	Ref	Unit	Qty	Hours	Hours £	Mat'ls £	O & P £	Total £
PART K **CEILING FINISHES**								
Plasterboard with taped butt joints fixed to joists	K1	m²	9.18	3.30	56.18	26.16	12.35	94.70
5mm skim coat to plasterboard ceilings	K2	m²	9.18	4.59	78.03	15.88	14.09	108.00
Two coats emulsion paint to ceilings	K3	m²	9.18	2.39	40.58	8.81	7.41	56.80
Carried to summary				10.28	174.79	50.86	33.85	259.50
PART L **ELECTRICAL WORK**								
13 amp double switched socket outlet with neon	L1	nr	3.00	2.40	45.60	26.43	10.80	82.83
Lighting point	L2	nr	2.00	1.40	26.60	14.68	6.19	47.47
Lighting switch	L3	nr	2.00	1.40	26.60	8.84	5.32	40.76
Lighting wiring	L4	m	6.00	1.20	22.80	11.16	5.09	39.05
Power cable	L5	m	16.00	4.80	91.20	43.36	20.18	154.74
Carried to summary				11.20	212.80	104.47	47.59	364.86
PART M **HEATING WORK**								
15mm copper pipe	M1	m	6.00	2.64	47.52	12.90	9.06	69.48
Elbow	M2	nr	4.00	2.24	40.32	9.76	7.51	57.59
Tee	M3	nr	1.00	0.56	10.08	2.44	1.88	14.40
Radiator, double convector size 1400 x 520mm	M4	nr	2.00	2.60	46.80	278.22	48.75	373.77
Break into existing pipe and insert tee	M5	nr	1.00	0.75	13.50	4.42	2.69	20.61
Carried to summary				8.79	158.22	307.74	69.89	535.85

	Ref	Unit	Qty	Hours	Hours £	Mat'ls £	O & P £	Total £
PART N **ALTERATION WORK**								
Take out existing window size 1500 x 1000mm and lintel over, adapt opening to receive 1770 x 2000mm patio door and insert new lintel over (both measured separately) and make good	N1	nr	1.00	20.00	340.00	32.10	55.82	427.92
Carried to summary				20.00	340.00	32.10	55.82	427.92

SUMMARY

	Hours	Hours £	Mat'ls £	O & P £	Total £
PART A **PRELIMINARIES**	0.00	0.00	0.00	0.00	1,788.00
PART B SUBSTRUCTURE TO **DPC LEVEL**	57.17	835.35	709.49	231.73	1,776.57
PART C **EXTERNAL WALLS**	82.31	1,399.32	1,827.77	484.06	3,711.15
PART D **FLAT ROOF**	0.00	0.00	0.00	0.00	0.00
PART E **PITCHED ROOF**	77.18	1,311.99	1,261.29	385.99	2,959.28
PART F WINDOWS AND **EXTERNAL DOORS**	18.86	320.62	993.08	197.05	1,510.75
PART G INTERNAL **PARTITIONS AND DOORS**	0.00	0.00	0.00	0.00	0.00
PART H **WALL FINISHES**	16.28	276.80	102.17	56.85	435.82
PART J **FLOOR FINISHES**	3.86	65.55	114.93	27.07	207.55
PART K **CEILING FINISHES**	10.28	174.79	50.86	33.85	259.50
PART L **ELECTRICAL WORK**	11.20	212.80	104.47	47.59	364.86
PART M **HEATING WORK**	8.79	158.22	307.74	69.89	535.85
PART N **ALTERATION WORK**	20.00	340.00	32.10	55.82	427.92
Final total	305.93	5,095.44	5,503.90	1,589.90	13,977.25

	Ref	Unit	Qty	Hours	Hours £	Mat'ls £	O & P £	Total £
PART A **PRELIMINARIES**								
Concrete mixer	A1	wks	8.00					400.00
Small tools	A2	wks	9.00					360.00
Scaffolding (m²/weeks)	A3		110					660.00
Skip	A4	wks	5.00					550.00
Clean up	A5	hrs	8.00					104.00
Carried to summary								2,074.00
PART B **SUBSTRUCTURE TO** **DPC LEVEL**								
Excavate topsoil 150mm thick by hand	B1	m²	17.50	5.25	68.25	0.00	10.24	78.49
Excavate to reduce levels	B2	m³	5.00	12.50	162.50	0.00	24.38	186.88
Excavate for trench foundations by hand	B3	m³	1.72	4.47	58.14	0.00	8.72	66.86
Earthwork support to sides of trenches	B4	m²	15.18	6.07	78.94	28.08	16.05	123.07
Backfilling with excavated material	B5	m³	0.55	0.33	4.29	0.00	0.64	4.93
Hardcore 225mm thick	B6	m²	11.88	2.38	30.89	74.73	15.84	121.46
Hardcore filling to trench	B7	m³	0.12	0.06	0.78	6.29	1.06	8.13
Concrete grade (1:3:6) in foundations	B8	m³	1.40	1.89	24.57	119.87	21.67	166.10
Concrete grade (1:2:4) in bed 150mm thick	B9	m²	11.88	3.56	46.33	182.00	34.25	262.58
Concrete (1:2:4) in cavity wall filling	B10	m²	4.70	0.94	11.97	11.19	3.47	26.63
Carried forward				37.45	486.65	422.15	136.32	1,045.13

150   One storey extension, size 3 x 5m, pitched roof

	Ref	Unit	Qty	Hours	Hours £	Mat'ls £	O & P £	Total £
Brought forward				37.45	486.65	422.15	136.32	1,045.13
Damp-proof membrane	B11	m²	12.38	0.50	6.44	28.72	5.27	40.43
Reinforcement ref A193 in foundation	B12	m²	5.20	0.62	8.11	10.50	2.79	21.41
Steel fabric reinforcement ref A193 in slab	B13	m²	11.88	1.78	23.17	23.76	7.04	53.96
Solid blockwork 140mm thick in cavity wall	B14	m²	6.76	8.79	149.40	112.76	39.32	301.48
Common bricks 112.5mm thick in cavity wall	B15	m²	5.20	8.84	150.28	85.54	35.37	271.19
Facing bricks in 112.5mm thick in skin of cavity wall	B16	m²	1.56	2.81	47.74	28.55	11.44	87.73
Form cavity 50mm wide in cavity wall	B17	m²	6.76	0.20	3.45	10.28	2.06	15.78
DPC 112mm wide	B18	m	10.40	0.52	8.84	10.61	2.92	22.37
DPC 140mm wide	B19	m	10.40	0.62	10.61	14.35	3.74	28.70
Bond in block wall	B20	m	1.30	0.57	9.72	3.78	2.03	15.53
Bond in half brick wall	B21	m	0.30	0.11	1.79	0.80	0.39	2.98
50mm thick insulation board	B22	m²	11.88	3.56	60.59	85.54	21.92	168.04
Carried to summary				66.38	966.77	837.34	270.62	2,074.73

PART C
EXTERNAL WALLS

	Ref	Unit	Qty	Hours	Hours £	Mat'ls £	O & P £	Total £
Solid blockwork 140mm thick in cavity wall	C1	m²	26.50	34.45	585.65	442.02	154.15	1,181.82
Facing brickwork 112.5mm thick in cavity wall	C2	m²	26.50	47.70	810.90	756.58	235.12	1,802.60
75mm thick insulation in cavity wall	C3	m²	26.50	5.83	99.11	141.51	36.09	276.71
Carried forward				87.98	1,495.66	1,340.11	425.36	3,261.13

	Ref	Unit	Qty	Hours	Hours £	Mat'ls £	O & P £	Total £
Brought forward				87.98	1,495.66	1,340.11	425.36	3,261.13
Steel lintel 2400mm long	C4	nr	2.00	0.50	8.50	296.58	45.76	350.84
Steel lintel 1500mm long	C5	nr	2.00	0.40	6.80	242.72	37.43	286.95
Close cavity wall at jambs	C7	m	8.96	0.45	7.62	25.18	4.92	37.71
Close cavity wall at cills	C8	m	3.67	0.18	3.12	10.31	2.01	15.45
Close cavity wall at top	C9	m	10.40	0.52	8.84	29.22	5.71	43.77
DPC 112mm wide at jambs	C10	m	8.96	0.45	7.62	9.14	2.51	19.27
DPC 112mm wide at cills	C11	m	3.87	0.19	3.29	3.95	1.09	8.32
Carried to summary				90.67	1,541.44	1,957.21	524.80	4,023.44
PART D FLAT ROOF				N/A	N/A	N/A	N/A	N/A
PART E PITCHED ROOF								
100 x 75mm sawn softwood wall plate	E1	m	5.00	1.50	25.50	13.05	5.78	44.33
200 x 50mm sawn softwood pole plate	E2	nr	5.30	1.59	27.03	17.60	6.69	51.32
100 x 50mm sawn softwood rafters	E3	m	31.50	6.30	107.10	78.12	27.78	213.00
100 x 50mm sawn softwood purlin	E4	m	5.30	1.06	18.02	15.37	5.01	38.40
150 x 50mm softwood joists	E5	m	24.50	4.90	83.30	76.44	23.96	183.70
150 x 50mm sawn softwood sprockets	E6	nr	18.00	2.16	36.72	28.08	9.72	74.52
100mm layer of insulation quilt laid over and between joists	E7	m²	15.00	7.20	122.40	174.00	44.46	340.86
Carried forward				24.71	420.07	402.66	123.41	946.13

	Ref	Unit	Qty	Hours	Hours £	Mat'ls £	O & P £	Total £
Brought forward				24.71	420.07	402.66	123.41	946.13
6mm softwood soffit 150mm wide	E8	m	12.30	4.92	83.64	25.95	16.44	126.03
19mm wrought softwood fascia/ barge board 200mm high	E9	m	5.30	2.65	45.05	14.36	8.91	68.32
Marley Plain roof tiles on felt and battens	E10	m²	23.85	45.32	770.36	864.32	245.20	1,879.88
Double eaves course	E11	m	5.30	1.86	31.54	22.31	8.08	61.93
Verge with tile undercloak	E12	m	9.00	2.25	38.25	59.49	14.66	112.40
Lead flashing code 5, 200mm girth	E13	m	5.30	3.18	54.06	45.69	14.96	114.71
Rake out joint for flashing	E14	m	5.30	1.86	31.54	1.27	4.92	37.73
112mm diameter PVC-U gutter	E15	m	5.30	1.38	23.43	17.17	6.09	46.69
Stop end	E16	nr	1.00	0.14	2.38	2.12	0.68	5.18
Stop end outlet	E17	nr	1.00	0.25	4.25	4.19	1.27	9.71
68mm diameter PVC-U down pipe	E18	m	2.50	0.63	10.63	14.25	3.73	28.61
Shoe	E19	nr	1.00	0.30	5.10	3.15	1.24	9.49
Paint fascia and soffit	E20	m²	3.69	0.74	12.55	5.90	2.77	21.22
Carried to summary				90.17	1,532.82	1,482.84	452.35	3,468.01
PART F **WINDOWS AND** **EXTERNAL DOORS**								
PVC-U door size 840 x 1980mm complete (B)	F1	0.00	0.00	0.00	0.00	0.00	0.00	0.00
PVC-U sliding patio door size 1700 x 2075mm (C)	F2	nr	2.00	14.00	238.00	616.94	128.24	983.18
Carried forward				14.00	238.00	616.94	128.24	983.18

	Ref	Unit	Qty	Hours	Hours £	Mat'ls £	O & P £	Total £
Brought forward				14.00	238.00	616.94	128.24	983.18
PVC-U window size 1200 x 1200mm complete (A)	F3	nr	2.00	4.00	68.00	364.28	64.84	497.12
25 x 225mm wrought softwood window board	F4	m	2.40	0.72	12.24	9.65	3.28	25.17
Paint window board	F5	m	2.40	0.14	2.38	2.21	0.69	5.28
Carried to summary				18.86	320.62	993.08	197.05	1,510.75
PART G INTERNAL PARTITIONS AND DOORS			N/A	N/A	N/A	N/A	N/A	N/A
PART H WALL FINISHES								
19 x 100mm wrought softwood skirting	H1	m	8.54	1.45	24.68	21.44	6.92	53.03
12mm plasterboard fixed to walls with dabs	H2	m²	19.00	7.41	125.97	54.15	27.02	207.14
12mm plasterboard fixed to walls less than 300mm wide	H3	m	13.47	2.42	41.22	12.39	8.04	61.65
Two coats emulsion paint to walls	H4	m²	21.02	5.47	92.91	20.18	16.96	130.05
Paint skirting	H5	m	8.54	1.71	29.04	7.86	5.53	42.43
Carried to summary				18.46	313.81	116.01	64.47	494.30
PART J FLOOR FINISHES								
Cement and sand floor screed 40mm thick	J1	m²	11.88	2.97	50.49	56.55	16.06	123.09
Vinyl floor tiles, size 300 x 300mm	J2	m²	11.88	2.02	34.33	92.19	18.98	145.50
Carried to summary				4.99	84.82	148.74	35.03	268.59

	Ref	Unit	Qty	Hours	Hours £	Mat'ls £	O & P £	Total £
PART K **CEILING FINISHES**								
Plasterboard with taped butt joints ixed to joists	K1	m²	11.88	4.28	72.71	33.86	15.98	122.55
5mm skim coat to plasterboard ceilings	K2	m²	11.88	5.94	100.98	20.55	18.23	139.76
Two coats emulsion paint to ceilings	K3	m²	11.88	3.09	52.51	11.40	9.59	73.50
Carried to summary				13.31	226.20	65.82	43.80	335.81
PART L **ELECTRICAL WORK**								
13 amp double switched socket outlet with neon	L1	nr	3.00	2.40	45.60	26.43	10.80	82.83
Lighting point	L2	nr	2.00	1.40	26.60	14.68	6.19	47.47
Lighting switch	L3	nr	2.00	1.40	26.60	8.84	5.32	40.76
Lighting wiring	L4	m	7.00	1.40	26.60	13.02	5.94	45.56
Power cable	L5	m	18.00	5.40	102.60	48.78	22.71	174.09
Carried to summary				12.00	228.00	111.75	50.96	390.71
PART M **HEATING WORK**								
15mm copper pipe	M1	m	7.00	3.08	55.44	15.05	10.57	81.06
Elbow	M2	nr	4.00	2.24	40.32	9.76	7.51	57.59
Tee	M3	nr	1.00	0.56	10.08	2.44	1.88	14.40
Radiator, double convector size 1400 x 520mm	M4	nr	2.00	2.60	46.80	278.22	48.75	373.77
Break into existing pipe and insert tee	M5	nr	1.00	0.75	13.50	4.42	2.69	20.61
Carried to summary				9.23	166.14	309.89	71.40	547.43

	Ref	Unit	Qty	Hours	Hours £	Mat'ls £	O & P £	Total £

PART N
ALTERATION WORK

	Ref	Unit	Qty	Hours	Hours £	Mat'ls £	O & P £	Total £
Take out existing window size 1500 x 1000mm and lintel over, adapt opening to receive 1770 x 2000mm patio door and insert new lintel over (both measured separately) and make good	N1	nr	1.00	20.00	340.00	32.10	55.82	427.92
Carried to summary				20.00	340.00	32.10	55.82	427.92

SUMMARY

	Hours	Hours £	Mat'ls £	O & P £	Total £
PART A PRELIMINARIES	0.00	0.00	0.00	0.00	2,074.00
PART B SUBSTRUCTURE TO DPC LEVEL	66.38	966.77	837.34	270.62	2,074.73
PART C EXTERNAL WALLS	90.67	1,541.44	1,957.21	524.80	4,023.45
PART D FLAT ROOF	0.00	0.00	0.00	0.00	0.00
PART E PITCHED ROOF	90.17	1,532.82	1,482.84	452.35	3,468.01
PART F WINDOWS AND EXTERNAL DOORS	18.86	320.62	993.08	197.05	1,510.75
PART G INTERNAL PARTITIONS AND DOORS	0.00	0.00	0.00	0.00	0.00
PART H WALL FINISHES	18.46	313.81	116.01	64.47	494.30
PART J FLOOR FINISHES	4.99	84.82	148.74	35.03	268.59
PART K CEILING FINISHES	13.31	226.20	65.82	43.80	335.81
PART L ELECTRICAL WORK	12.00	228.00	111.75	50.96	390.71
PART M HEATING WORK	9.23	166.14	309.89	71.40	547.43
PART N ALTERATION WORK	20.00	340.00	32.10	55.82	427.92
Final total	344.07	5,720.62	6,054.78	1,766.29	15,615.69

	Ref	Unit	Qty	Hours	Hours £	Mat'ls £	O & P £	Total £
PART A **PRELIMINARIES**								
Concrete mixer	A1	wks	9.00					450.00
Small tools	A2	wks	10.00					400.00
Scaffolding (m²/weeks)	A3		130					780.00
Skip	A4	wks	6.00					660.00
Clean up	A5	hrs	10.00					130.00
Carried to summary								2,420.00
PART B **SUBSTRUCTURE TO** **DPC LEVEL**								
Excavate topsoil 150mm thick by hand	B1	m²	19.85	5.96	77.42	0.00	11.61	89.03
Excavate to reduce levels	B2	m³	5.95	14.88	193.38	0.00	29.01	222.38
Excavate for trench foundations by hand	B3	m³	1.88	4.89	63.54	0.00	9.53	73.08
Earthwork support to sides of trenches	B4	m²	16.64	6.66	86.53	30.78	17.60	134.91
Backfilling with excavated material	B5	m³	0.70	0.42	5.46	0.00	0.82	6.28
Hardcore 225mm thick	B6	m²	14.58	2.92	37.91	91.71	19.44	149.06
Hardcore filling to trench	B7	m³	0.15	0.08	0.98	6.29	1.09	8.35
Concrete grade (1:3:6) in foundations	B8	m³	1.54	2.08	27.03	131.85	23.83	182.71
Concrete grade (1:2:4) in bed 150mm thick	B9	m²	14.58	4.37	56.86	223.37	42.03	322.26
Concrete (1:2:4) in cavity wall filling	B10	m²	5.70	1.14	11.97	13.57	3.83	29.37
Carried forward				43.38	561.06	497.57	158.79	1,217.43

	Ref	Unit	Qty	Hours	Hours £	Mat'ls £	O & P £	Total £
Brought forward				43.38	561.06	497.57	158.79	1,217.43
Damp-proof membrane	B11	m²	15.13	0.61	7.87	35.10	6.45	49.41
Reinforcement ref A193 in foundation	B12	m²	5.70	0.68	8.89	11.51	3.06	23.47
Steel fabric reinforcement ref A193 in slab	B13	m²	14.58	2.19	28.43	29.16	8.64	66.23
Solid blockwork 140mm thick in cavity wall	B14	m²	7.40	9.62	163.54	123.43	43.05	330.02
Common bricks 112.5mm thick in cavity wall	B15	m²	5.70	9.69	164.73	93.77	38.77	297.27
Facing bricks in 112.5mm thick in skin of cavity wall	B16	m²	1.71	3.08	52.33	28.55	12.13	93.01
Form cavity 50mm wide in cavity wall	B17	m²	7.41	0.22	3.78	11.26	2.26	17.30
DPC 112mm wide	B18	m	11.40	0.57	9.69	11.63	3.20	24.52
DPC 140mm wide	B19	m	11.40	0.68	11.63	15.73	4.10	31.46
Bond in block wall	B20	m	1.30	0.57	9.72	3.78	2.03	15.53
Bond in half brick wall	B21	m	0.30	0.11	1.79	0.80	0.39	2.98
50mm thick insulation board	B22	m²	14.58	4.37	74.36	104.98	26.90	206.23
Carried to summary				75.77	1,097.81	967.28	309.76	2,374.86

PART C
EXTERNAL WALLS

	Ref	Unit	Qty	Hours	Hours £	Mat'ls £	O & P £	Total £
Solid blockwork 140mm thick in cavity wall	C1	m²	29.00	37.70	640.90	483.72	168.69	1,293.31
Facing brickwork 112.5mm thick in cavity wall	C2	m²	29.00	52.20	887.40	827.95	257.30	1,972.65
75mm thick insulation in cavity wall	C3	m²	29.00	6.38	108.46	154.86	39.50	302.82
Carried forward				96.28	1,636.76	1,466.53	465.49	3,568.78

	Ref	Unit	Qty	Hours	Hours £	Mat'ls £	O & P £	Total £
Brought forward				96.28	1,636.76	1,466.53	465.49	3,568.78
Steel lintel 2400mm long	C4	nr	2.00	0.50	8.50	296.58	45.76	350.84
Steel lintel 1500mm long	C5	nr	2.00	0.40	6.80	242.72	37.43	286.95
Steel lintel 1150mm long	C6	nr	1.00	0.15	2.55	98.72	15.19	116.46
Close cavity wall at jambs	C7	m	17.32	0.87	14.72	48.67	9.51	72.90
Close cavity wall at cills	C8	m	5.71	0.29	4.85	16.05	3.13	24.03
Close cavity wall at top	C9	m	11.40	0.57	9.69	32.03	6.26	47.98
DPC 112mm wide at jambs	C10	m	17.32	0.87	14.72	17.67	4.86	37.25
DPC 112mm wide at cills	C11	m	5.71	0.29	4.85	5.82	1.60	12.28
Carried to summary				100.05	1,700.90	2,126.07	574.05	4,401.02
PART D FLAT ROOF				N/A	N/A	N/A	N/A	N/A
PART E PITCHED ROOF								
100 x 75mm sawn softwood wall plate	E1	m	6.00	1.80	30.60	15.66	6.94	53.20
200 x 50mm sawn softwood pole plate	E2	nr	6.30	1.89	32.13	20.92	7.96	61.00
100 x 50mm sawn softwood rafters	E3	m	31.50	6.30	107.10	78.12	27.78	213.00
100 x 50mm sawn softwood purlin	E4	m	6.30	1.26	21.42	18.27	5.95	45.64
150 x 50mm softwood joists	E5	m	24.50	4.90	83.30	76.44	23.96	183.70
150 x 50mm sawn softwood sprockets	E6	nr	18.00	2.16	36.72	28.08	9.72	74.52
100mm layer of insulation quilt laid over and between between joists	E7	m²	18.00	8.64	146.88	208.80	53.35	409.03
Carried forward				26.95	548.59	474.35	153.44	1,176.38

	Ref	Unit	Qty	Hours	Hours £	Mat'ls £	O & P £	Total £
Brought forward				26.95	548.59	474.35	153.44	1,176.38
6mm softwood soffit 150mm wide	E8	m	13.30	5.32	90.44	28.06	17.78	136.28
19mm wrought softwood fascia/ barge board 200mm high	E9	m	6.30	3.15	53.55	17.07	10.59	81.22
Marley Plain roof tiles on felt and battens	E10	m²	28.35	53.87	915.71	1,027.40	291.47	2,234.58
Double eaves course	E11	m	6.30	2.21	37.49	26.52	9.60	73.61
Verge with tile undercloak	E12	m	9.00	2.25	38.25	59.49	14.66	112.40
Lead flashing code 5, 200mm girth	E13	m	6.30	3.78	64.26	54.31	17.78	136.35
Rake out joint for flashing	E14	m	6.30	2.21	37.49	1.51	5.85	44.85
112mm diameter PVC-U gutter	E15	m	6.30	1.64	27.85	20.41	7.24	55.50
Stop end	E16	nr	1.00	0.14	2.38	2.12	0.68	5.18
Stop end outlet	E17	nr	1.00	0.25	4.25	4.19	1.27	9.71
68mm diameter PVC-U down pipe	E18	m	2.50	0.63	10.63	14.25	3.73	28.61
Shoe	E19	nr	1.00	0.30	5.10	3.15	1.24	9.49
Paint fascia and soffit	E20	m²	3.99	0.80	13.57	6.38	2.99	22.94
Carried to summary				103.50	1,759.09	1,711.16	520.54	3,773.23

**PART F
WINDOWS AND
EXTERNAL DOORS**

	Ref	Unit	Qty	Hours	Hours £	Mat'ls £	O & P £	Total £
PVC-U door size 840 x 1980mm complete (B)	F1	nr	1.00	2.50	42.50	278.67	48.18	369.35
PVC-U sliding patio door size 1700 x 2075mm (C)	F2	nr	2.00	14.00	238.00	616.94	128.24	983.18
Carried forward				16.50	280.50	895.61	176.42	1,352.53

	Ref	Unit	Qty	Hours	Hours £	Mat'ls £	O & P £	Total £
Brought forward				16.50	280.50	895.61	176.42	1,352.53
PVC-U window size 1200 x 1200mm complete (A)	F3	nr	3.00	6.00	102.00	546.42	97.26	745.68
25 x 225mm wrought softwood window board	F4	m	3.60	1.08	18.36	14.47	4.92	37.76
Paint window board	F5	m	3.60	0.14	2.38	3.31	0.85	6.55
Carried to summary				23.72	403.24	1,459.81	279.46	2,142.51
PART G INTERNAL PARTITIONS AND DOORS			N/A	N/A	N/A	N/A	N/A	N/A
PART H WALL FINISHES								
19 x 100mm wrought softwood skirting	H1	m	8.70	1.48	25.14	21.84	7.05	54.03
12mm plasterboard fixed to walls with dabs	H2	m²	18.40	7.18	121.99	52.44	26.16	200.60
12mm plasterboard fixed to walls less than 300mm wide	H3	m	23.03	4.15	70.47	21.19	13.75	105.41
Two coats emulsion paint to walls	H4	m²	21.85	5.68	96.58	20.98	17.63	135.19
Paint skirting	H5	m	8.70	1.74	29.58	8.00	5.64	43.22
Carried to summary				20.22	343.76	124.44	70.23	538.44
PART J FLOOR FINISHES								
Cement and sand floor screed 40mm thick	J1	m²	14.58	3.65	61.97	69.40	19.70	151.07
Vinyl floor tiles, size 300 x 300mm	J2	m²	14.58	2.48	42.14	113.14	23.29	178.57
Carried to summary				6.12	104.10	182.54	43.00	329.64

	Ref	Unit	Qty	Hours	Hours £	Mat'ls £	O & P £	Total £
PART K **CEILING FINISHES**								
Plasterboard with taped butt joints fixed to joists	K1	m²	14.58	5.25	89.23	41.55	19.62	150.40
5mm skim coat to plasterboard ceilings	K2	m²	14.58	7.29	123.93	25.22	22.37	171.53
Two coats emulsion paint to ceilings	K3	m²	14.58	3.79	64.44	14.00	11.77	90.21
Carried to summary				16.33	277.60	80.77	53.76	412.13
PART L **ELECTRICAL WORK**								
13 amp double switched socket outlet with neon	L1	nr	4.00	3.20	60.80	35.24	14.41	110.45
Lighting point	L2	nr	2.00	1.40	26.60	14.68	6.19	47.47
Lighting switch	L3	nr	3.00	2.10	39.90	13.26	7.97	61.13
Lighting wiring	L4	m	8.00	1.60	30.40	14.88	6.79	52.07
Power cable	L5	m	20.00	6.00	114.00	54.20	25.23	193.43
Carried to summary				14.30	271.70	132.26	60.59	464.55
PART M **HEATING WORK**								
15mm copper pipe	M1	m	8.00	3.52	63.36	17.20	12.08	92.64
Elbow	M2	nr	4.00	2.24	40.32	9.76	7.51	57.59
Tee	M3	nr	1.00	0.56	10.08	2.44	1.88	14.40
Radiator, double convector size 1400 x 520mm	M4	nr	2.00	2.60	46.80	278.22	48.75	373.77
Break into existing pipe and insert tee	M5	nr	1.00	0.75	13.50	4.42	2.69	20.61
Carried to summary				9.67	174.06	312.04	72.92	559.02

	Ref	Unit	Qty	Hours	Hours £	Mat'ls £	O & P £	Total £
PART N								
ALTERATION WORK								
Take out existing window size 1500 x 1000mm and lintel over, adapt opening to receive 1770 x 2000mm patio door and insert new lintel over (both measured separately) and make good	N1	nr	1.00	20.00	340.00	32.10	55.82	427.92
Carried to summary				20.00	340.00	32.10	55.82	427.92

SUMMARY

	Hours	Hours £	Mat'ls £	O & P £	Total £
PART A **PRELIMINARIES**	0.00	0.00	0.00	0.00	2,420.00
PART B SUBSTRUCTURE TO **DPC LEVEL**	75.77	1,097.81	967.28	309.76	2,374.86
PART C **EXTERNAL WALLS**	100.05	1,700.90	2,126.07	574.05	4,401.02
PART D **FLAT ROOF**	0.00	0.00	0.00	0.00	0.00
PART E **PITCHED ROOF**	103.50	1,759.09	1,711.16	520.54	3,773.23
PART F WINDOWS AND **EXTERNAL DOORS**	23.72	403.24	1,459.81	279.46	2,142.51
PART G INTERNAL **PARTITIONS AND DOORS**	0.00	0.00	0.00	0.00	0.00
PART H **WALL FINISHES**	20.22	343.76	124.44	70.23	538.44
PART J **FLOOR FINISHES**	6.12	104.10	182.54	43.00	329.64
PART K **CEILING FINISHES**	16.33	277.60	80.77	53.76	412.13
PART L **ELECTRICAL WORK**	14.30	271.70	132.26	60.59	464.55
PART M **HEATING WORK**	9.67	174.06	312.04	72.92	559.02
PART N **ALTERATION WORK**	20.00	340.00	32.10	55.82	427.92
Final total	389.68	6,472.26	7,128.47	2,040.12	17,843.31

	Ref	Unit	Qty	Hours	Hours £	Mat'ls £	O & P £	Total £
PART A **PRELIMINARIES**								
Concrete mixer	A1	wks	8.00					400.00
Small tools	A2	wks	9.00					360.00
Scaffolding (m²/weeks)	A3		130					780.00
Skip	A4	wks	5.00					550.00
Clean up	A5	hrs	10.00					130.00
Carried to summary								2,220.00
PART B **SUBSTRUCTURE TO** **DPC LEVEL**								
Excavate topsoil 150mm thick by hand	B1	m²	17.85	5.36	69.62	0.00	10.44	80.06
Excavate to reduce levels	B2	m³	5.35	13.38	173.88	0.00	26.08	199.96
Excavate for trench foundations by hand	B3	m³	1.88	4.89	63.54	0.00	9.53	73.08
Earthwork support to sides of trenches	B4	m²	16.64	6.66	86.53	30.78	17.60	134.91
Backfilling with excavated material	B5	m³	0.63	0.38	4.91	0.00	0.74	5.65
Hardcore 225mm thick	B6	m²	13.69	2.74	35.59	86.11	18.26	139.96
Hardcore filling to trench	B7	m³	0.14	0.07	0.91	6.29	1.08	8.28
Concrete grade (1:3:6) in foundations	B8	m³	1.54	2.08	27.03	131.85	23.83	182.71
Concrete grade (1:2:4) in bed 150mm thick	B9	m²	13.69	4.11	53.39	209.73	39.47	302.59
Concrete (1:2:4) in cavity wall filling	B10	m²	5.70	1.14	11.97	13.57	3.83	29.37
Carried forward				40.79	527.37	478.34	150.86	1,156.56

	Ref	Unit	Qty	Hours	Hours £	Mat'ls £	O & P £	Total £
Brought forward				40.79	527.37	478.34	150.86	1,156.56
Damp-proof membrane	B11	m²	13.13	0.53	6.83	30.46	5.59	42.88
Reinforcement ref A193 in foundation	B12	m²	5.70	0.68	8.89	11.51	3.06	23.47
Steel fabric reinforcement ref A193 in slab	B13	m²	13.69	2.05	26.70	27.38	8.11	62.19
Solid blockwork 140mm thick in cavity wall	B14	m²	7.41	9.63	163.76	123.60	43.10	330.46
Common bricks 112.5mm thick in cavity wall	B15	m²	5.70	9.69	164.73	93.77	38.77	297.27
Facing bricks in 112.5mm thick in skin of cavity wall	B16	m²	1.71	3.08	52.33	28.55	12.13	93.01
Form cavity 50mm wide in cavity wall	B17	m²	7.41	0.22	3.78	11.26	2.26	17.30
DPC 112mm wide	B18	m	11.40	0.57	9.69	11.63	3.20	24.52
DPC 140mm wide	B19	m	11.40	0.68	11.63	15.73	4.10	31.46
Bond in block wall	B20	m	1.30	0.57	9.72	3.78	2.03	15.53
Bond in half brick wall	B21	m	0.30	0.11	1.79	0.80	0.39	2.98
50mm thick insulation board	B22	m²	13.69	4.11	69.82	98.57	25.26	193.65
Carried to summary				72.71	1,057.03	935.38	298.86	2,291.27
PART C **EXTERNAL WALLS**								
Solid blockwork 140mm thick in cavity wall	C1	m²	31.90	41.47	704.99	532.09	185.56	1,422.64
Facing brickwork 112.5mm thick in cavity wall	C2	m²	31.90	57.42	976.14	910.75	283.03	2,169.92
75mm thick insulation in cavity wall	C3	m²	31.90	7.02	119.31	170.35	43.45	333.10
Carried forward				105.91	1,800.44	1,613.18	512.04	3,925.66

	Ref	Unit	Qty	Hours	Hours £	Mat'ls £	O & P £	Total £
Brought forward				105.91	1,800.44	1,613.18	512.04	3,925.66
Steel lintel 2400mm long	C4	nr	2.00	0.50	8.50	296.58	45.76	350.84
Steel lintel 1500mm long	C5	nr	3.00	0.60	10.20	364.08	56.14	430.42
Steel lintel 1150mm long	C6	nr	1.00	0.15	2.55	98.72	15.19	116.46
Close cavity wall at jambs	C7	m	17.32	0.87	14.72	48.67	9.51	72.90
Close cavity wall at cills	C8	m	5.71	0.29	4.85	16.05	3.13	24.03
Close cavity wall at top	C9	m	11.40	0.57	9.69	32.03	6.26	47.98
DPC 112mm wide at jambs	C10	m	17.32	0.87	14.72	17.67	4.86	37.25
DPC 112mm wide at cills	C11	m	5.71	0.29	4.85	5.82	1.60	12.28
Carried to summary				109.88	1,867.98	2,394.08	639.31	4,901.37
PART D FLAT ROOF				N/A	N/A	N/A	N/A	N/A
PART E PITCHED ROOF								
100 x 75mm sawn softwood wall plate	E1	m	4.00	1.20	20.40	10.44	4.63	35.47
200 x 50mm sawn softwood pole plate	E2	nr	4.30	1.29	21.93	14.28	5.43	41.64
100 x 50mm sawn softwood rafters	E3	m	49.50	9.90	168.30	122.76	43.66	334.72
100 x 50mm sawn softwood purlin	E4	m	4.30	0.86	14.62	12.47	4.06	31.15
150 x 50mm softwood joists	E5	m	40.50	8.10	137.70	126.36	39.61	303.67
150 x 50mm sawn softwood sprockets	E6	nr	20.00	2.40	40.80	31.20	10.80	82.80
100mm layer of insulation quilt laid over and between joists	E7	m²	16.00	7.68	130.56	185.60	47.42	363.58
Carried forward				36.75	624.75	531.17	173.39	1,329.31

	Ref	Unit	Qty	Hours	Hours £	Mat'ls £	O & P £	Total £
Brought forward				36.75	624.75	531.17	173.39	1,329.31
6mm softwood soffit 150mm wide	E8	m	13.30	5.32	90.44	28.06	17.78	136.28
19mm wrought softwood fascia/ barge board 200mm high	E9	m	4.30	2.15	36.55	11.65	7.23	55.43
Marley Plain roof tiles on felt and battens	E10	m²	23.65	44.94	763.90	857.08	243.15	1,864.12
Double eaves course	E11	m	4.30	1.51	25.59	18.10	6.55	50.24
Verge with tile undercloak	E12	m	11.00	2.75	46.75	72.71	17.92	137.38
Lead flashing code 5, 200mm girth	E13	m	4.30	2.58	43.86	37.07	12.14	93.06
Rake out joint for flashing	E14	m	4.30	1.51	25.59	1.03	3.99	30.61
112mm diameter PVC-U gutter	E15	m	4.30	1.12	19.01	13.93	4.94	37.88
Stop end	E16	nr	1.00	0.14	2.38	2.12	0.68	5.18
Stop end outlet	E17	nr	1.00	0.25	4.25	4.19	1.27	9.71
68mm diameter PVC-U down pipe	E18	m	2.50	0.63	10.63	14.25	3.73	28.61
Shoe	E19	nr	1.00	0.30	5.10	3.15	1.24	9.49
Paint fascia and soffit	E20	m²	3.99	0.80	13.57	6.38	2.99	22.94
Carried to summary				95.41	1,621.90	1,572.84	479.21	3,525.45

**PART F
WINDOWS AND
EXTERNAL DOORS**

	Ref	Unit	Qty	Hours	Hours £	Mat'ls £	O & P £	Total £
PVC-U door size 840 x 1980mm complete (B)	F1	nr	1.00	2.50	42.50	278.67	48.18	369.35
PVC-U sliding patio door size 1700 x 2075mm (C)	F2	nr	2.00	14.00	238.00	616.94	128.24	983.18
Carried forward				16.50	280.50	895.61	176.42	1,352.53

	Ref	Unit	Qty	Hours	Hours £	Mat'ls £	O & P £	Total £
Brought forward				16.50	280.50	895.61	176.42	1,352.53
PVC-U window size 1200 x 1200mm complete (A)	F3	nr	3.00	6.00	102.00	546.42	97.26	745.68
25 x 225mm wrought softwood window board	F4	m	3.60	1.08	18.36	14.47	4.92	37.76
Paint window board	F5	m	3.60	0.14	2.38	3.31	0.85	6.55
Carried to summary				23.72	403.24	1,459.81	279.46	2,142.51
PART G INTERNAL PARTITIONS AND DOORS			N/A	N/A	N/A	N/A	N/A	N/A
PART H WALL FINISHES								
19 x 100mm wrought softwood skirting	H1	m	8.70	1.48	25.14	21.84	7.05	54.03
12mm plasterboard fixed to walls with dabs	H2	m²	18.40	7.18	121.99	52.44	26.16	200.60
12mm plasterboard fixed to walls ess than 300mm wide	H3	m	23.03	4.15	70.47	21.19	13.75	105.41
Two coats emulsion paint to walls	H4	m²	21.85	5.68	96.58	20.98	17.63	135.19
Paint skirting	H5	m	8.70	1.74	29.58	8.00	5.64	43.22
Carried to summary				20.22	343.76	124.44	70.23	538.44
PART J FLOOR FINISHES								
Cement and sand floor screed 40mm thick	J1	m²	12.58	3.15	53.47	59.88	17.00	130.35
Vinyl floor tiles, size 300 x 300mm	J2	m²	12.58	2.14	36.36	97.62	20.10	154.07
Carried to summary				5.28	89.82	157.50	37.10	284.42

	Ref	Unit	Qty	Hours	Hours £	Mat'ls £	O & P £	Total £
PART K								
CEILING FINISHES								
Plasterboard with taped butt joints fixed to joists	K1	m²	12.58	4.53	76.99	35.85	16.93	129.77
5mm skim coat to plasterboard ceilings	K2	m²	12.58	6.29	106.93	21.76	19.30	148.00
Two coats emulsion paint to ceilings	K3	m²	12.58	3.27	55.60	12.08	10.15	77.83
Carried to summary				14.09	239.52	69.69	46.38	355.60
PART L								
ELECTRICAL WORK								
13 amp double switched socket outlet with neon	L1	nr	3.00	2.40	45.60	26.43	10.80	82.83
Lighting point	L2	nr	2.00	1.40	26.60	14.68	6.19	47.47
Lighting switch	L3	nr	3.00	2.10	39.90	13.26	7.97	61.13
Lighting wiring	L4	m	8.00	1.60	30.40	14.88	6.79	52.07
Power cable	L5	m	18.00	5.40	102.60	48.78	22.71	174.09
Carried to summary				12.90	245.10	118.03	54.47	417.60
PART M								
HEATING WORK								
15mm copper pipe	M1	m	9.00	3.96	71.28	19.35	13.59	104.22
Elbow	M2	nr	4.00	2.24	40.32	9.76	7.51	57.59
Tee	M3	nr	1.00	0.56	10.08	2.44	1.88	14.40
Radiator, double convector size 1400 x 520mm	M4	nr	2.00	2.60	46.80	278.22	48.75	373.77
Break into existing pipe and insert tee	M5	nr	1.00	0.75	13.50	4.42	2.69	20.61
Carried to summary				10.11	181.98	314.19	74.43	570.60

	Ref	Unit	Qty	Hours	Hours £	Mat'ls £	O & P £	Total £

PART N
ALTERATION WORK

	Ref	Unit	Qty	Hours	Hours £	Mat'ls £	O & P £	Total £
Take out existing window size 1500 x 1000mm and lintel over, adapt opening to receive 1770 x 2000mm patio door and insert new lintel over (both measured separately) and make good	N1	nr	1.00	20.00	340.00	32.10	55.82	427.92
Carried to summary				20.00	340.00	32.10	55.82	427.92

SUMMARY

	Hours	Hours £	Mat'ls £	O & P £	Total £
PART A **PRELIMINARIES**	0.00	0.00	0.00	0.00	2,220.00
PART B SUBSTRUCTURE TO **DPC LEVEL**	72.71	1,057.03	935.38	298.86	2,291.27
PART C **EXTERNAL WALLS**	109.88	1,867.98	2,394.08	639.31	4,901.37
PART D **FLAT ROOF**	0.00	0.00	0.00	0.00	0.00
PART E **PITCHED ROOF**	95.41	1,621.90	1,572.84	479.21	3,525.45
PART F WINDOWS AND **EXTERNAL DOORS**	23.72	403.24	1,459.81	279.46	2,142.51
PART G INTERNAL **PARTITIONS AND DOORS**	0.00	0.00	0.00	0.00	0.00
PART H **WALL FINISHES**	20.22	343.76	124.44	70.23	538.44
PART J **FLOOR FINISHES**	5.28	89.82	157.50	37.10	284.42
PART K **CEILING FINISHES**	14.09	239.52	69.69	46.38	355.60
PART L **ELECTRICAL WORK**	12.90	245.10	118.03	54.47	417.60
PART M **HEATING WORK**	10.11	181.98	314.19	74.43	570.60
PART N **ALTERATION WORK**	20.00	340.00	32.10	55.82	427.92
Final total	384.32	6,390.33	7,178.06	2,035.27	17,675.18

	Ref	Unit	Qty	Hours	Hours £	Mat'ls £	O & P £	Total £
PART A **PRELIMINARIES**								
Concrete mixer	A1	wks	9.00					450.00
Small tools	A2	wks	10.00					400.00
Scaffolding (m²/weeks)	A3		140					840.00
Skip	A4	wks	6.00					660.00
Clean up	A5	hrs	10.00					130.00
Carried to summary								2,480.00
PART B **SUBSTRUCTURE TO** **DPC LEVEL**								
Excavate topsoil 150mm thick by hand	B1	m²	22.01	6.60	85.84	0.00	12.88	98.71
Excavate to reduce levels	B2	m³	6.60	16.50	214.50	0.00	32.18	246.68
Excavate for trench foundations by hand	B3	m³	2.05	5.33	69.29	0.00	10.39	79.68
Earthwork support to sides of trenches	B4	m²	18.10	7.24	94.12	33.49	19.14	146.75
Backfilling with excavated material	B5	m³	0.63	0.38	4.91	0.00	0.74	5.65
Hardcore 225mm thick	B6	m²	16.28	3.26	42.33	102.40	21.71	166.44
Hardcore filling to trench	B7	m³	0.15	0.08	0.98	6.29	1.09	8.35
Concrete grade (1:3:6) in foundations	B8	m³	1.67	2.25	29.31	142.99	25.84	198.14
Concrete grade (1:2:4) in bed 150mm thick	B9	m²	16.28	4.88	63.49	249.41	46.94	359.84
Concrete (1:2:4) in cavity wall filling	B10	m²	6.20	1.24	11.97	14.76	4.01	30.73
Carried forward				47.76	616.74	549.33	174.91	1,340.97

	Ref	Unit	Qty	Hours	Hours £	Mat'ls £	O & P £	Total £
Brought forward				47.76	616.74	549.33	174.91	1,340.97
Damp-proof membrane	B11	m²	16.88	0.68	8.78	39.16	7.19	55.13
Reinforcement ref A193 in foundation	B12	m²	6.20	0.74	9.67	12.52	3.33	25.53
Steel fabric reinforcement ref A193 in slab	B13	m²	16.28	2.44	31.75	32.56	9.65	73.95
Solid blockwork 140mm thick in cavity wall	B14	m²	8.06	10.48	178.13	134.44	46.89	359.45
Common bricks 112.5mm thick in cavity wall	B15	m²	6.20	10.54	179.18	101.99	42.18	323.35
Facing bricks in 112.5mm thick in skin of cavity wall	B16	m²	1.86	3.35	56.92	28.55	12.82	98.29
Form cavity 50mm wide in cavity wall	B17	m²	8.06	0.24	4.11	12.25	2.45	18.82
DPC 112mm wide	B18	m	12.40	0.62	10.54	12.65	3.48	26.67
DPC 140mm wide	B19	m	12.40	0.74	12.65	17.11	4.46	34.22
Bond in block wall	B20	m	1.30	0.57	9.72	3.78	2.03	15.53
Bond in half brick wall	B21	m	0.30	0.11	1.79	0.80	0.39	2.98
50mm thick insulation board	B22	m²	16.28	4.88	83.03	117.22	30.04	230.28
Carried to summary				83.15	1,202.99	1,062.37	339.80	2,605.16

**PART C
EXTERNAL WALLS**

	Ref	Unit	Qty	Hours	Hours £	Mat'ls £	O & P £	Total £
Solid blockwork 140mm thick in cavity wall	C1	m²	32.96	42.85	728.42	549.77	191.73	1,469.92
Facing brickwork 112.5mm thick in cavity wall	C2	m²	32.96	59.33	1,008.58	941.01	292.44	2,242.02
75mm thick insulation in cavity wall	C3	m²	32.96	7.25	123.27	176.01	44.89	344.17
Carried forward				109.43	1,860.26	1,666.79	529.06	4,056.11

	Ref	Unit	Qty	Hours	Hours £	Mat'ls £	O & P £	Total £
Brought forward				109.43	1,860.26	1,666.79	529.06	4,056.11
Steel lintel 2400mm long	C4	nr	2.00	0.50	8.50	296.58	45.76	350.84
Steel lintel 1500mm long	C5	nr	4.00	0.80	13.60	485.44	74.86	573.90
Steel lintel 1150mm long	C6	nr	1.00	0.15	2.55	98.72	15.19	116.46
Close cavity wall at jambs	C7	m	19.72	0.99	16.76	55.41	10.83	83.00
Close cavity wall at cills	C8	m	6.91	0.35	5.87	19.42	3.79	29.08
Close cavity wall at top	C9	m	12.40	0.62	10.54	34.84	6.81	52.19
DPC 112mm wide at jambs	C10	m	19.72	0.99	16.76	20.11	5.53	42.41
DPC 112mm wide at cills	C11	m	6.91	0.35	5.87	7.05	1.94	14.86
Carried to summary				114.01	1,938.17	2,585.64	678.57	5,202.39
PART D **FLAT ROOF**				N/A	N/A	N/A	N/A	N/A
PART E **PITCHED ROOF**								
100 x 75mm sawn softwood wall plate	E1	m	5.00	1.50	25.50	13.05	5.78	44.33
200 x 50mm sawn softwood pole plate	E2	nr	5.30	1.59	27.03	17.60	6.69	51.32
100 x 50mm sawn softwood rafters	E3	m	49.50	9.90	168.30	122.76	43.66	334.72
100 x 50mm sawn softwood purlin	E4	m	5.30	1.06	18.02	15.37	5.01	38.40
150 x 50mm softwood joists	E5	m	40.50	8.10	137.70	126.36	39.61	303.67
150 x 50mm sawn softwood sprockets	E6	nr	20.00	2.40	40.80	31.20	10.80	82.80
100mm layer of insulation quilt laid over and between joists	E7	m²	20.00	9.60	163.20	232.00	59.28	454.48
Carried forward				40.67	691.39	592.73	192.62	1,476.74

	Ref	Unit	Qty	Hours	Hours £	Mat'ls £	O & P £	Total £
Brought forward				40.67	691.39	592.73	192.62	1,476.74
6mm softwood soffit 150mm wide	E8	m	16.30	6.52	110.84	34.39	21.78	167.02
19mm wrought softwood fascia/ barge board 200mm high	E9	m	5.30	2.65	45.05	14.36	8.91	68.32
Marley Plain roof tiles on felt and battens	E10	m²	29.15	55.39	941.55	1,056.40	299.69	2,297.63
Double eaves course	E11	m	5.30	1.86	31.54	22.31	8.08	61.93
Verge with tile undercloak	E12	m	11.00	2.75	46.75	72.71	17.92	137.38
Lead flashing code 5, 200mm girth	E13	m	5.30	3.18	54.06	45.69	14.96	114.71
Rake out joint for flashing	E14	m	5.30	1.86	31.54	1.27	4.92	37.73
112mm diameter PVC-U gutter	E15	m	5.30	1.38	23.43	17.17	6.09	46.69
Stop end	E16	nr	1.00	0.14	2.38	2.12	0.68	5.18
Stop end outlet	E17	nr	1.00	0.25	4.25	4.19	1.27	9.71
68mm diameter PVC-U down pipe	E18	m	2.50	0.63	10.63	14.25	3.73	28.61
Shoe	E19	nr	1.00	0.30	5.10	3.15	1.24	9.49
Paint fascia and soffit	E20	m²	4.29	0.86	14.59	6.86	3.22	24.67
Carried to summary				111.90	1,902.23	1,853.22	563.32	4,135.73

PART F
WINDOWS AND
EXTERNAL DOORS

	Ref	Unit	Qty	Hours	Hours £	Mat'ls £	O & P £	Total £
PVC-U door size 840 x 1980mm complete (B)	F1	nr	1.00	2.50	42.50	278.67	48.18	369.35
PVC-U sliding patio door size 1700 x 2075mm (C)	F2	nr	2.00	14.00	238.00	616.94	128.24	983.18
Carried forward				16.50	280.50	895.61	176.42	1,352.53

	Ref	Unit	Qty	Hours	Hours £	Mat'ls £	O & P £	Total £
Brought forward				16.50	280.50	895.61	176.42	1,352.53
PVC-U window size 1200 x 1200mm complete (A)	F3	nr	4.00	8.00	136.00	728.56	129.68	994.24
25 x 225mm wrought softwood window board	F4	m	4.80	1.44	24.48	19.30	6.57	50.34
Paint window board	F5	m	4.80	0.14	2.38	4.42	1.02	7.82
Carried to summary				26.08	443.36	1,647.88	313.69	2,404.93
PART G INTERNAL PARTITIONS AND DOORS			N/A	N/A	N/A	N/A	N/A	N/A
PART H WALL FINISHES								
19 x 100mm wrought softwood skirting	H1	m	9.70	1.65	28.03	24.35	7.86	60.24
12mm plasterboard fixed to walls with dabs	H2	m²	19.60	7.64	129.95	55.86	27.87	213.68
12mm plasterboard fixed to walls less than 300mm wide	H3	m	26.63	4.79	81.49	24.50	15.90	121.89
Two coats emulsion paint to walls	H4	m²	23.46	6.10	103.69	22.52	18.93	145.15
Paint skirting	H5	m	19.70	3.94	66.98	18.12	12.77	97.87
Carried to summary				24.13	410.14	145.35	83.32	638.82
PART J FLOOR FINISHES								
Cement and sand floor screed 40mm thick	J1	m²	16.28	4.07	69.19	77.49	22.00	168.69
Vinyl floor tiles, size 300 x 300mm	J2	m²	16.28	2.77	47.05	126.33	26.01	199.39
Carried to summary				6.84	116.24	203.83	48.01	368.07

	Ref	Unit	Qty	Hours	Hours £	Mat'ls £	O & P £	Total £
PART K **CEILING FINISHES**								
Plasterboard with taped butt joints fixed to joists	K1	m²	16.28	5.86	99.63	46.40	21.90	167.94
5mm skim coat to plasterboard ceilings	K2	m²	16.28	8.14	138.38	28.16	24.98	191.53
Two coats emulsion paint to ceilings	K3	m²	16.28	4.23	71.96	15.63	13.14	100.72
Carried to summary				18.23	309.97	90.19	60.02	460.19
PART L **ELECTRICAL WORK**								
13 amp double switched socket outlet with neon	L1	nr	4.00	3.20	60.80	35.24	14.41	110.45
Lighting point	L2	nr	3.00	2.10	39.90	22.02	9.29	71.21
Lighting switch	L3	nr	3.00	2.10	39.90	13.26	7.97	61.13
Lighting wiring	L4	m	9.00	1.80	34.20	16.74	7.64	58.58
Power cable	L5	m	20.00	6.00	114.00	54.20	25.23	193.43
Carried to summary				15.20	288.80	141.46	64.54	494.80
PART M **HEATING WORK**								
15mm copper pipe	M1	m	10.00	4.40	79.20	21.50	15.11	115.81
Elbow	M2	nr	4.00	2.24	40.32	9.76	7.51	57.59
Tee	M3	nr	1.00	0.56	10.08	2.44	1.88	14.40
Radiator, double convector size 1400 x 520mm	M4	nr	3.00	3.90	70.20	417.33	73.13	560.66
Break into existing pipe and insert tee	M5	nr	1.00	0.75	13.50	4.42	2.69	20.61
Carried to summary				11.85	213.30	455.45	100.31	769.06

	Ref	Unit	Qty	Hours	Hours £	Mat'ls £	O & P £	Total £

PART N
ALTERATION WORK

	Ref	Unit	Qty	Hours	Hours £	Mat'ls £	O & P £	Total £
Take out existing window size 1500 x 1000mm and lintel over, adapt opening to receive 1770 x 2000mm patio door and insert new lintel over (both measured separately) and make good	N1	nr	1.00	20.00	340.00	32.10	55.82	427.92
Carried to summary				20.00	340.00	32.10	55.82	427.92

SUMMARY

	Hours	Hours £	Mat'ls £	O & P £	Total £
PART A **PRELIMINARIES**	0.00	0.00	0.00	0.00	2,480.00
PART B SUBSTRUCTURE TO **DPC LEVEL**	83.15	1,202.99	1,062.37	339.80	2,605.16
PART C **EXTERNAL WALLS**	114.01	1,938.17	2,585.64	678.57	5,202.39
PART D **FLAT ROOF**	0.00	0.00	0.00	0.00	0.00
PART E **PITCHED ROOF**	111.90	1,902.23	1,853.22	563.32	4,135.73
PART F WINDOWS AND **EXTERNAL DOORS**	26.08	443.36	1,647.88	313.69	2,404.93
PART G INTERNAL **PARTITIONS AND DOORS**	0.00	0.00	0.00	0.00	0.00
PART H **WALL FINISHES**	24.13	410.14	145.35	83.32	638.82
PART J **FLOOR FINISHES**	6.84	116.24	203.83	48.01	368.07
PART K **CEILING FINISHES**	18.23	309.97	90.19	60.02	460.19
PART L **ELECTRICAL WORK**	15.20	288.80	141.46	64.54	494.80
PART M **HEATING WORK**	11.85	213.30	455.45	100.31	769.06
PART N **ALTERATION WORK**	20.00	340.00	32.10	55.82	427.92
Final total	431.39	7,165.20	8,217.49	2,307.40	19,987.07

	Ref	Unit	Qty	Hours	Hours £	Mat'ls £	O & P £	Total £

PART A
PRELIMINARIES

	Ref	Unit	Qty	Hours	Hours £	Mat'ls £	O & P £	Total £
Concrete mixer	A1	wks	10.00					500.00
Small tools	A2	wks	11.00					440.00
Scaffolding (m²/weeks)	A3		150					900.00
Skip	A4	wks	7.00					770.00
Clean up	A5	hrs	10.00					130.00
Carried to summary								2,740.00

PART B
SUBSTRUCTURE TO
DPC LEVEL

	Ref	Unit	Qty	Hours	Hours £	Mat'ls £	O & P £	Total £
Excavate topsoil 150mm thick by hand	B1	m²	26.15	7.85	101.99	0.00	15.30	117.28
Excavate to reduce levels	B2	m³	7.85	19.63	255.13	0.00	38.27	293.39
Excavate for trench foundations by hand	B3	m³	2.21	5.75	74.70	0.00	11.20	85.90
Earthwork support to sides of trenches	B4	m²	19.56	7.82	101.71	36.19	20.68	158.58
Backfilling with excavated material	B5	m³	0.77	0.46	6.01	0.00	0.90	6.91
Hardcore 225mm thick	B6	m²	19.98	4.00	51.95	125.67	26.64	204.27
Hardcore filling to trench	B7	m³	0.17	0.09	1.11	6.29	1.11	8.50
Concrete grade (1:3:6) in foundations	B8	m³	1.81	2.44	31.77	154.97	28.01	214.75
Concrete grade (1:2:4) in bed 150mm thick	B9	m²	19.98	5.99	77.92	306.09	57.60	441.62
Concrete (1:2:4) in cavity wall filling	B10	m²	6.70	1.34	11.97	15.95	4.19	32.10
Carried forward				55.36	714.24	645.16	203.91	1,563.31

	Ref	Unit	Qty	Hours	Hours £	Mat'ls £	O & P £	Total £
Brought forward				55.36	714.24	645.16	203.91	1,563.31
Damp-proof membrane	B11	m²	20.63	0.83	10.73	47.86	8.79	67.38
Reinforcement ref A193 in foundation	B12	m²	6.70	0.80	10.45	13.53	3.60	27.58
Steel fabric reinforcement ref A193 in slab	B13	m²	19.98	3.00	38.96	39.96	11.84	90.76
Solid blockwork 140mm thick in cavity wall	B14	m²	8.71	11.32	192.49	145.28	50.67	388.44
Common bricks 112.5mm thick in cavity wall	B15	m²	6.70	11.39	193.63	110.22	45.58	349.42
Facing bricks in 112.5mm thick in skin of cavity wall	B16	m²	2.01	3.62	61.51	28.55	13.51	103.56
Form cavity 50mm wide in cavity wall	B17	m²	8.71	0.26	4.44	13.24	2.65	20.33
DPC 112mm wide	B18	m	13.40	0.67	11.39	13.67	3.76	28.82
DPC 140mm wide	B19	m	13.40	0.80	13.67	18.49	4.82	36.98
Bond in block wall	B20	m	1.30	0.57	9.72	3.78	2.03	15.53
Bond in half brick wall	B21	m	0.30	0.11	1.79	0.80	0.39	2.98
50mm thick insulation board	B22	m²	19.98	5.99	101.90	143.86	36.86	282.62
Carried to summary				94.72	1,364.91	1,224.41	388.40	2,977.72

PART C
EXTERNAL WALLS

	Ref	Unit	Qty	Hours	Hours £	Mat'ls £	O & P £	Total £
Solid blockwork 140mm thick in cavity wall	C1	m²	34.46	44.80	761.57	574.79	200.45	1,536.81
Facing brickwork 112.5mm thick in cavity wall	C2	m²	34.46	62.03	1,054.48	983.83	305.75	2,344.06
75mm thick insulation in cavity wall	C3	m²	34.36	7.56	128.51	183.48	46.80	358.79
Carried forward				114.39	1,944.55	1,742.11	553.00	4,239.66

	Ref	Unit	Qty	Hours	Hours £	Mat'ls £	O & P £	Total £
Brought forward				114.39	1,944.55	1,742.11	553.00	4,239.66
Steel lintel 2400mm long	C4	nr	2.00	0.50	8.50	296.58	45.76	350.84
Steel lintel 1500mm long	C5	nr	4.00	0.80	13.60	485.44	74.86	573.90
Steel lintel 1150mm long	C6	nr	1.00	0.15	2.55	98.72	15.19	116.46
Close cavity wall at jambs	C7	m	19.72	0.99	16.76	55.41	10.83	83.00
Close cavity wall at cills	C8	m	6.91	0.35	5.87	19.42	3.79	29.08
Close cavity wall at top	C9	m	13.04	0.65	11.08	36.64	7.16	54.89
DPC 112mm wide at jambs	C10	m	19.72	0.99	16.76	20.11	5.53	42.41
DPC 112mm wide at cills	C11	m	6.91	0.35	5.87	7.05	1.94	14.86
Carried to summary				119.00	2,023.00	2,662.76	702.87	5,388.63
PART D **FLAT ROOF**				N/A	N/A	N/A	N/A	N/A
PART E **PITCHED ROOF**								
100 x 75mm sawn softwood wall plate	E1	m	6.00	1.80	30.60	15.66	6.94	53.20
200 x 50mm sawn softwood pole plate	E2	nr	6.30	1.89	32.13	20.92	7.96	61.00
100 x 50mm sawn softwood rafters	E3	m	49.50	9.90	168.30	122.76	43.66	334.72
100 x 50mm sawn softwood purlin	E4	m	6.30	1.26	21.42	18.27	5.95	45.64
150 x 50mm softwood joists	E5	m	40.50	8.10	137.70	126.36	39.61	303.67
150 x 50mm sawn softwood sprockets	E6	nr	20.00	2.40	40.80	31.20	10.80	82.80
100mm layer of insulation quilt laid over and between joists	E7	m²	24.00	11.52	195.84	278.40	71.14	545.38
Carried forward				42.99	730.83	645.85	206.50	1,583.18

	Ref	Unit	Qty	Hours	Hours £	Mat'ls £	O & P £	Total £
Brought forward				42.99	730.83	645.85	206.50	1,583.18
6mm softwood soffit 150mm wide	E8	m	15.30	6.12	104.04	32.28	20.45	156.77
19mm wrought softwood fascia/ barge board 200mm high	E9	m	6.30	3.15	53.55	17.07	10.59	81.22
Marley Plain roof tiles on felt and battens	E10	m²	36.65	69.64	1,183.80	1,328.20	376.80	2,888.79
Double eaves course	E11	m	6.30	2.21	37.49	26.52	9.60	73.61
Verge with tile undercloak	E12	m	11.00	2.75	46.75	72.71	17.92	137.38
Lead flashing code 5, 200mm girth	E13	m	6.30	3.78	64.26	54.31	17.78	136.35
Rake out joint for flashing	E14	m	6.30	2.21	37.49	1.51	5.85	44.85
112mm diameter PVC-U gutter	E15	m	6.30	1.64	27.85	20.41	7.24	55.50
Stop end	E16	nr	1.00	0.14	2.38	2.12	0.68	5.18
Stop end outlet	E17	nr	1.00	0.25	4.25	4.19	1.27	9.71
68mm diameter PVC-U down pipe	E18	m	2.50	0.63	10.63	14.25	3.73	28.61
Shoe	E19	nr	1.00	0.30	5.10	3.15	1.24	9.49
Paint fascia and soffit	E20	m²	4.59	0.92	15.61	7.34	3.44	26.39
Carried to summary				130.59	2,219.96	2,197.64	662.64	4,862.67

PART F
WINDOWS AND
EXTERNAL DOORS

	Ref	Unit	Qty	Hours	Hours £	Mat'ls £	O & P £	Total £
PVC-U door size 840 x 1980mm complete (B)	F1	nr	1.00	2.50	42.50	278.67	48.18	369.35
PVC-U sliding patio door size 1700 x 2075mm (C)	F2	nr	2.00	14.00	238.00	616.94	128.24	983.18
Carried forward				16.50	280.50	895.61	176.42	1,352.53

	Ref	Unit	Qty	Hours	Hours £	Mat'ls £	O & P £	Total £
Brought forward				16.50	280.50	895.61	176.42	1,352.53
PVC-U window size 1200 x 1200mm complete (A)	F3	nr	4.00	8.00	136.00	728.56	129.68	994.24
25 x 225mm wrought softwood window board	F4	m	4.80	1.44	24.48	19.30	6.57	50.34
Paint window board	F5	m	4.80	0.14	2.38	4.42	1.02	7.82
Carried to summary				26.08	443.36	1,647.88	313.69	2,404.93
PART G INTERNAL PARTITIONS AND DOORS			N/A	N/A	N/A	N/A	N/A	N/A
PART H WALL FINISHES								
19 x 100mm wrought softwood skirting	H1	m	10.07	1.71	29.10	25.28	8.16	62.53
12mm plasterboard fixed to walls with dabs	H2	m²	21.96	8.56	145.59	62.59	31.23	239.41
12mm plasterboard fixed to walls less than 300mm wide	H3	m	26.63	4.79	81.49	24.50	15.90	121.89
Two coats emulsion paint to walls	H4	m²	25.96	6.75	114.74	24.92	20.95	160.61
Paint skirting	H5	m	10.70	2.14	36.38	9.84	6.93	53.16
Carried to summary				23.96	407.31	147.13	83.17	637.60
PART J FLOOR FINISHES								
Cement and sand floor screed 40mm thick	J1	m²	19.88	4.97	84.49	94.63	26.87	205.99
Vinyl floor tiles, size 300 x 300mm	J2	m²	19.88	3.38	57.45	154.27	31.76	243.48
Carried to summary				8.35	141.94	248.90	58.63	449.47

	Ref	Unit	Qty	Hours	Hours £	Mat'ls £	O & P £	Total £
PART K **CEILING FINISHES**								
Plasterboard with taped butt joints fixed to joists	K1	m²	19.88	7.16	121.67	56.66	26.75	205.07
5mm skim coat to plasterboard ceilings	K2	m²	19.88	9.94	168.98	34.39	30.51	233.88
Two coats emulsion paint to ceilings	K3	m²	19.88	5.17	87.87	19.08	16.04	123.00
Carried to summary				22.27	378.52	110.14	73.30	561.95
PART L **ELECTRICAL WORK**								
13 amp double switched socket outlet with neon	L1	nr	5.00	4.00	76.00	44.05	18.01	138.06
Lighting point	L2	nr	3.00	2.10	39.90	22.02	9.29	71.21
Lighting switch	L3	nr	3.00	2.10	39.90	13.26	7.97	61.13
Lighting wiring	L4	m	11.00	2.20	41.80	20.46	9.34	71.60
Power cable	L5	m	22.00	6.60	125.40	59.62	27.75	212.77
Carried to summary				17.00	323.00	159.41	72.36	554.77
PART M **HEATING WORK**								
15mm copper pipe	M1	m	11.00	4.84	87.12	23.65	16.62	127.39
Elbow	M2	nr	4.00	2.24	40.32	9.76	7.51	57.59
Tee	M3	nr	1.00	0.56	10.08	2.44	1.88	14.40
Radiator, double convector size 1400 x 520mm	M4	nr	3.00	3.90	70.20	417.33	73.13	560.66
Break into existing pipe and insert tee	M5	nr	1.00	0.75	13.50	4.42	2.69	20.61
Carried to summary				12.29	221.22	457.60	101.82	780.64

	Ref	Unit	Qty	Hours	Hours £	Mat'ls £	O & P £	Total £
PART N **ALTERATION WORK**								
Take out existing window size 1500 x 1000mm and lintel over, adapt opening to receive 1770 x 2000mm patio door and insert new lintel over (both measured separately) and make good	N1	nr	1.00	20.00	340.00	32.10	55.82	427.92
Carried to summary				20.00	340.00	32.10	55.82	427.92

SUMMARY

	Hours	Hours £	Mat'ls £	O & P £	Total £
PART A **PRELIMINARIES**	0.00	0.00	0.00	0.00	2,740.00
PART B SUBSTRUCTURE TO **DPC LEVEL**	94.72	1,364.91	1,224.41	388.40	2,977.72
PART C **EXTERNAL WALLS**	119.00	2,023.00	2,662.76	702.87	5,388.63
PART D **FLAT ROOF**	0.00	0.00	0.00	0.00	0.00
PART E **PITCHED ROOF**	130.59	2,219.96	2,197.64	662.64	4,862.67
PART F WINDOWS AND **EXTERNAL DOORS**	26.08	443.36	1,647.88	313.69	2,404.93
PART G INTERNAL **PARTITIONS AND DOORS**	0.00	0.00	0.00	0.00	0.00
PART H **WALL FINISHES**	23.96	407.31	147.13	83.17	637.60
PART J **FLOOR FINISHES**	8.35	141.94	248.90	58.63	449.47
PART K **CEILING FINISHES**	22.27	378.52	110.14	73.30	561.95
PART L **ELECTRICAL WORK**	17.00	323.00	159.41	72.36	554.77
PART M **HEATING WORK**	12.29	221.22	457.60	101.82	780.64
PART N **ALTERATION WORK**	20.00	340.00	32.10	55.82	427.92
Final total	474.26	7,863.22	8,887.97	2,512.70	21,786.30

	Ref	Unit	Qty	Hours	Hours £	Mat'ls £	O & P £	Total £

PART A
PRELIMINARIES

	Ref	Unit	Qty	Hours	Hours £	Mat'ls £	O & P £	Total £
Concrete mixer	A1	wks	7.00					350.00
Small tools	A2	wks	8.00					320.00
Scaffolding (m²/weeks)	A3		140					840.00
Skip	A4	wks	4.00					440.00
Clean up	A5	hrs	8.00					104.00
Carried to summary								2,054.00

PART B
SUBSTRUCTURE TO
DPC LEVEL

	Ref	Unit	Qty	Hours	Hours £	Mat'ls £	O & P £	Total £
Excavate topsoil 150mm thick by hand	B1	m²	7.10	2.13	27.69	0.00	4.15	31.84
Excavate to reduce levels	B2	m³	2.13	5.33	69.23	0.00	10.38	79.61
Excavate for trench foundations by hand	B3	m³	1.06	2.76	35.83	0.00	5.37	41.20
Earthwork support to sides of trenches	B4	m²	9.34	3.74	48.57	17.28	9.88	75.72
Backfilling with excavated material	B5	m³	0.40	0.24	3.12	0.00	0.47	3.59
Hardcore 225mm thick	B6	m²	4.08	0.82	10.61	25.66	5.44	41.71
Hardcore filling to trench	B7	m³	0.08	0.04	0.52	6.29	1.02	7.83
Concrete grade (1:3:6) in foundations	B8	m³	0.86	1.16	15.09	73.63	13.31	102.04
Concrete grade (1:2:4) in bed 150mm thick	B9	m²	4.08	1.22	15.91	62.51	11.76	90.18
Concrete (1:2:4) in cavity wall filling	B10	m²	3.20	0.64	11.97	7.62	2.94	22.52
Carried forward				18.07	238.53	192.99	64.73	496.25

	Ref	Unit	Qty	Hours	Hours £	Mat'ls £	O & P £	Total £
Brought forward				18.07	238.53	192.99	64.73	496.25
Damp-proof membrane	B11	m²	4.38	0.18	2.28	10.16	1.87	14.31
Reinforcement ref A193 in foundation	B12	m²	3.22	0.39	5.02	6.50	1.73	13.26
Steel fabric reinforcement ref A193 in slab	B13	m²	4.08	0.61	7.96	8.16	2.42	18.53
Solid blockwork 140mm thick in cavity wall	B14	m²	4.16	5.41	91.94	69.39	24.20	185.52
Common bricks 112.5mm thick in cavity wall	B15	m²	3.20	5.44	92.48	52.64	21.77	166.89
Facing bricks in 112.5mm thick in skin of cavity wall	B16	m²	0.96	1.73	29.38	28.55	8.69	66.61
Form cavity 50mm wide in cavity wall	B17	m²	4.16	0.12	2.12	6.32	1.27	9.71
DPC 112mm wide	B18	m	6.40	0.32	5.44	6.53	1.80	13.76
DPC 140mm wide	B19	m	6.40	0.38	6.53	8.83	2.30	17.66
Bond in block wall	B20	m	1.30	0.57	9.72	3.78	2.03	15.53
Bond in half brick wall	B21	m	0.30	0.11	1.79	0.80	0.39	2.98
50mm thick insulation board	B22	m²	4.08	1.22	20.81	29.38	7.53	57.71
Carried to summary				34.55	513.99	424.04	140.70	1,078.73

PART C
EXTERNAL WALLS

	Ref	Unit	Qty	Hours	Hours £	Mat'ls £	O & P £	Total £
Solid blockwork 140mm thick in cavity wall	C1	m²	23.94	31.12	529.07	399.32	139.26	1,067.65
Facing brickwork 112.5mm thick n cavity wall	C2	m²	23.94	43.09	732.56	683.49	212.41	1,628.46
75mm thick insulation in cavity wall	C3	m²	23.94	5.27	89.54	127.84	32.61	249.98
Carried forward				79.48	1,351.17	1,210.65	384.27	2,946.09

	Ref	Unit	Qty	Hours	Hours £	Mat'ls £	O & P £	Total £
Brought forward				79.48	1,351.17	1,210.65	384.27	2,946.09
Steel lintel 2400mm long	C4	nr	2.00	0.50	8.50	296.58	45.76	350.84
Steel lintel 1500mm long	C5	nr	4.00	0.80	13.60	485.44	74.86	573.90
Close cavity wall at jambs	C7	m	13.76	0.69	11.70	38.67	7.55	57.92
Close cavity wall at cills	C8	m	5.07	0.25	4.31	14.25	2.78	21.34
Close cavity wall at top	C9	m	6.40	0.32	5.44	17.98	3.51	26.94
DPC 112mm wide at jambs	C10	m	13.76	0.69	11.70	14.04	3.86	29.59
DPC 112mm wide at cills	C11	m	5.07	0.25	4.31	5.17	1.42	10.90
Carried to summary				82.98	1,410.72	2,082.77	524.02	4,017.52

PART D
FLAT ROOF

	Ref	Unit	Qty	Hours	Hours £	Mat'ls £	O & P £	Total £
200 x 50mm sawn softwood joists	D1	m	15.05	3.76	63.96	53.88	17.68	135.52
200 x 50mm sawn softwood sprocket pieces	D2	nr	8.00	1.12	19.04	14.32	5.00	38.36
18mm thick WPB grade decking	D3	m²	7.10	6.39	108.63	81.44	28.51	218.58
50 x 50mm (avg) wide sawn softwood firrings	D4	m	15.05	2.71	46.05	28.75	11.22	86.02
High density polyethylene vapour barrier 150mm thick	D5	m²	6.00	1.20	20.40	67.14	13.13	100.67
100 x 75mm sawn softwood wall plate	D6	m	7.00	2.10	35.70	19.18	8.23	63.11
100 x 75mm sawn softwood tilt fillet	D7	m	3.30	0.83	14.03	8.61	3.40	26.03
Build in ends of 200 x 50mm joists	D8	nr	7.00	2.45	41.65	1.54	6.48	49.67
Carried forward				20.56	349.46	274.85	93.65	717.96

	Ref	Unit	Qty	Hours	Hours £	Mat'ls £	O & P £	Total £
Brought forward				20.56	349.46	274.85	93.65	717.96
Rake out joint for flashing	D9	m	3.30	1.16	19.64	0.79	3.06	23.49
6mm thick soffit 150mm wide	D10	m	7.30	2.92	49.64	15.40	9.76	74.80
19mm wrought softwood fascia 200mm high	D11	m	7.30	3.65	62.05	19.78	12.27	94.11
Three layer fibre-based roofing felt	D12	m²	7.10	3.91	66.39	92.58	23.85	182.81
Felt turn-down 100mm girth	D13	m	7.30	0.73	12.41	10.22	3.39	26.02
Felt flashing 150mm girth	D14	m	3.30	0.33	5.61	5.68	1.69	12.98
112mm diameter PVC-U gutter	D15	m	3.30	0.86	14.59	10.69	3.79	29.07
Stop end	D16	nr	1.00	0.14	2.38	2.12	0.68	5.18
Stop end outlet	D17	nr	1.00	0.25	4.25	4.19	1.27	9.71
68mm diameter PVC-U down pipe	D18	m	4.80	1.20	20.40	27.36	7.16	54.92
Shoe	D19	nr	1.00	0.30	5.10	3.15	1.24	9.49
Paint fascia and soffit	D20	m²	2.19	0.44	7.45	3.50	1.64	12.59
Carried to summary				36.43	619.35	470.33	163.45	1,253.13
PART E **PITCHED ROOF**				N/A	N/A	N/A	N/A	N/A
PART F **WINDOWS AND** **EXTERNAL DOORS**								
PVC-U door size 840 x 1980mm complete (B)	F1	nr	1.00	2.50	42.50	278.67	48.18	369.35
PVC-U sliding patio door size 1700 x 2075mm (C)	F2	nr	2.00	14.00	238.00	616.94	128.24	983.18
Carried forward				16.50	280.50	895.61	176.42	1,352.53

	Ref	Unit	Qty	Hours	Hours £	Mat'ls £	O & P £	Total £
Brought forward				16.50	280.50	895.61	176.42	1,352.53
PVC-U window size 1200 x 1200mm complete (A)	F3	nr	4.00	8.00	136.00	728.56	129.68	994.24
25 x 225mm wrought softwood window board	F4	m	4.80	1.44	24.48	19.30	6.57	50.34
Paint window board	F5	m	4.80	0.14	2.38	4.42	1.02	7.82
Carried to summary				26.08	443.36	1,647.88	313.69	2,404.93
PART G INTERNAL PARTITIONS AND DOORS				N/A	N/A	N/A	N/A	N/A
PART H WALL FINISHES								
19 x 100mm wrought softwood skirting	H1	m	10.34	1.76	29.88	25.95	8.38	64.21
12mm plasterboard fixed to walls with dabs	H2	m²	20.04	7.82	132.87	57.11	28.50	218.48
12mm plasterboard fixed to walls less than 300mm wide	H3	m	18.33	3.30	56.09	16.86	10.94	83.90
Two coats emulsion paint to walls	H4	m²	22.86	5.94	101.04	21.95	18.45	141.43
Paint skirting	H5	m	10.34	2.07	35.16	9.51	6.70	51.37
Carried to summary				20.88	355.03	131.39	72.96	559.39
PART J FLOOR FINISHES								
Cement and sand floor screed 40mm thick	J1	m²	4.08	1.02	17.34	19.42	5.51	42.27
Vinyl floor tiles, size 300 x 300mm	J2	m²	4.08	0.69	11.79	31.66	6.52	49.97
Carried forward				1.71	29.13	51.08	12.03	92.24

	Ref	Unit	Qty	Hours	Hours £	Mat'ls £	O & P £	Total £
Brought forward				1.71	29.13	51.08	12.03	92.24
25mm thick tongued and grooved boarding	J3	m²	4.08	3.02	51.33	50.92	15.34	117.58
150 x 50mm softwood joists	J4	m	9.50	2.09	35.53	29.64	9.78	74.95
Cut and pin end of joists to existing brick wall	J5	nr	5.00	0.90	15.30	1.50	2.52	19.32
Build in ends of joists to blockwork	J6	nr	5.00	0.50	8.50	1.20	1.46	11.16
Carried to summary				8.22	139.79	134.34	41.12	315.25

PART K
CEILING FINISHES

	Ref	Unit	Qty	Hours	Hours £	Mat'ls £	O & P £	Total £
Plasterboard with taped butt joints fixed to joists	K1	m²	8.16	2.94	49.94	23.26	10.98	84.17
5mm skim coat to plasterboard ceilings	K2	m²	8.16	4.08	69.36	14.12	12.52	96.00
Two coats emulsion paint to ceilings	K3	m²	8.16	2.12	36.07	7.83	6.59	50.49
Carried to summary				9.14	155.37	45.21	30.09	230.66

PART L
ELECTRICAL WORK

	Ref	Unit	Qty	Hours	Hours £	Mat'ls £	O & P £	Total £
13 amp double switched socket outlet with neon	L1	nr	4.00	3.20	60.80	35.24	14.41	110.45
Lighting point	L2	nr	2.00	1.40	26.60	14.68	6.19	47.47
Lighting switch	L3	nr	4.00	2.80	53.20	17.68	10.63	81.51
Lighting wiring	L4	m	10.00	2.00	38.00	18.60	8.49	65.09
Power cable	L5	m	24.00	7.20	136.80	65.04	30.28	232.12
Carried to summary				16.60	315.40	151.24	70.00	536.64

	Ref	Unit	Qty	Hours	Hours £	Mat'ls £	O & P £	Total £
PART M **HEATING WORK**								
15mm copper pipe	M1	m	9.00	3.96	71.28	19.35	13.59	104.22
Elbow	M2	nr	8.00	4.48	80.64	19.52	15.02	115.18
Tee	M3	nr	2.00	1.12	20.16	4.88	3.76	28.80
Radiator, double convector size 1400 x 520mm	M4	nr	4.00	5.20	93.60	556.44	97.51	747.55
Break into existing pipe and insert tee	M5	nr	1.00	0.75	13.50	4.42	2.69	20.61
Carried to summary				15.51	279.18	604.61	132.57	1,016.36
PART N **ALTERATION WORK**								
Take out existing window size 1500 x 1000mm and lintel over, adapt opening to receive 1770 x 2000mm patio door and insert new lintel over (both measured separately) and make good	N1	nr	1.00	20.00	340.00	32.10	55.82	427.92
Take out existing window size 1500 x 1000mm, enlarge opening to receive new PVC–U door (measured separately)	N2	nr	1.00	24.00	408.00	58.16	69.92	536.08
Carried to summary				44.00	748.00	90.26	125.74	964.00

SUMMARY

	Hours	Hours £	Mat'ls £	O & P £	Total £
PART A **PRELIMINARIES**	0.00	0.00	0.00	0.00	2,054.00
PART B SUBSTRUCTURE TO **DPC LEVEL**	34.55	513.99	424.04	140.70	1,078.73
PART C **EXTERNAL WALLS**	82.98	1,410.72	2,082.77	524.02	4,017.51
PART D **FLAT ROOF**	36.43	619.35	470.33	163.45	1,253.13
PART E **PITCHED ROOF**	0.00	0.00	0.00	0.00	0.00
PART F WINDOWS AND **EXTERNAL DOORS**	26.08	443.36	1,647.88	313.69	2,404.93
PART G INTERNAL **PARTITIONS AND DOORS**	0.00	0.00	0.00	0.00	0.00
PART H **WALL FINISHES**	20.88	355.03	131.39	72.96	559.39
PART J **FLOOR FINISHES**	8.22	139.79	134.34	41.12	315.25
PART K **CEILING FINISHES**	9.14	155.37	45.21	30.09	230.66
PART L **ELECTRICAL WORK**	16.60	315.40	151.24	70.00	536.64
PART M **HEATING WORK**	15.51	279.18	604.61	132.57	1,016.36
PART N **ALTERATION WORK**	44.00	748.00	90.26	125.74	964.00
Final total	294.39	4,980.19	5,782.07	1,614.34	14,430.60

	Ref	Unit	Qty	Hours	Hours £	Mat'ls £	O & P £	Total £
PART A **PRELIMINARIES**								
Concrete mixer	A1	wks	8.00					400.00
Small tools	A2	wks	9.00					360.00
Scaffolding (m²/weeks)	A3		160					960.00
Skip	A4	wks	5.00					550.00
Clean up	A5	hrs	10.00					130.00
Carried to summary								2,400.00
PART B **SUBSTRUCTURE TO** **DPC LEVEL**								
Excavate topsoil 150mm thick by hand	B1	m²	9.25	2.78	36.08	0.00	5.41	41.49
Excavate to reduce levels	B2	m³	2.77	6.93	90.03	0.00	13.50	103.53
Excavate for trench foundations by hand	B3	m³	1.22	3.17	41.24	0.00	6.19	47.42
Earthwork support to sides of trenches	B4	m²	10.80	4.32	56.16	19.98	11.42	87.56
Backfilling with excavated material	B5	m³	0.47	0.28	3.67	0.00	0.55	4.22
Hardcore 225mm thick	B6	m²	5.78	1.16	15.03	36.36	7.71	59.09
Hardcore filling to trench	B7	m³	0.09	0.05	0.59	6.29	1.03	7.91
Concrete grade (1:3:6) in foundations	B8	m³	1.00	1.35	17.55	85.62	15.48	118.65
Concrete grade (1:2:4) in bed 150mm thick	B9	m²	5.78	1.73	22.54	88.55	16.66	127.76
Concrete (1:2:4) in cavity wall filling	B10	m²	3.70	0.74	11.97	8.81	3.12	23.89
Carried forward				22.50	294.84	245.60	81.07	621.50

	Ref	Unit	Qty	Hours	Hours £	Mat'ls £	O & P £	Total £
Brought forward				22.50	294.84	245.60	81.07	621.50
Damp-proof membrane	B11	m²	6.13	0.25	3.19	14.22	2.61	20.02
Reinforcement ref A193 in foundation	B12	m²	3.70	0.44	5.77	7.47	1.99	15.23
Steel fabric reinforcement ref A193 in slab	B13	m²	5.78	0.87	11.27	11.56	3.42	26.26
Solid blockwork 140mm thick in cavity wall	B14	m²	4.81	6.25	106.30	80.23	27.98	214.51
Common bricks 112.5mm thick in cavity wall	B15	m²	3.70	6.29	106.93	60.87	25.17	192.96
Facing bricks in 112.5mm thick in skin of cavity wall	B16	m²	1.11	2.00	33.97	28.55	9.38	71.89
Form cavity 50mm wide in cavity wall	B17	m²	4.81	0.14	2.45	7.31	1.46	11.23
DPC 112mm wide	B18	m	7.40	0.37	6.29	7.55	2.08	15.91
DPC 140mm wide	B19	m	7.40	0.44	7.55	10.21	2.66	20.42
Bond in block wall	B20	m	1.30	0.57	9.72	3.78	2.03	15.53
Bond in half brick wall	B21	m	0.30	0.11	1.79	0.80	0.39	2.98
50mm thick insulation board	B22	m²	5.78	1.73	29.48	41.62	10.66	81.76
Carried to summary				41.97	619.54	519.78	170.90	1,310.22

PART C
EXTERNAL WALLS

	Ref	Unit	Qty	Hours	Hours £	Mat'ls £	O & P £	Total £
Solid blockwork 140mm thick in cavity wall	C1	m²	28.99	37.69	640.68	483.55	168.63	1,292.87
Facing brickwork 112.5mm thick in cavity wall	C2	m²	28.99	52.18	887.09	827.66	257.21	1,971.97
75mm thick insulation in cavity wall	C3	m²	28.99	6.38	108.42	154.81	39.48	302.71
Carried forward				96.25	1,636.20	1,466.02	465.33	3,567.55

	Ref	Unit	Qty	Hours	Hours £	Mat'ls £	O & P £	Total £
Brought forward				96.25	1,636.20	1,466.02	465.33	3,567.55
Steel lintel 2400mm long	C4	nr	2.00	0.50	8.50	296.58	45.76	350.84
Steel lintel 1500mm long	C5	nr	4.00	0.80	13.60	485.44	74.86	573.90
Close cavity wall at jambs	C7	m	13.76	0.69	11.70	38.67	7.55	57.92
Close cavity wall at cills	C8	m	5.07	0.25	4.31	14.25	2.78	21.34
Close cavity wall at top	C9	m	7.60	0.38	6.46	21.36	4.17	31.99
DPC 112mm wide at jambs	C10	m	13.76	0.69	11.70	14.04	3.86	29.59
DPC 112mm wide at cills	C11	m	5.07	0.25	4.31	5.17	1.42	10.90
Carried to summary				99.81	1,696.77	2,341.52	605.74	4,644.03

PART D
FLAT ROOF

	Ref	Unit	Qty	Hours	Hours £	Mat'ls £	O & P £	Total £
200 x 50mm sawn softwood joists	D1	m	19.35	4.84	82.24	69.27	22.73	174.24
200 x 50mm sawn softwood sprocket pieces	D2	nr	8.00	1.12	19.04	14.32	5.00	38.36
18mm thick WPB grade decking	D3	m²	9.25	8.33	141.53	106.10	37.14	284.77
50 x 50mm (avg) wide sawn softwood firrings	D4	m	19.25	3.47	58.91	36.77	14.35	110.02
High density polyethylene vapour barrier 150mm thick	D5	m²	8.00	1.60	27.20	89.52	17.51	134.23
100 x 75mm sawn softwood wall plate	D6	m	8.00	2.40	40.80	21.92	9.41	72.13
100 x 75mm sawn softwood tilt fillet	D7	m	4.30	1.08	18.28	11.22	4.42	33.92
Build in ends of 200 x 50mm joists	D8	nr	9.00	3.15	53.55	1.98	8.33	63.86
Carried forward				25.97	441.53	351.10	118.90	911.53

	Ref	Unit	Qty	Hours	Hours £	Mat'ls £	O & P £	Total £
Brought forward				25.97	441.53	351.10	118.90	911.53
Rake out joint for flashing	D9	m	4.30	1.51	25.59	1.03	3.99	30.61
6mm thick soffit 150mm wide	D10	m	8.30	3.32	56.44	17.51	11.09	85.05
19mm wrought softwood fascia 200mm high	D11	m	8.30	4.15	70.55	22.49	13.96	107.00
Three layer fibre-based roofing felt	D12	m²	9.25	5.09	86.49	120.62	31.07	238.17
Felt turn-down 100mm girth	D13	m	8.30	0.83	14.11	11.62	3.86	29.59
Felt flashing 150mm girth	D14	m	4.30	0.43	7.31	7.40	2.21	16.91
112mm diameter PVC-U gutter	D15	m	4.30	1.12	19.01	13.93	4.94	37.88
Stop end	D16	nr	1.00	0.14	2.38	2.12	0.68	5.18
Stop end outlet	D17	nr	1.00	0.25	4.25	4.19	1.27	9.71
68mm diameter PVC-U down pipe	D18	m	4.80	1.20	20.40	27.36	7.16	54.92
Shoe	D19	nr	1.00	0.30	5.10	3.15	1.24	9.49
Paint fascia and soffit	D20	m²	2.49	0.50	8.47	3.98	1.87	14.32
Carried to summary				44.80	761.62	586.51	202.22	1,550.35
PART E **PITCHED ROOF**				N/A	N/A	N/A	N/A	N/A
PART F **WINDOWS AND** **EXTERNAL DOORS**								
PVC-U door size 840 x 1980mm complete (B)	F1	nr	1.00	2.50	42.50	278.67	48.18	369.35
PVC-U sliding patio door size 1700 x 2075mm (C)	F2	nr	2.00	14.00	238.00	616.94	128.24	983.18
Carried forward				16.50	280.50	895.61	176.42	1,352.53

	Ref	Unit	Qty	Hours	Hours £	Mat'ls £	O & P £	Total £
Brought forward				16.50	280.50	895.61	176.42	1,352.53
PVC-U window size 1200 x 1200mm complete (A)	F3	nr	4.00	8.00	136.00	728.56	129.68	994.24
25 x 225mm wrought softwood window board	F4	m	4.80	1.44	24.48	19.30	6.57	50.34
Paint window board	F5	m	4.80	0.14	2.38	4.42	1.02	7.82
Carried to summary				26.08	443.36	1,647.88	313.69	2,404.93
PART G INTERNAL PARTITIONS AND DOORS				N/A	N/A	N/A	N/A	N/A
PART H WALL FINISHES								
19 x 100mm wrought softwood skirting	H1	m	12.34	2.10	35.66	30.97	10.00	76.63
12mm plasterboard fixed to walls with dabs	H2	m²	24.94	9.73	165.35	71.08	35.46	271.90
12mm plasterboard fixed to walls less than 300mm wide	H3	m	18.33	3.30	56.09	16.86	10.94	83.90
Two coats emulsion paint to walls	H4	m²	26.76	6.96	118.28	25.69	21.60	165.56
Paint skirting	H5	m	12.34	2.47	41.96	11.35	8.00	61.31
Carried to summary				24.55	417.34	155.96	85.99	659.29
PART J FLOOR FINISHES								
Cement and sand floor screed 40mm thick	J1	m²	5.78	1.45	24.57	27.51	7.81	59.89
Vinyl floor tiles, size 300 x 300mm	J2	m²	5.78	0.98	16.70	44.85	9.23	70.79
Carried forward				2.43	41.27	72.37	17.05	130.68

	Ref	Unit	Qty	Hours	Hours £	Mat'ls £	O & P £	Total £
Brought forward				2.43	41.27	72.37	17.05	130.68
25mm thick tongued and grooved boarding	J3	m²	5.78	4.28	72.71	72.13	21.73	166.57
150 x 50mm softwood joists	J4	m	13.30	2.93	49.74	41.50	13.69	104.92
Cut and pin end of joists to existing brick wall	J5	nr	7.00	1.26	21.42	2.10	3.53	27.05
Build in ends of joists to blockwork	J6	nr	7.00	0.70	11.90	1.68	2.04	15.62
Carried to summary				11.59	197.04	189.78	58.02	444.84

PART K
CEILING FINISHES

	Ref	Unit	Qty	Hours	Hours £	Mat'ls £	O & P £	Total £
Plasterboard with taped butt joints fixed to joists	K1	m²	11.56	4.16	70.75	32.95	15.55	119.25
5mm skim coat to plasterboard ceilings	K2	m²	11.56	5.78	98.26	20.00	17.74	136.00
Two coats emulsion paint to ceilings	K3	m²	11.56	3.01	51.10	11.10	9.33	71.52
Carried to summary				12.95	220.10	64.04	42.62	326.77

PART L
ELECTRICAL WORK

	Ref	Unit	Qty	Hours	Hours £	Mat'ls £	O & P £	Total £
13 amp double switched socket outlet with neon	L1	nr	4.00	3.20	60.80	35.24	14.41	110.45
Lighting point	L2	nr	4.00	2.80	53.20	29.36	12.38	94.94
Lighting switch	L3	nr	4.00	2.80	53.20	17.68	10.63	81.51
Lighting wiring	L4	m	12.00	2.40	45.60	22.32	10.19	78.11
Power cable	L5	m	28.00	8.40	159.60	75.88	35.32	270.80
Carried to summary				19.60	372.40	180.48	82.93	635.81

	Ref	Unit	Qty	Hours	Hours £	Mat'ls £	O & P £	Total £
PART M **HEATING WORK**								
15mm copper pipe	M1	m	11.00	4.84	87.12	23.65	16.62	127.39
Elbow	M2	nr	8.00	4.48	80.64	19.52	15.02	115.18
Tee	M3	nr	2.00	1.12	20.16	4.88	3.76	28.80
Radiator, double convector size 1400 x 520mm	M4	nr	4.00	5.20	93.60	556.44	97.51	747.55
Break into existing pipe and insert tee	M5	nr	1.00	0.75	13.50	4.42	2.69	20.61
Carried to summary				16.39	295.02	608.91	135.59	1,039.52
PART N **ALTERATION WORK**								
Take out existing window size 1500 x 1000mm and lintel over, adapt opening to receive 1770 x 2000mm patio door and insert new lintel over (both measured separately) and make good	N1	nr	1.00	20.00	340.00	32.10	55.82	427.92
Take out existing window size 1500 x 1000mm, enlarge opening to receive new PVC–U door (measured separately)	N2	nr	1.00	24.00	408.00	58.16	69.92	536.08
Carried to summary				44.00	748.00	90.26	125.74	964.00

SUMMARY

	Hours	Hours £	Mat'ls £	O & P £	Total £
PART A **PRELIMINARIES**	0.00	0.00	0.00	0.00	2,400.00
PART B SUBSTRUCTURE TO **DPC LEVEL**	41.97	619.54	519.78	170.90	1,310.22
PART C **EXTERNAL WALLS**	99.81	1,696.77	2,341.52	605.74	4,644.03
PART D **FLAT ROOF**	44.80	761.62	586.51	202.22	1,550.35
PART E **PITCHED ROOF**	0.00	0.00	0.00	0.00	0.00
PART F WINDOWS AND **EXTERNAL DOORS**	26.08	443.36	1,647.88	313.69	2,404.93
PART G INTERNAL **PARTITIONS AND DOORS**	0.00	0.00	0.00	0.00	0.00
PART H **WALL FINISHES**	24.55	417.34	155.96	85.99	659.29
PART J **FLOOR FINISHES**	11.59	197.04	189.78	58.02	444.84
PART K **CEILING FINISHES**	12.95	220.10	64.04	42.62	326.76
PART L **ELECTRICAL WORK**	19.60	372.40	180.48	82.93	635.81
PART M **HEATING WORK**	16.39	295.02	608.91	135.59	1,039.52
PART N **ALTERATION WORK**	44.00	748.00	90.26	125.74	964.00
Final total	341.74	5,771.19	6,385.12	1,823.44	16,379.75

	Ref	Unit	Qty	Hours	Hours £	Mat'ls £	O & P £	Total £
PART A **PRELIMINARIES**								
Concrete mixer	A1	wks	9.00					450.00
Small tools	A2	wks	10.00					400.00
Scaffolding (m²/weeks)	A3		180					1,080.00
Skip	A4	wks	6.00					660.00
Clean up	A5	hrs	12.00					156.00
Carried to summary								2,746.00
PART B **SUBSTRUCTURE TO** **DPC LEVEL**								
Excavate topsoil 150mm thick by hand	B1	m²	11.40	3.42	44.46	0.00	6.67	51.13
Excavate to reduce levels	B2	m³	3.42	8.55	111.15	0.00	16.67	127.82
Excavate for trench foundations by hand	B3	m³	1.39	3.61	46.98	0.00	7.05	54.03
Earthwork support to sides of trenches	B4	m²	12.26	4.90	63.75	22.68	12.96	99.40
Backfilling with excavated material	B5	m³	0.55	0.33	4.29	0.00	0.64	4.93
Hardcore 225mm thick	B6	m²	7.48	1.50	19.45	47.05	9.97	76.47
Hardcore filling to trench	B7	m³	0.12	0.06	0.78	6.29	1.06	8.13
Concrete grade (1:3:6) in foundations	B8	m³	1.13	1.53	19.83	96.75	17.49	134.07
Concrete grade (1:2:4) in bed 150mm thick	B9	m²	7.48	2.24	29.17	114.59	21.56	165.33
Concrete (1:2:4) in cavity wall filling	B10	m²	4.20	0.84	11.97	10.00	3.29	25.26
Carried forward				26.98	351.84	297.36	97.38	746.58

	Ref	Unit	Qty	Hours	Hours £	Mat'ls £	O & P £	Total £
Brought forward				26.98	351.84	297.36	97.38	746.58
Damp-proof membrane	B11	m²	7.88	0.32	4.10	18.28	3.36	25.74
Reinforcement ref A193 in foundation	B12	m²	4.20	0.50	6.55	8.48	2.26	17.29
Steel fabric reinforcement ref A193 in slab	B13	m²	7.48	1.12	14.59	14.96	4.43	33.98
Solid blockwork 140mm thick in cavity wall	B14	m²	5.46	7.10	120.67	91.07	31.76	243.50
Common bricks 112.5mm thick in cavity wall	B15	m²	4.20	7.14	121.38	69.09	28.57	219.04
Facing bricks in 112.5mm thick in skin of cavity wall	B16	m²	1.26	2.27	38.56	28.55	10.07	77.17
Form cavity 50mm wide in cavity wall	B17	m²	5.46	0.16	2.78	8.30	1.66	12.75
DPC 112mm wide	B18	m	8.40	0.42	7.14	8.57	2.36	18.06
DPC 140mm wide	B19	m	8.40	0.50	8.57	11.59	3.02	23.18
Bond in block wall	B20	m	1.30	0.57	9.72	3.78	2.03	15.53
Bond in half brick wall	B21	m	0.30	0.11	1.79	0.80	0.39	2.98
50mm thick insulation board	B22	m²	7.48	2.24	38.15	53.86	13.80	105.80
Carried to summary				49.44	725.82	614.70	201.08	1,541.60

**PART C
EXTERNAL WALLS**

	Ref	Unit	Qty	Hours	Hours £	Mat'ls £	O & P £	Total £
Solid blockwork 140mm thick in cavity wall	C1	m²	34.04	44.25	752.28	567.79	198.01	1,518.08
Facing brickwork 112.5mm thick in cavity wall	C2	m²	34.04	61.27	1,041.62	971.84	302.02	2,315.49
75mm thick insulation in cavity wall	C3	m²	34.04	7.49	127.31	181.77	46.36	355.45
Carried forward				113.01	1,921.22	1,721.40	546.39	4,189.01

	Ref	Unit	Qty	Hours	Hours £	Mat'ls £	O & P £	Total £
Brought forward				113.01	1,921.22	1,721.40	546.39	4,189.01
Steel lintel 2400mm long	C4	nr	2.00	0.50	8.50	296.58	45.76	350.84
Steel lintel 1500mm long	C5	nr	4.00	0.80	13.60	485.44	74.86	573.90
Close cavity wall at jambs	C7	m	13.76	0.69	11.70	38.67	7.55	57.92
Close cavity wall at cills	C8	m	5.07	0.25	4.31	14.25	2.78	21.34
Close cavity wall at top	C9	m	8.40	0.42	7.14	23.60	4.61	35.36
DPC 112mm wide at jambs	C10	m	13.76	0.69	11.70	14.04	3.86	29.59
DPC 112mm wide at cills	C11	m	5.07	0.25	4.31	5.17	1.42	10.90
Carried to summary				116.62	1,982.47	2,599.15	687.24	5,268.86

PART D
FLAT ROOF

	Ref	Unit	Qty	Hours	Hours £	Mat'ls £	O & P £	Total £
200 x 50mm sawn softwood joists	D1	m	23.65	5.91	100.51	84.67	27.78	212.96
200 x 50mm sawn softwood sprocket pieces	D2	nr	8.00	1.12	19.04	14.32	5.00	38.36
18mm thick WPB grade decking	D3	m²	11.40	10.26	174.42	130.76	45.78	350.95
50 x 50mm (avg) wide sawn softwood firrings	D4	m	23.65	4.26	72.37	45.17	17.63	135.17
High density polyethylene vapour barrier 150mm thick	D5	m²	10.00	2.00	34.00	111.90	21.89	167.79
100 x 75mm sawn softwood wall plate	D6	m	9.00	2.70	45.90	24.66	10.58	81.14
100 x 75mm sawn softwood tilt fillet	D7	m	5.30	1.33	22.53	13.83	5.45	41.81
Build in ends of 200 x 50mm joists	D8	nr	11.00	3.85	65.45	2.42	10.18	78.05
Carried forward				31.42	534.22	427.73	144.29	1,106.24

	Ref	Unit	Qty	Hours	Hours £	Mat'ls £	O & P £	Total £
Brought forward				31.42	534.22	427.73	144.29	1,106.24
Rake out joint for flashing	D9	m	5.30	1.86	31.54	1.27	4.92	37.73
6mm thick soffit 150mm wide	D10	m	9.30	3.72	63.24	19.62	12.43	95.29
19mm wrought softwood fascia 200mm high	D11	m	9.30	4.65	79.05	25.20	15.64	119.89
Three layer fibre-based roofing felt	D12	m²	11.40	6.27	106.59	148.66	38.29	293.53
Felt turn-down 100mm girth	D13	m	9.30	0.93	15.81	13.02	4.32	33.15
Felt flashing 150mm girth	D14	m	5.30	0.53	9.01	9.12	2.72	20.84
112mm diameter PVC-U gutter	D15	m	5.30	1.38	23.43	17.17	6.09	46.69
Stop end	D16	nr	1.00	0.14	2.38	2.12	0.68	5.18
Stop end outlet	D17	nr	1.00	0.25	4.25	4.19	1.27	9.71
68mm diameter PVC-U down pipe	D18	m	4.80	1.20	20.40	27.36	7.16	54.92
Shoe	D19	nr	1.00	0.30	5.10	3.15	1.24	9.49
Paint fascia and soffit	D20	m²	2.79	0.56	9.49	4.46	2.09	16.04
Carried to summary				53.21	904.49	703.08	241.14	1,848.70
PART E PITCHED ROOF				N/A	N/A	N/A	N/A	N/A
PART F WINDOWS AND EXTERNAL DOORS								
PVC-U door size 840 x 1980mm complete (B)	F1	nr	1.00	2.50	42.50	278.67	48.18	369.35
PVC-U sliding patio door size 1700 x 2075mm (C)	F2	nr	2.00	14.00	238.00	616.94	128.24	983.18
Carried forward				16.50	280.50	895.61	176.42	1,352.53

	Ref	Unit	Qty	Hours	Hours £	Mat'ls £	O & P £	Total £
Brought forward				16.50	280.50	895.61	176.42	1,352.53
PVC-U window size 1200 x 1200mm complete (A)	F3	nr	4.00	8.00	136.00	728.56	129.68	994.24
25 x 225mm wrought softwood window board	F4	m	4.80	1.44	24.48	19.30	6.57	50.34
Paint window board	F5	m	4.80	0.14	2.38	4.42	1.02	7.82
Carried to summary				26.08	443.36	1,647.88	313.69	2,404.93
PART G INTERNAL PARTITIONS AND DOORS				N/A	N/A	N/A	N/A	N/A
PART H WALL FINISHES								
19 x 100mm wrought softwood skirting	H1	m	14.34	2.44	41.44	35.99	11.62	89.05
12mm plasterboard fixed to walls with dabs	H2	m²	29.84	11.64	197.84	85.04	42.43	325.32
12mm plasterboard fixed to walls less than 300mm wide	H3	m	18.33	3.30	56.09	16.86	10.94	83.90
Two coats emulsion paint to walls	H4	m²	32.66	8.49	144.36	31.35	26.36	202.07
Paint skirting	H5	m	14.34	2.87	48.76	13.19	9.29	71.24
Carried to summary				28.73	488.48	182.45	100.64	771.57
PART J FLOOR FINISHES								
Cement and sand floor screed 40mm thick	J1	m²	7.48	1.87	31.79	35.60	10.11	77.50
Vinyl floor tiles, size 300 x 300mm	J2	m²	7.48	1.27	21.62	58.04	11.95	91.61
Carried forward				3.14	53.41	93.65	22.06	169.12

	Ref	Unit	Qty	Hours	Hours £	Mat'ls £	O & P £	Total £
Brought forward				3.14	53.41	93.65	22.06	169.12
25mm thick tongued and grooved boarding	J3	m²	7.48	5.54	94.10	93.35	28.12	215.57
150 x 50mm softwood joists	J4	m	15.20	3.34	56.85	47.42	15.64	119.91
Cut and pin end of joists to existing brick wall	J5	nr	8.00	1.44	24.48	2.40	4.03	30.91
Build in ends of joists to blockwork	J6	nr	8.00	0.80	13.60	1.92	2.33	17.85
Carried to summary				14.26	242.43	238.74	72.18	553.35

PART K
CEILING FINISHES

	Ref	Unit	Qty	Hours	Hours £	Mat'ls £	O & P £	Total £
Plasterboard with taped butt joints fixed to joists	K1	m²	14.96	5.39	91.56	42.64	20.13	154.32
5mm skim coat to plasterboard ceilings	K2	m²	14.96	7.48	127.16	25.88	22.96	176.00
Two coats emulsion paint to ceilings	K3	m²	14.96	3.89	66.12	14.36	12.07	92.56
Carried to summary				16.76	284.84	82.88	55.16	422.87

PART L
ELECTRICAL WORK

	Ref	Unit	Qty	Hours	Hours £	Mat'ls £	O & P £	Total £
13 amp double switched socket outlet with neon	L1	nr	6.00	4.80	91.20	52.86	21.61	165.67
Lighting point	L2	nr	4.00	2.80	53.20	29.36	12.38	94.94
Lighting switch	L3	nr	4.00	2.80	53.20	17.68	10.63	81.51
Lighting wiring	L4	m	14.00	2.80	53.20	26.04	11.89	91.13
Power cable	L5	m	32.00	9.60	182.40	86.72	40.37	309.49
Carried to summary				22.80	433.20	212.66	96.88	742.74

	Ref	Unit	Qty	Hours	Hours £	Mat'ls £	O & P £	Total £
PART M **HEATING WORK**								
15mm copper pipe	M1	m	13.00	5.72	102.96	27.95	19.64	150.55
Elbow	M2	nr	8.00	4.48	80.64	19.52	15.02	115.18
Tee	M3	nr	2.00	1.12	20.16	4.88	3.76	28.80
Radiator, double convector size 1400 x 520mm	M4	nr	4.00	5.20	93.60	556.44	97.51	747.55
Break into existing pipe and insert tee	M5	nr	1.00	0.75	13.50	4.42	2.69	20.61
Carried to summary				17.27	310.86	613.21	138.61	1,062.68
PART N **ALTERATION WORK**								
Take out existing window size 1500 x 1000mm and lintel over, adapt opening to receive 1770 x 2000mm patio door and insert new lintel over (both measured separately) and make good	N1	nr	1.00	20.00	340.00	32.10	55.82	427.92
Take out existing window size 1500 x 1000mm, enlarge opening to receive new PVC–U door (measured separately)	N2	nr	1.00	24.00	408.00	58.16	69.92	536.08
Carried to summary				44.00	748.00	90.26	125.74	964.00

SUMMARY

	Hours	Hours £	Mat'ls £	O & P £	Total £
PART A **PRELIMINARIES**	0.00	0.00	0.00	0.00	2,746.00
PART B SUBSTRUCTURE TO **DPC LEVEL**	49.44	725.82	614.70	201.08	1,541.60
PART C **EXTERNAL WALLS**	116.62	1,982.47	2,599.15	687.24	5,268.86
PART D **FLAT ROOF**	53.21	904.49	703.08	241.14	1,848.70
PART E **PITCHED ROOF**	0.00	0.00	0.00	0.00	0.00
PART F WINDOWS AND **EXTERNAL DOORS**	26.08	443.36	1,647.88	313.69	2,404.93
PART G INTERNAL **PARTITIONS AND DOORS**	0.00	0.00	0.00	0.00	0.00
PART H **WALL FINISHES**	28.73	488.48	182.45	100.64	771.57
PART J **FLOOR FINISHES**	14.26	242.43	238.74	72.18	553.35
PART K **CEILING FINISHES**	16.76	284.84	82.88	55.16	422.88
PART L **ELECTRICAL WORK**	22.80	433.20	212.66	96.88	742.74
PART M **HEATING WORK**	17.27	310.86	613.21	138.61	1,062.68
PART N **ALTERATION WORK**	44.00	748.00	90.26	125.74	964.00
Final total	389.17	6,563.95	6,985.01	2,032.36	18,327.31

	Ref	Unit	Qty	Hours	Hours £	Mat'ls £	O & P £	Total £
PART A PRELIMINARIES								
Concrete mixer	A1	wks	9.00					450.00
Small tools	A2	wks	9.00					360.00
Scaffolding (m²/weeks)	A3		180					1,080.00
Skip	A4	wks	5.00					550.00
Clean up	A5	hrs	10.00					130.00
Carried to summary								2,570.00
PART B SUBSTRUCTURE TO DPC LEVEL								
Excavate topsoil 150mm thick by hand	B1	m²	10.40	3.12	40.56	0.00	6.08	46.64
Excavate to reduce levels	B2	m³	3.12	7.80	101.40	0.00	15.21	116.61
Excavate for trench foundations by hand	B3	m³	1.39	3.61	46.98	0.00	7.05	54.03
Earthwork support to sides of trenches	B4	m²	12.26	4.90	63.75	22.68	12.96	99.40
Backfilling with excavated material	B5	m³	0.47	0.28	3.67	0.00	0.55	4.22
Hardcore 225mm thick	B6	m²	6.48	1.30	16.85	40.76	8.64	66.25
Hardcore filling to trench	B7	m³	0.09	0.05	0.59	6.29	1.03	7.91
Concrete grade (1:3:6) in foundations	B8	m³	1.13	1.53	19.83	96.75	17.49	134.07
Concrete grade (1:2:4) in bed 150mm thick	B9	m²	6.48	1.94	25.27	99.27	18.68	143.23
Concrete (1:2:4) in cavity wall filling	B10	m²	4.20	0.84	11.97	10.00	3.29	25.26
Carried forward				25.37	330.87	275.75	90.99	697.61

	Ref	Unit	Qty	Hours	Hours £	Mat'ls £	O & P £	Total £
Brought forward				25.37	330.87	275.75	90.99	697.61
Damp-proof membrane	B11	m²	6.88	0.28	3.58	15.96	2.93	22.47
Reinforcement ref A193 in foundation	B12	m²	4.20	0.50	6.55	8.48	2.26	17.29
Steel fabric reinforcement ref A193 in slab	B13	m²	6.48	0.97	12.64	12.96	3.84	29.44
Solid blockwork 140mm thick in cavity wall	B14	m²	5.46	7.10	120.67	91.07	31.76	243.50
Common bricks 112.5mm thick in cavity wall	B15	m²	4.20	7.14	121.38	69.09	28.57	219.04
Facing bricks in 112.5mm thick in skin of cavity wall	B16	m²	1.26	2.27	38.56	28.55	10.07	77.17
Form cavity 50mm wide in cavity wall	B17	m²	5.46	0.16	2.78	8.30	1.66	12.75
DPC 112mm wide	B18	m	8.40	0.42	7.14	8.57	2.36	18.06
DPC 140mm wide	B19	m	8.40	0.50	8.57	11.59	3.02	23.18
Bond in block wall	B20	m	1.30	0.57	9.72	3.78	2.03	15.53
Bond in half brick wall	B21	m	0.30	0.11	1.79	0.80	0.39	2.98
50mm thick insulation board	B22	m²	6.48	1.94	33.05	46.66	11.96	91.66
Carried to summary				47.34	697.28	581.57	191.83	1,470.68

**PART C
EXTERNAL WALLS**

	Ref	Unit	Qty	Hours	Hours £	Mat'ls £	O & P £	Total £
Solid blockwork 140mm thick in cavity wall	C1	m²	34.04	44.25	752.28	567.79	198.01	1,518.08
Facing brickwork 112.5mm thick in cavity wall	C2	m²	34.04	61.27	1,041.62	971.84	302.02	2,315.49
75mm thick insulation in cavity wall	C3	m²	34.04	7.49	127.31	181.77	46.36	355.45
Carried forward				113.01	1,921.22	1,721.40	546.39	4,189.01

	Ref	Unit	Qty	Hours	Hours £	Mat'ls £	O & P £	Total £
Brought forward				113.01	1,921.22	1,721.40	546.39	4,189.01
Steel lintel 2400mm long	C4	nr	2.00	0.50	8.50	296.58	45.76	350.84
Steel lintel 1500mm long	C5	nr	4.00	0.80	13.60	485.44	74.86	573.90
Close cavity wall at jambs	C7	m	13.76	0.69	11.70	38.67	7.55	57.92
Close cavity wall at cills	C8	m	5.07	0.25	4.31	14.25	2.78	21.34
Close cavity wall at top	C9	m	8.40	0.42	7.14	23.60	4.61	35.36
DPC 112mm wide at jambs	C10	m	13.76	0.69	11.70	14.04	3.86	29.59
DPC 112mm wide at cills	C11	m	5.07	0.25	4.31	5.17	1.42	10.90
Carried to summary				116.62	1,982.47	2,599.15	687.24	5,268.86

PART D
FLAT ROOF

	Ref	Unit	Qty	Hours	Hours £	Mat'ls £	O & P £	Total £
200 x 50mm sawn softwood joists	D1	m	22.05	5.51	93.71	78.94	25.90	198.55
200 x 50mm sawn softwood sprocket pieces	D2	nr	12.00	1.68	28.56	21.48	7.51	57.55
18mm thick WPB grade decking	D3	m²	10.40	9.36	159.12	119.29	41.76	320.17
50 x 50mm (avg) wide sawn softwood firrings	D4	m	22.05	3.97	67.47	42.12	16.44	126.03
High density polyethylene vapour barrier 150mm thick	D5	m²	9.00	1.80	30.60	100.71	19.70	151.01
100 x 75mm sawn softwood wall plate	D6	m	9.00	2.70	45.90	24.66	10.58	81.14
100 x 75mm sawn softwood tilt fillet	D7	m	3.30	0.83	14.03	8.61	3.40	26.03
Build in ends of 200 x 50mm joists	D8	nr	7.00	2.45	41.65	1.54	6.48	49.67
Carried forward				28.30	481.04	397.35	131.76	1,010.14

	Ref	Unit	Qty	Hours	Hours £	Mat'ls £	O & P £	Total £
Brought forward				28.30	481.04	397.35	131.76	1,010.14
Rake out joint for flashing	D9	m	3.30	1.16	19.64	0.79	3.06	23.49
6mm thick soffit 150mm wide	D10	m	9.30	3.72	63.24	19.62	12.43	95.29
19mm wrought softwood fascia 200mm high	D11	m	9.30	4.65	79.05	25.20	15.64	119.89
Three layer fibre-based roofing felt	D12	m²	10.40	5.72	97.24	135.62	34.93	267.78
Felt turn-down 100mm girth	D13	m	9.30	0.93	15.81	13.02	4.32	33.15
Felt flashing 150mm girth	D14	m	3.30	0.33	5.61	5.68	1.69	12.98
112mm diameter PVC-U gutter	D15	m	3.30	0.86	14.59	10.69	3.79	29.07
Stop end	D16	nr	1.00	0.14	2.38	2.12	0.68	5.18
Stop end outlet	D17	nr	1.00	0.25	4.25	4.19	1.27	9.71
68mm diameter PVC-U down pipe	D18	m	4.80	1.20	20.40	27.36	7.16	54.92
Shoe	D19	nr	1.00	0.30	5.10	3.15	1.24	9.49
Paint fascia and soffit	D20	m²	2.79	0.56	9.49	4.46	2.09	16.04
Carried to summary				48.11	817.83	649.25	220.06	1,687.14
PART E **PITCHED ROOF**				N/A	N/A	N/A	N/A	N/A
PART F **WINDOWS AND** **EXTERNAL DOORS**								
PVC-U door size 840 x 1980mm complete (B)	F1	nr	1.00	2.50	42.50	278.67	48.18	369.35
PVC-U sliding patio door size 1700 x 2075mm (C)	F2	nr	2.00	14.00	238.00	616.94	128.24	983.18
Carried forward				16.50	280.50	895.61	176.42	1,352.53

	Ref	Unit	Qty	Hours	Hours £	Mat'ls £	O & P £	Total £
Brought forward				16.50	280.50	895.61	176.42	1,352.53
PVC-U window size 1200 x 1200mm complete (A)	F3	nr	4.00	8.00	136.00	728.56	129.68	994.24
25 x 225mm wrought softwood window board	F4	m	4.80	1.44	24.48	19.30	6.57	50.34
Paint window board	F5	m	4.80	0.14	2.38	4.42	1.02	7.82
Carried to summary				26.08	443.36	1,647.88	313.69	2,404.93
PART G **INTERNAL** **PARTITIONS AND** **DOORS**				N/A	N/A	N/A	N/A	N/A
PART H **WALL FINISHES**								
19 x 100mm wrought softwood skirting	H1	m	14.34	2.44	41.44	35.99	11.62	89.05
12mm plasterboard fixed to walls with dabs	H2	m²	29.84	11.64	197.84	85.04	42.43	325.32
12mm plasterboard fixed to walls less than 300mm wide	H3	m	18.33	3.30	56.09	16.86	10.94	83.90
Two coats emulsion paint to walls	H4	m²	32.66	8.49	144.36	31.35	26.36	202.07
Paint skirting	H5	m	14.34	2.87	48.76	13.19	9.29	71.24
Carried to summary				28.73	488.48	182.45	100.64	771.57
PART J **FLOOR FINISHES**								
Cement and sand floor screed 40mm thick	J1	m²	6.48	1.62	27.54	30.84	8.76	67.14
Vinyl floor tiles, size 300 x 300mm	J2	m²	6.48	1.10	18.73	50.28	10.35	79.36
Carried forward				2.72	46.27	81.13	19.11	146.51

218 Two storey extension, size 3 x 3m, flat roof

	Ref	Unit	Qty	Hours	Hours £	Mat'ls £	O & P £	Total £
Brought forward				2.72	46.27	81.13	19.11	146.51
25mm thick tongued and grooved boarding	J3	m²	6.48	4.80	81.52	80.87	24.36	186.75
150 x 50mm softwood joists	J4	m	14.50	3.19	54.23	45.24	14.92	114.39
Cut and pin end of joists to excisting brick wall	J5	nr	5.00	0.90	15.30	1.50	2.52	19.32
Build in ends of joists to blockwork	J6	nr	5.00	0.50	8.50	1.20	1.46	11.16
Carried to summary				12.11	205.82	209.94	62.36	478.12

PART K
CEILING FINISHES

	Ref	Unit	Qty	Hours	Hours £	Mat'ls £	O & P £	Total £
Plasterboard with taped butt joints fixed to joists	K1	m²	12.96	4.67	79.32	36.94	17.44	133.69
5mm skim coat to plasterboard ceilings	K2	m²	12.96	6.48	110.16	22.42	19.89	152.47
Two coats emulsion paint to ceilings	K3	m²	12.96	3.37	57.28	12.44	10.46	80.18
Carried to summary				14.52	246.76	71.80	47.78	366.34

PART L
ELECTRICAL WORK

	Ref	Unit	Qty	Hours	Hours £	Mat'ls £	O & P £	Total £
13 amp double switched socket outlet with neon	L1	nr	4.00	3.20	60.80	35.24	14.41	110.45
Lighting point	L2	nr	2.00	1.40	26.60	14.68	6.19	47.47
Lighting switch	L3	nr	4.00	2.80	53.20	17.68	10.63	81.51
Lighting wiring	L4	m	12.00	2.40	45.60	22.32	10.19	78.11
Power cable	L5	m	28.00	8.40	159.60	75.88	35.32	270.80
Carried to summary				18.20	345.80	165.80	76.74	588.34

	Ref	Unit	Qty	Hours	Hours £	Mat'ls £	O & P £	Total £
PART M **HEATING WORK**								
15mm copper pipe	M1	m	11.00	4.84	87.12	23.65	16.62	127.39
Elbow	M2	nr	8.00	4.48	80.64	19.52	15.02	115.18
Tee	M3	nr	2.00	1.12	20.16	4.88	3.76	28.80
Radiator, double convector size 1400 x 520mm	M4	nr	4.00	5.20	93.60	556.44	97.51	747.55
Break into existing pipe and insert tee	M5	nr	1.00	0.75	13.50	4.42	2.69	20.61
Carried to summary				16.39	295.02	608.91	135.59	1,039.52
PART N **ALTERATION WORK**								
Take out existing window size 1500 x 1000mm and lintel over, adapt opening to receive 1770 x 2000mm patio door and insert new lintel over (both measured separately) and make good	N1	nr	1.00	20.00	340.00	32.10	55.82	427.92
Take out existing window size 1500 x 1000mm, enlarge opening to receive new PVC–U door (measured separately)	N2	nr	1.00	24.00	408.00	58.16	69.92	536.08
Carried to summary				44.00	748.00	90.26	125.74	964.00

SUMMARY

	Hours	Hours £	Mat'ls £	O & P £	Total £
PART A **PRELIMINARIES**	0.00	0.00	0.00	0.00	2,570.00
PART B SUBSTRUCTURE TO **DPC LEVEL**	47.34	697.28	581.57	191.83	1,470.68
PART C **EXTERNAL WALLS**	116.62	1,982.47	2,599.15	687.24	5,268.86
PART D **FLAT ROOF**	48.11	817.83	649.25	220.06	1,687.14
PART E **PITCHED ROOF**	0.00	0.00	0.00	0.00	0.00
PART F WINDOWS AND **EXTERNAL DOORS**	26.08	443.36	1,647.88	313.69	2,404.93
PART G INTERNAL **PARTITIONS AND DOORS**	0.00	0.00	0.00	0.00	0.00
PART H **WALL FINISHES**	28.73	488.48	182.45	100.64	771.57
PART J **FLOOR FINISHES**	12.11	205.82	209.94	62.36	478.12
PART K **CEILING FINISHES**	14.52	246.76	71.80	47.78	366.34
PART L **ELECTRICAL WORK**	18.20	345.80	165.80	76.74	588.34
PART M **HEATING WORK**	16.39	295.02	608.91	135.59	1,039.52
PART N **ALTERATION WORK**	44.00	748.00	90.26	125.74	964.00
Final total	372.10	6,270.82	6,807.01	1,961.68	17,609.50

	Ref	Unit	Qty	Hours	Hours £	Mat'ls £	O & P £	Total £
PART A **PRELIMINARIES**								
Concrete mixer	A1	wks	10.00					500.00
Small tools	A2	wks	10.00					400.00
Scaffolding (m²/weeks)	A3		200					1,200.00
Skip	A4	wks	6.00					660.00
Clean up	A5	hrs	12.00					156.00
Carried to summary								2,916.00
PART B **SUBSTRUCTURE TO** **DPC LEVEL**								
Excavate topsoil 150mm thick by hand	B1	m²	14.19	4.26	55.34	0.00	8.30	63.64
Excavate to reduce levels	B2	m³	4.06	10.15	131.95	0.00	19.79	151.74
Excavate for trench foundations by hand	B3	m³	1.39	3.61	46.98	0.00	7.05	54.03
Earthwork support to sides of trenches	B4	m²	13.72	5.49	71.34	25.38	14.51	111.23
Backfilling with excavated material	B5	m³	0.55	0.33	4.29	0.00	0.64	4.93
Hardcore 225mm thick	B6	m²	9.18	1.84	23.87	57.74	12.24	93.85
Hardcore filling to trench	B7	m³	0.12	0.06	0.78	6.29	1.06	8.13
Concrete grade (1:3:6) in foundations	B8	m³	1.27	1.71	22.29	108.74	19.65	150.68
Concrete grade (1:2:4) in bed 150mm thick	B9	m²	9.18	2.75	35.80	140.64	26.47	202.91
Concrete (1:2:4) in cavity wall filling	B10	m²	4.70	0.94	11.97	11.19	3.47	26.63
Carried forward				31.14	404.62	349.98	113.19	867.78

	Ref	Unit	Qty	Hours	Hours £	Mat'ls £	O & P £	Total £
Brought forward				31.14	404.62	349.98	113.19	867.78
Damp-proof membrane	B11	m²	9.63	0.39	5.01	22.34	4.10	31.45
Reinforcement ref A193 in foundation	B12	m²	4.70	0.56	7.33	9.49	2.52	19.35
Steel fabric reinforcement ref A193 in slab	B13	m²	9.18	1.38	17.90	18.36	5.44	41.70
Solid blockwork 140mm thick in cavity wall	B14	m²	6.11	7.94	135.03	101.91	35.54	272.49
Common bricks 112.5mm thick in cavity wall	B15	m²	4.70	7.99	135.83	77.32	31.97	245.12
Facing bricks in 112.5mm thick in skin of cavity wall	B16	m²	1.41	2.54	43.15	28.55	10.75	82.45
Form cavity 50mm wide in cavity wall	B17	m²	1.41	0.04	0.72	2.14	0.43	3.29
DPC 112mm wide	B18	m	9.40	0.47	7.99	9.59	2.64	20.21
DPC 140mm wide	B19	m	9.40	0.56	9.59	12.97	3.38	25.94
Bond in block wall	B20	m	1.30	0.57	9.72	3.78	2.03	15.53
Bond in half brick wall	B21	m	0.30	0.11	1.79	0.80	0.39	2.98
50mm thick insulation board	B22	m²	9.18	2.75	46.82	66.10	16.94	129.85
Carried to summary				56.45	825.49	703.34	229.32	1,758.15

**PART C
EXTERNAL WALLS**

	Ref	Unit	Qty	Hours	Hours £	Mat'ls £	O & P £	Total £
Solid blockwork 140mm thick in cavity wall	C1	m²	39.09	50.82	863.89	652.02	227.39	1,743.30
Facing brickwork 112.5mm thick in cavity wall	C2	m²	39.09	70.36	1,196.15	1,116.02	346.83	2,659.00
75mm thick insulation in cavity wall	C3	m²	39.09	8.60	146.20	208.74	53.24	408.18
Carried forward				129.78	2,206.24	1,976.78	627.45	4,810.47

	Ref	Unit	Qty	Hours	Hours £	Mat'ls £	O & P £	Total £
Brought forward				129.78	2,206.24	1,976.78	627.45	4,810.47
Steel lintel 2400mm long	C4	nr	2.00	0.50	8.50	296.58	45.76	350.84
Steel lintel 1500mm long	C5	nr	4.00	0.80	13.60	485.44	74.86	573.90
Close cavity wall at jambs	C7	m	13.76	0.69	11.70	38.67	7.55	57.92
Close cavity wall at cills	C8	m	5.07	0.25	4.31	14.25	2.78	21.34
Close cavity wall at top	C9	m	9.40	0.47	7.99	26.41	5.16	39.56
DPC 112mm wide at jambs	C10	m	13.76	0.69	11.70	14.04	3.86	29.59
DPC 112mm wide at cills	C11	m	5.07	0.25	4.31	5.17	1.42	10.90
Carried to summary				133.43	2,268.34	2,857.33	768.85	5,894.53

PART D
FLAT ROOF

	Ref	Unit	Qty	Hours	Hours £	Mat'ls £	O & P £	Total £
200 x 50mm sawn softwood joists	D1	m	28.35	7.09	120.49	101.49	33.30	255.28
200 x 50mm sawn softwood sprocket pieces	D2	nr	12.00	1.68	28.56	21.48	7.51	57.55
18mm thick WPB grade decking	D3	m²	13.55	12.20	207.32	155.42	54.41	417.14
50 x 50mm (avg) wide sawn softwood firrings	D4	m	28.35	5.10	86.75	54.15	21.13	162.03
High density polyethylene vapour barrier 150mm thick	D5	m²	12.00	2.40	40.80	134.28	26.26	201.34
100 x 75mm sawn softwood wall plate	D6	m	10.00	3.00	51.00	27.40	11.76	90.16
100 x 75mm sawn softwood tilt fillet	D7	m	4.30	1.08	18.28	11.22	4.42	33.92
Build in ends of 200 x 50mm joists	D8	nr	9.00	3.15	53.55	1.98	8.33	63.86
Carried forward				35.69	606.74	507.42	167.12	1,281.29

	Ref	Unit	Qty	Hours	Hours £	Mat'ls £	O & P £	Total £
Brought forward				35.69	606.74	507.42	167.12	1,281.29
Rake out joint for flashing	D9	m	4.30	1.51	25.59	1.03	3.99	30.61
6mm thick soffit 150mm wide	D10	m	40.30	16.12	274.04	85.03	53.86	412.93
19mm wrought softwood fascia 200mm high	D11	m	10.30	5.15	87.55	27.91	17.32	132.78
Three layer fibre-based roofing felt	D12	m²	13.55	7.45	126.69	176.69	45.51	348.89
Felt turn-down 100mm girth	D13	m	10.30	1.03	17.51	14.42	4.79	36.72
Felt flashing 150mm girth	D14	m	4.30	0.43	7.31	7.40	2.21	16.91
112mm diameter PVC-U gutter	D15	m	4.30	1.12	19.01	13.93	4.94	37.88
Stop end	D16	nr	1.00	0.14	2.38	2.12	0.68	5.18
Stop end outlet	D17	nr	1.00	0.25	4.25	4.19	1.27	9.71
68mm diameter PVC-U down pipe	D18	m	4.80	1.20	20.40	27.36	7.16	54.92
Shoe	D19	nr	1.00	0.30	5.10	3.15	1.24	9.49
Paint fascia and soffit	D20	m²	3.09	0.62	10.51	4.94	2.32	17.77
Carried to summary				71.00	1,207.07	875.61	312.40	2,395.07
PART E **PITCHED ROOF**				N/A	N/A	N/A	N/A	N/A
PART F **WINDOWS AND** **EXTERNAL DOORS**								
PVC-U door size 840 x 1980mm complete (B)	F1	nr	1.00	2.50	42.50	278.67	48.18	369.35
PVC-U sliding patio door size 1700 x 2075mm (C)	F2	nr	2.00	14.00	238.00	616.94	128.24	983.18
Carried forward				16.50	280.50	895.61	176.42	1,352.53

	Ref	Unit	Qty	Hours	Hours £	Mat'ls £	O & P £	Total £
Brought forward				16.50	280.50	895.61	176.42	1,352.53
PVC-U window size 1200 x 1200mm complete (A)	F3	nr	4.00	8.00	136.00	728.56	129.68	994.24
25 x 225mm wrought softwood window board	F4	m	4.80	1.44	24.48	19.30	6.57	50.34
Paint window board	F5	m	4.80	0.14	2.38	4.42	1.02	7.82
Carried to summary				26.08	443.36	1,647.88	313.69	2,404.93
PART G INTERNAL PARTITIONS AND DOORS				N/A	N/A	N/A	N/A	N/A
PART H WALL FINISHES								
19 x 100mm wrought softwood skirting	H1	m	16.34	2.78	47.22	41.01	13.24	101.47
12mm plasterboard fixed to walls with dabs	H2	m²	34.74	13.55	230.33	99.01	49.40	378.74
12mm plasterboard fixed to walls less than 300mm wide	H3	m	18.33	3.30	56.09	16.86	10.94	83.90
Two coats emulsion paint to walls	H4	m²	38.56	10.03	170.44	37.02	31.12	238.57
Paint skirting	H5	m	16.34	3.27	55.56	15.03	10.59	81.18
Carried to summary				32.92	559.63	208.94	115.28	883.85
PART J FLOOR FINISHES								
Cement and sand floor screed 40mm thick	J1	m²	9.18	2.30	39.02	43.70	12.41	95.12
Vinyl floor tiles, size 300 x 300mm	J2	m²	9.18	1.56	26.53	71.24	14.67	112.43
Carried forward				3.86	65.55	114.93	27.07	207.55

226 *Two storey extension, size 3 x 4m, flat roof*

	Ref	Unit	Qty	Hours	Hours £	Mat'ls £	O & P £	Total £
Brought forward				3.86	65.55	114.93	27.07	207.55
25mm thick tongued and grooved boarding	J3	m²	9.18	6.79	115.48	114.57	34.51	264.56
150 x 50mm softwood joists	J4	m	20.30	4.47	75.92	63.34	20.89	160.15
Cut and pin end of joists to exising brick wall	J5	nr	7.00	1.26	21.42	2.10	3.53	27.05
Build in ends of joists to blockwork	J6	nr	7.00	0.70	11.90	1.68	2.04	15.62
Carried to summary				17.07	290.27	296.62	88.03	674.92

PART K
CEILING FINISHES

	Ref	Unit	Qty	Hours	Hours £	Mat'ls £	O & P £	Total £
Plasterboard with taped butt joints fixed to joists	K1	m²	18.36	6.61	112.36	52.33	24.70	189.39
5mm skim coat to plasterboard ceilings	K2	m²	18.36	9.18	156.06	31.76	28.17	216.00
Two coats emulsion paint to ceilings	K3	m²	18.36	4.77	81.15	17.63	14.82	113.59
Carried to summary				20.56	349.57	101.71	67.69	518.98

PART L
ELECTRICAL WORK

	Ref	Unit	Qty	Hours	Hours £	Mat'ls £	O & P £	Total £
13 amp double switched socket outlet with neon	L1	nr	6.00	4.80	91.20	52.86	21.61	165.67
Lighting point	L2	nr	4.00	2.80	53.20	29.36	12.38	94.94
Lighting switch	L3	nr	4.00	2.80	53.20	17.68	10.63	81.51
Lighting wiring	L4	m	12.00	2.40	45.60	22.32	10.19	78.11
Power cable	L5	m	32.00	9.60	182.40	86.72	40.37	309.49
Carried to summary				22.40	425.60	208.94	95.18	729.72

	Ref	Unit	Qty	Hours	Hours £	Mat'ls £	O & P £	Total £
PART M **HEATING WORK**								
15mm copper pipe	M1	m	13.00	5.72	102.96	27.95	19.64	150.55
Elbow	M2	nr	8.00	4.48	80.64	19.52	15.02	115.18
Tee	M3	nr	2.00	1.12	20.16	4.88	3.76	28.80
Radiator, double convector size 1400 x 520mm	M4	nr	4.00	5.20	93.60	556.44	97.51	747.55
Break into existing pipe and insert tee	M5	nr	1.00	0.75	13.50	4.42	2.69	20.61
Carried to summary				17.27	310.86	613.21	138.61	1,062.68
PART N **ALTERATION WORK**								
Take out existing window size 1500 x 1000mm and lintel over, adapt opening to receive 1770 x 2000mm patio door and insert new lintel over (both measured separately) and make good	N1	nr	1.00	20.00	340.00	32.10	55.82	427.92
Take out existing window size 1500 x 1000mm, enlarge opening to receive new PVC–U door (measured separately)	N2	nr	1.00	24.00	408.00	58.16	69.92	536.08
Carried to summary				44.00	748.00	90.26	125.74	964.00

SUMMARY

	Hours	Hours £	Mat'ls £	O & P £	Total £
PART A **PRELIMINARIES**	0.00	0.00	0.00	0.00	2,916.00
PART B SUBSTRUCTURE TO **DPC LEVEL**	56.40	825.49	703.34	229.32	1,758.15
PART C **EXTERNAL WALLS**	133.43	2,268.34	2,857.33	768.85	5,894.33
PART D **FLAT ROOF**	71.00	1,207.07	875.61	312.40	2,395.07
PART E **PITCHED ROOF**	0.00	0.00	0.00	0.00	0.00
PART F WINDOWS AND **EXTERNAL DOORS**	26.08	443.36	1,647.88	313.69	2,404.93
PART G INTERNAL **PARTITIONS AND DOORS**	0.00	0.00	0.00	0.00	0.00
PART H **WALL FINISHES**	32.92	559.63	208.94	115.28	883.85
PART J **FLOOR FINISHES**	17.07	290.27	296.62	88.03	674.92
PART K **CEILING FINISHES**	20.56	349.57	101.71	67.69	518.98
PART L **ELECTRICAL WORK**	22.40	425.60	208.94	95.18	729.72
PART M **HEATING WORK**	17.27	310.86	613.21	138.61	1,062.68
PART N **ALTERATION WORK**	44.00	748.00	90.26	125.74	964.00
Final total	441.13	7,428.19	7,603.84	2,254.79	20,202.63

	Ref	Unit	Qty	Hours	Hours £	Mat'ls £	O & P £	Total £
PART A PRELIMINARIES								
Concrete mixer	A1	wks	11.00					550.00
Small tools	A2	wks	11.00					440.00
Scaffolding (m²/weeks)	A3		220					1,320.00
Skip	A4	wks	7.00					770.00
Clean up	A5	hrs	12.00					156.00
Carried to summary								3,236.00
PART B SUBSTRUCTURE TO DPC LEVEL								
Excavate topsoil 150mm thick by hand	B1	m²	17.50	5.25	68.25	0.00	10.24	78.49
Excavate to reduce levels	B2	m³	5.00	12.50	162.50	0.00	24.38	186.88
Excavate for trench foundations by hand	B3	m³	1.72	4.47	58.14	0.00	8.72	66.86
Earthwork support to sides of trenches	B4	m²	15.18	6.07	78.94	28.08	16.05	123.07
Backfilling with excavated material	B5	m³	0.63	0.38	4.91	0.00	0.74	5.65
Hardcore 225mm thick	B6	m²	11.88	2.38	30.89	74.73	15.84	121.46
Hardcore filling to trench	B7	m³	0.14	0.07	0.91	6.29	1.08	8.28
Concrete grade (1:3:6) in foundations	B8	m³	1.40	1.89	24.57	119.87	21.67	166.10
Concrete grade (1:2:4) in bed 150mm thick	B9	m²	11.88	3.56	46.33	182.00	34.25	262.58
Concrete (1:2:4) in cavity wall filling	B10	m²	5.20	1.04	11.97	12.38	3.65	28.00
Carried forward				37.61	487.41	423.34	136.61	1,047.37

	Ref	Unit	Qty	Hours	Hours £	Mat'ls £	O & P £	Total £
Brought forward				37.61	487.41	423.34	136.61	1,047.37
Damp-proof membrane	B11	m²	12.38	0.50	6.44	28.72	5.27	40.43
Reinforcement ref A193 in foundation	B12	m²	5.20	0.62	8.11	10.50	2.79	21.41
Steel fabric reinforcement ref A193 in slab	B13	m²	11.38	1.71	22.19	22.76	6.74	51.69
Solid blockwork 140mm thick in cavity wall	B14	m²	6.76	8.79	149.40	112.76	39.32	301.48
Common bricks 112.5mm thick in cavity wall	B15	m²	5.20	8.84	150.28	85.54	35.37	271.19
Facing bricks in 112.5mm thick in skin of cavity wall	B16	m²	1.56	2.81	47.74	28.55	11.44	87.73
Form cavity 50mm wide in cavity wall	B17	m²	6.76	0.20	3.45	10.28	2.06	15.78
DPC 112mm wide	B18	m	10.40	0.52	8.84	10.61	2.92	22.37
DPC 140mm wide	B19	m	10.40	0.62	10.61	14.35	3.74	28.70
Bond in block wall	B20	m	1.30	0.57	9.72	3.78	2.03	15.53
Bond in half brick wall	B21	m	0.30	0.11	1.79	0.80	0.39	2.98
50mm thick insulation board	B22	m²	11.88	3.56	60.59	85.54	21.92	168.04
Carried to summary				66.46	966.55	837.53	270.61	2,074.71

**PART C
EXTERNAL WALLS**

	Ref	Unit	Qty	Hours	Hours £	Mat'ls £	O & P £	Total £
Solid blockwork 140mm thick in cavity wall	C1	m²	44.14	57.38	975.49	736.26	256.76	1,968.51
Facing brickwork 112.5mm thick in cavity wall	C2	m²	44.14	79.45	1,350.68	1,260.20	391.63	3,002.51
75mm thick insulation in cavity wall	C3	m²	44.14	9.71	165.08	235.71	60.12	460.91
Carried forward				146.54	2,491.26	2,232.16	708.51	5,431.93

	Ref	Unit	Qty	Hours	Hours £	Mat'ls £	O & P £	Total £
Brought forward				146.54	2,491.26	2,232.16	708.51	5,431.93
Steel lintel 2400mm long	C4	nr	2.00	0.50	8.50	296.58	45.76	350.84
Steel lintel 1500mm long	C5	nr	4.00	0.80	13.60	485.44	74.86	573.90
Close cavity wall at jambs	C7	m	13.76	0.69	11.70	38.67	7.55	57.92
Close cavity wall at cills	C8	m	5.07	0.25	4.31	14.25	2.78	21.34
Close cavity wall at top	C9	m	10.40	0.52	8.84	29.22	5.71	43.77
DPC 112mm wide at jambs	C10	m	13.76	0.69	11.70	14.04	3.86	29.59
DPC 112mm wide at cills	C11	m	5.07	0.25	4.31	5.17	1.42	10.90
Carried to summary				150.25	2,554.21	3,115.52	850.46	6,520.20

**PART D
FLAT ROOF**

	Ref	Unit	Qty	Hours	Hours £	Mat'ls £	O & P £	Total £
200 x 50mm sawn softwood joists	D1	m	34.65	8.66	147.26	124.05	40.70	312.01
200 x 50mm sawn softwood sprocket pieces	D2	nr	12.00	1.68	28.56	21.48	7.51	57.55
18mm thick WPB grade decking	D3	m²	16.70	15.03	255.51	191.55	67.06	514.12
50 x 50mm (avg) wide sawn softwood firrings	D4	m	34.65	6.24	106.03	66.18	25.83	198.04
High density polyethylene vapour barrier 150mm thick	D5	m²	15.00	3.00	51.00	167.85	32.83	251.68
100 x 75mm sawn softwood wall plate	D6	m	11.00	3.30	56.10	30.14	12.94	99.18
100 x 75mm sawn softwood tilt fillet	D7	m	5.30	1.33	22.53	13.83	5.45	41.81
Build in ends of 200 x 50mm joists	D8	nr	11.00	3.85	65.45	2.42	10.18	78.05
Carried forward				43.08	732.44	617.50	202.49	1,552.43

	Ref	Unit	Qty	Hours	Hours £	Mat'ls £	O & P £	Total £
Brought forward				43.08	732.44	617.50	202.49	1,552.43
Rake out joint for flashing	D9	m	5.30	1.86	31.54	1.27	4.92	37.73
6mm thick soffit 150mm wide	D10	m	11.30	4.52	76.84	23.84	15.10	115.79
19mm wrought softwood fascia 200mm high	D11	m	11.30	5.65	96.05	30.62	19.00	145.67
Three layer fibre-based roofing felt	D12	m²	16.70	9.19	156.15	217.77	56.09	430.00
Felt turn-down 100mm girth	D13	m	11.30	1.13	19.21	15.82	5.25	40.28
Felt flashing 150mm girth	D14	m	5.30	0.53	9.01	9.12	2.72	20.84
112mm diameter PVC-U gutter	D15	m	5.30	1.38	23.43	17.17	6.09	46.69
Stop end	D16	nr	1.00	0.14	2.38	2.12	0.68	5.18
Stop end outlet	D17	nr	1.00	0.25	4.25	4.19	1.27	9.71
68mm diameter PVC-U down pipe	D18	m	4.80	1.20	20.40	27.36	7.16	54.92
Shoe	D19	nr	1.00	0.30	5.10	3.15	1.24	9.49
Paint fascia and soffit	D20	m²	3.39	0.68	11.53	5.42	2.54	19.49
Carried to summary				69.90	1,188.31	975.36	324.55	2,488.22
PART E PITCHED ROOF				N/A	N/A	N/A	N/A	N/A
PART F WINDOWS AND EXTERNAL DOORS								
PVC-U door size 840 x 1980mm complete (B)	F1	nr	1.00	2.50	42.50	278.67	48.18	369.35
PVC-U sliding patio door size 1700 x 2075mm (C)	F2	nr	2.00	14.00	238.00	616.94	128.24	983.18
Carried forward				16.50	280.50	895.61	176.42	1,352.53

	Ref	Unit	Qty	Hours	Hours £	Mat'ls £	O & P £	Total £
Brought forward				16.50	280.50	895.61	176.42	1,352.53
PVC-U window size 1200 x 1200mm complete (A)	F3	nr	4.00	8.00	136.00	728.56	129.68	994.24
25 x 225mm wrought softwood window board	F4	m	4.80	1.44	24.48	19.30	6.57	50.34
Paint window board	F5	m	4.80	0.14	2.38	4.42	1.02	7.82
Carried to summary				26.08	443.36	1,647.88	313.69	2,404.93
PART G INTERNAL PARTITIONS AND DOORS				N/A	N/A	N/A	N/A	N/A
PART H WALL FINISHES								
19 x 100mm wrought softwood skirting	H1	m	18.34	3.12	53.00	46.03	14.86	113.89
12mm plasterboard fixed to walls with dabs	H2	m²	39.64	15.46	262.81	112.97	56.37	432.16
12mm plasterboard fixed to walls less than 300mm wide	H3	m	18.33	3.30	56.09	16.86	10.94	83.90
Two coats emulsion paint to walls	H4	m²	42.46	11.04	187.67	40.76	34.27	262.70
Paint skirting	H5	m	18.34	3.67	62.36	16.87	11.88	91.11
Carried to summary				36.58	621.93	233.51	128.32	983.76
PART J FLOOR FINISHES								
Cement and sand floor screed 40mm thick	J1	m²	11.88	2.97	50.49	56.55	16.06	123.09
Vinyl floor tiles, size 300 x 300mm	J2	m²	11.88	2.02	34.33	92.19	18.98	145.50
Carried forward				4.99	84.82	148.74	35.03	268.59

	Ref	Unit	Qty	Hours	Hours £	Mat'ls £	O & P £	Total £
Brought forward				4.99	84.82	148.74	35.03	268.59
25mm thick tongued and grooved boarding	J3	m²	11.88	8.79	149.45	148.26	44.66	342.37
150 x 50mm softwood joists	J4	m	23.20	5.10	86.77	72.38	23.87	183.02
Cut and pin end of joists to existing brick wall	J5	nr	8.00	1.44	24.48	2.40	4.03	30.91
Build in ends of joists to blockwork	J6	nr	8.00	0.80	13.60	1.92	2.33	17.85
Carried to summary				21.12	359.12	373.70	109.92	842.75

PART K
CEILING FINISHES

	Ref	Unit	Qty	Hours	Hours £	Mat'ls £	O & P £	Total £
Plasterboard with taped butt joints fixed to joists	K1	m²	23.76	8.55	145.41	67.72	31.97	245.10
5mm skim coat to plasterboard ceilings	K2	m²	23.76	11.88	201.96	41.10	36.46	279.52
Two coats emulsion paint to ceilings	K3	m²	23.76	6.18	105.02	22.81	19.17	147.00
Carried to summary				26.61	452.39	131.63	87.60	671.62

PART L
ELECTRICAL WORK

	Ref	Unit	Qty	Hours	Hours £	Mat'ls £	O & P £	Total £
13 amp double switched socket outlet with neon	L1	nr	6.00	4.80	91.20	52.86	21.61	165.67
Lighting point	L2	nr	4.00	2.80	53.20	29.36	12.38	94.94
Lighting switch	L3	nr	4.00	2.80	53.20	17.68	10.63	81.51
Lighting wiring	L4	m	14.00	2.80	53.20	26.04	11.89	91.13
Power cable	L5	m	36.00	10.80	205.20	97.56	45.41	348.17
Carried to summary				24.00	456.00	223.50	101.93	781.43

	Ref	Unit	Qty	Hours	Hours £	Mat'ls £	O & P £	Total £
PART M **HEATING WORK**								
15mm copper pipe	M1	m	15.00	6.60	118.80	32.25	22.66	173.71
Elbow	M2	nr	8.00	4.48	80.64	19.52	15.02	115.18
Tee	M3	nr	2.00	1.12	20.16	4.88	3.76	28.80
Radiator, double convector size 1400 x 520mm	M4	nr	4.00	5.20	93.60	556.44	97.51	747.55
Break into existing pipe and insert tee	M5	nr	1.00	0.75	13.50	4.42	2.69	20.61
Carried to summary				18.15	326.70	617.51	141.63	1,085.84
PART N **ALTERATION WORK**								
Take out existing window size 1500 x 1000mm and lintel over, adapt opening to receive 1770 x 2000mm patio door and insert new lintel over (both measured separately) and make good	N1	nr	1.00	20.00	340.00	32.10	55.82	427.92
Take out existing window size 1500 x 1000mm, enlarge opening to receive new PVC–U door (measured separately)	N2	nr	1.00	24.00	408.00	58.16	69.92	536.08
Carried to summary				44.00	748.00	90.26	125.74	964.00

SUMMARY

	Hours	Hours £	Mat'ls £	O & P £	Total £
PART A **PRELIMINARIES**	0.00	0.00	0.00	0.00	3,236.00
PART B SUBSTRUCTURE TO **DPC LEVEL**	66.46	966.55	837.53	270.61	2,074.70
PART C **EXTERNAL WALLS**	150.25	2,554.21	3,115.52	850.46	6,520.20
PART D **FLAT ROOF**	69.90	1,188.31	975.36	324.55	2,488.22
PART E **PITCHED ROOF**	0.00	0.00	0.00	0.00	0.00
PART F WINDOWS AND **EXTERNAL DOORS**	26.08	443.36	1,647.88	313.69	2,404.93
PART G INTERNAL **PARTITIONS AND DOORS**	0.00	0.00	0.00	0.00	0.00
PART H **WALL FINISHES**	36.58	621.93	233.51	128.32	983.76
PART J **FLOOR FINISHES**	21.12	359.12	373.70	109.92	842.75
PART K **CEILING FINISHES**	26.61	452.39	131.63	87.60	671.62
PART L **ELECTRICAL WORK**	24.00	456.00	223.50	101.93	781.43
PART M **HEATING WORK**	18.15	326.70	617.51	141.63	1,085.84
PART N **ALTERATION WORK**	44.00	748.00	90.26	125.74	964.00
Final total	483.15	8,116.57	8,246.40	2,454.45	22,053.45

	Ref	Unit	Qty	Hours	Hours £	Mat'ls £	O & P £	Total £
PART A **PRELIMINARIES**								
Concrete mixer	A1	wks	12.00					600.00
Small tools	A2	wks	12.00					480.00
Scaffolding (m²/weeks)	A3		260					1,560.00
Skip	A4	wks	8.00					880.00
Clean up	A5	hrs	12.00					156.00
Carried to summary								3,676.00
PART B **SUBSTRUCTURE TO** **DPC LEVEL**								
Excavate topsoil 150mm thick by hand	B1	m²	19.85	5.96	77.42	0.00	11.61	89.03
Excavate to reduce levels	B2	m³	5.95	14.88	193.38	0.00	29.01	222.38
Excavate for trench foundations by hand	B3	m³	1.88	4.89	63.54	0.00	9.53	73.08
Earthwork support to sides of trenches	B4	m²	16.64	6.66	86.53	30.78	17.60	134.91
Backfilling with excavated material	B5	m³	0.70	0.42	5.46	0.00	0.82	6.28
Hardcore 225mm thick	B6	m²	14.58	2.92	37.91	91.71	19.44	149.06
Hardcore filling to trench	B7	m³	0.15	0.08	0.98	6.29	1.09	8.35
Concrete grade (1:3:6) in foundations	B8	m³	1.54	2.08	27.03	131.85	23.83	182.71
Concrete grade (1:2:4) in bed 150mm thick	B9	m²	14.58	4.37	56.86	223.37	42.03	322.26
Concrete (1:2:4) in cavity wall filling	B10	m²	5.70	1.14	11.97	13.57	3.83	29.37
Carried forward				43.38	561.06	497.57	158.79	1,217.43

	Ref	Unit	Qty	Hours	Hours £	Mat'ls £	O & P £	Total £
Brought forward				43.38	561.06	497.57	158.79	1,217.43
Damp-proof membrane	B11	m²	15.13	0.61	7.87	35.10	6.45	49.41
Reinforcement ref A193 in foundation	B12	m²	5.70	0.68	8.89	11.51	3.06	23.47
Steel fabric reinforcement ref A193 in slab	B13	m²	14.58	2.19	28.43	29.16	8.64	66.23
Solid blockwork 140mm thick in cavity wall	B14	m²	7.41	9.63	163.76	123.60	43.10	330.46
Common bricks 112.5mm thick in cavity wall	B15	m²	5.70	9.69	164.73	93.77	38.77	297.27
Facing bricks in 112.5mm thick in skin of cavity wall	B16	m²	1.71	3.08	52.33	28.55	12.13	93.01
Form cavity 50mm wide in cavity wall	B17	m²	7.41	0.22	3.78	11.26	2.26	17.30
DPC 112mm wide	B18	m	11.40	0.57	9.69	11.63	3.20	24.52
DPC 140mm wide	B19	m	11.40	0.68	11.63	15.73	4.10	31.46
Bond in block wall	B20	m	1.30	0.57	9.72	3.78	2.03	15.53
Bond in half brick wall	B21	m	0.30	0.11	1.79	0.80	0.39	2.98
50mm thick insulation board	B22	m²	14.58	4.37	74.36	104.98	26.90	206.23
Carried to summary				75.78	1,098.04	967.44	309.82	2,375.30

**PART C
EXTERNAL WALLS**

	Ref	Unit	Qty	Hours	Hours £	Mat'ls £	O & P £	Total £
Solid blockwork 140mm thick in cavity wall	C1	m²	46.65	60.65	1,030.97	778.12	271.36	2,080.45
Facing brickwork 112.5mm thick in cavity wall	C2	m²	46.65	83.97	1,427.49	1,331.86	413.90	3,173.25
75mm thick insulation in cavity wall	C3	m²	46.65	10.26	174.47	249.11	63.54	487.12
Carried forward				154.88	2,632.93	2,359.09	748.80	5,740.82

	Ref	Unit	Qty	Hours	Hours £	Mat'ls £	O & P £	Total £
Brought forward				154.88	2,632.93	2,359.09	748.80	5,740.82
Steel lintel 2400mm long	C4	nr	2.00	0.50	8.50	296.58	45.76	350.84
Steel lintel 1500mm long	C5	nr	6.00	1.20	20.40	728.16	112.28	860.84
Steel lintel 1500mm long	C6	nr	1.00	0.15	2.55	98.72	15.19	116.46
Close cavity wall at jambs	C7	m	24.52	1.23	20.84	68.90	13.46	103.20
Close cavity wall at cills	C8	m	9.31	0.47	7.91	26.16	5.11	39.19
Close cavity wall at top	C9	m	11.40	0.57	9.69	32.03	6.26	47.98
DPC 112mm wide at jambs	C10	m	24.52	1.23	20.84	25.01	6.88	52.73
DPC 112mm wide at cills	C11	m	9.31	0.47	7.91	9.50	2.61	20.02
Carried to summary				160.68	2,731.58	3,644.15	956.36	7,332.09

PART D
FLAT ROOF

	Ref	Unit	Qty	Hours	Hours £	Mat'ls £	O & P £	Total £
200 x 50mm sawn softwood joists	D1	m	40.95	10.24	174.04	146.60	48.10	368.73
200 x 50mm sawn softwood sprocket pieces	D2	nr	12.00	1.68	28.56	21.48	7.51	57.55
18mm thick WPB grade decking	D3	m²	19.85	17.87	303.71	227.68	79.71	611.09
50 x 50mm (avg) wide sawn softwood firrings	D4	m	40.95	7.37	125.31	78.21	30.53	234.05
High density polyethylene vapour barrier 150mm thick	D5	m²	18.00	3.60	61.20	201.42	39.39	302.01
100 x 75mm sawn softwood wall plate	D6	m	12.00	3.60	61.20	32.88	14.11	108.19
100 x 75mm sawn softwood tilt fillet	D7	m	6.30	1.58	26.78	16.44	6.48	49.70
Build in ends of 200 x 50mm joists	D8	nr	13.00	4.55	77.35	2.86	12.03	92.24
Carried forward				50.48	858.13	727.58	237.86	1,823.57

	Ref	Unit	Qty	Hours	Hours £	Mat'ls £	O & P £	Total £
Brought forward				50.48	858.13	727.58	237.86	1,823.57
Rake out joint for flashing	D9	m	6.30	2.21	37.49	1.51	5.85	44.85
6mm thick soffit 150mm wide	D10	m	12.30	4.92	83.64	25.95	16.44	126.03
19mm wrought softwood fascia 200mm high	D11	m	12.30	6.15	104.55	33.33	20.68	158.57
Three layer fibre-based roofing felt	D12	m²	19.85	10.92	185.60	258.84	66.67	511.11
Felt turn-down 100mm girth	D13	m	12.30	1.23	20.91	17.22	5.72	43.85
Felt flashing 150mm girth	D14	m	6.30	0.63	10.71	10.84	3.23	24.78
112mm diameter PVC-U gutter	D15	m	6.30	1.64	27.85	20.41	7.24	55.50
Stop end	D16	nr	1.00	0.14	2.38	2.12	0.68	5.18
Stop end outlet	D17	nr	1.00	0.25	4.25	4.19	1.27	9.71
68mm diameter PVC-U down pipe	D18	m	4.80	1.20	20.40	27.36	7.16	54.92
Shoe	D19	nr	1.00	0.30	5.10	3.15	1.24	9.49
Paint fascia and soffit	D20	m²	3.69	0.74	12.55	5.90	2.77	21.22
Carried to summary				80.80	1,373.55	1,138.41	376.79	2,888.76
PART E PITCHED ROOF				N/A	N/A	N/A	N/A	N/A
PART F WINDOWS AND EXTERNAL DOORS								
PVC-U door size 840 x 1980mm complete (B)	F1	nr	2.00	5.00	85.00	278.67	54.55	418.22
PVC-U sliding patio door size 1700 x 2075mm (C)	F2	nr	2.00	14.00	238.00	616.94	128.24	983.18
Carried forward				19.00	323.00	895.61	182.79	1,401.40

	Ref	Unit	Qty	Hours	Hours £	Mat'ls £	O & P £	Total £
Brought forward				19.00	323.00	895.61	182.79	1,401.40
PVC-U window size 1200 x 1200mm complete (A)	F3	nr	6.00	12.00	204.00	1,092.84	194.53	1,491.37
25 x 225mm wrought softwood window board	F4	m	7.20	2.16	36.72	28.94	9.85	75.51
Paint window board	F5	m	7.20	0.14	2.38	6.62	1.35	10.35
Carried to summary				33.30	566.10	2,024.02	388.52	2,978.64
PART G INTERNAL PARTITIONS AND DOORS				N/A	N/A	N/A	N/A	N/A
PART H WALL FINISHES								
19 x 100mm wrought softwood skirting	H1	m	19.56	3.33	56.53	49.10	15.84	121.47
12mm plasterboard fixed to walls with dabs	H2	m²	40.00	15.60	265.20	114.00	56.88	436.08
12mm plasterboard fixed to walls less than 300mm wide	H3	m	13.83	2.49	42.32	12.72	8.26	63.30
Two coats emulsion paint to walls	H4	m²	25.07	6.52	110.81	24.07	20.23	155.11
Paint skirting	H5	m	19.56	3.91	66.50	18.00	12.67	97.17
Carried to summary				31.84	541.36	217.88	113.89	873.13
PART J FLOOR FINISHES								
Cement and sand floor screed 40mm thick	J1	m²	14.58	3.65	61.97	69.40	19.70	151.07
Vinyl floor tiles, size 300 x 300mm	J2	m²	14.58	2.48	42.14	113.14	23.29	178.57
Carried forward				6.12	104.10	182.54	43.00	329.64

	Ref	Unit	Qty	Hours	Hours £	Mat'ls £	O & P £	Total £
Brought forward				6.12	104.10	182.54	43.00	329.64
25mm thick tongued and grooved boarding	J3	m²	14.58	10.79	183.42	181.96	54.81	420.18
150 x 50mm softwood joists	J4	m	29.00	6.38	108.46	90.48	29.84	228.78
Cut and pin end of joists to existing brick wall	J5	nr	10.00	1.80	30.60	3.00	5.04	38.64
Build in ends of joists to blockwork	J6	nr	10.00	1.00	17.00	2.40	2.91	22.31
Carried to summary				26.09	443.58	460.38	135.59	1,039.55

PART K
CEILING FINISHES

	Ref	Unit	Qty	Hours	Hours £	Mat'ls £	O & P £	Total £
Plasterboard with taped butt joints fixed to joists	K1	m²	29.18	10.50	178.58	83.16	39.26	301.01
5mm skim coat to plasterboard ceilings	K2	m²	29.18	14.59	248.03	50.48	44.78	343.29
Two coats emulsion paint to ceilings	K3	m²	29.18	7.59	128.98	28.01	23.55	180.54
Carried to summary				32.68	555.59	161.66	107.59	824.83

PART L
ELECTRICAL WORK

	Ref	Unit	Qty	Hours	Hours £	Mat'ls £	O & P £	Total £
13 amp double switched socket outlet with neon	L1	nr	8.00	6.40	121.60	70.48	28.81	220.89
Lighting point	L2	nr	4.00	2.80	53.20	29.36	12.38	94.94
Lighting switch	L3	nr	5.00	3.50	66.50	22.10	13.29	101.89
Lighting wiring	L4	m	16.00	3.20	60.80	29.76	13.58	104.14
Power cable	L5	m	40.00	12.00	228.00	108.40	50.46	386.86
Carried to summary				27.90	530.10	260.10	118.53	908.73

	Ref	Unit	Qty	Hours	Hours £	Mat'ls £	O & P £	Total £
PART M								
HEATING WORK								
15mm copper pipe	M1	m	18.00	7.92	142.56	38.70	27.19	208.45
Elbow	M2	nr	8.00	4.48	80.64	19.52	15.02	115.18
Tee	M3	nr	2.00	1.12	20.16	4.88	3.76	28.80
Radiator, double convector size 1400 x 520mm	M4	nr	4.00	5.20	93.60	556.44	97.51	747.55
Break into existing pipe and insert tee	M5	nr	1.00	0.75	13.50	4.42	2.69	20.61
Carried to summary				19.47	350.46	623.96	146.16	1,120.58
PART N								
ALTERATION WORK								
Take out existing window size 1500 x 1000mm and lintel over, adapt opening to receive 1770 x 2000mm patio door and insert new lintel over (both measured separately) and make good	N1	nr	1.00	20.00	340.00	32.10	55.82	427.92
Take out existing window size 1500 x 1000mm, enlarge opening to receive new PVC–U door (measured separately)	N2	nr	1.00	24.00	408.00	58.16	69.92	536.08
Carried to summary				44.00	748.00	90.26	125.74	964.00

SUMMARY

	Hours	Hours £	Mat'ls £	O & P £	Total £
PART A **PRELIMINARIES**	0.00	0.00	0.00	0.00	3,676.00
PART B SUBSTRUCTURE TO **DPC LEVEL**	75.78	1,098.04	967.44	309.82	2,375.30
PART C **EXTERNAL WALLS**	160.68	2,731.58	3,644.15	956.36	7,332.09
PART D **FLAT ROOF**	80.80	1,373.55	1,138.41	376.79	2,888.76
PART E **PITCHED ROOF**	0.00	0.00	0.00	0.00	0.00
PART F WINDOWS AND **EXTERNAL DOORS**	33.30	566.10	2,024.02	388.52	2,978.64
PART G INTERNAL **PARTITIONS AND DOORS**	0.00	0.00	0.00	0.00	0.00
PART H **WALL FINISHES**	31.84	541.36	217.88	113.89	873.13
PART J **FLOOR FINISHES**	26.09	443.58	460.38	135.59	1,039.55
PART K **CEILING FINISHES**	32.68	555.59	161.66	107.59	824.83
PART L **ELECTRICAL WORK**	27.90	530.10	260.10	118.53	908.73
PART M **HEATING WORK**	19.47	350.46	623.96	146.16	1,120.58
PART N **ALTERATION WORK**	44.00	748.00	90.26	125.74	964.00
Final total	532.54	8,938.36	9,588.26	2,778.99	24,981.61

	Ref	Unit	Qty	Hours	Hours £	Mat'ls £	O & P £	Total £
PART A								
PRELIMINARIES								
Concrete mixer	A1	wks	11.00					550.00
Small tools	A2	wks	11.00					440.00
Scaffolding (m²/weeks)	A3		260					1,560.00
Skip	A4	wks	7.00					770.00
Clean up	A5	hrs	12.00					156.00
Carried to summary								3,476.00
PART B								
SUBSTRUCTURE TO								
DPC LEVEL								
Excavate topsoil 150mm thick by hand	B1	m²	17.85	5.36	69.62	0.00	10.44	80.06
Excavate to reduce levels	B2	m³	5.35	13.38	173.88	0.00	26.08	199.96
Excavate for trench foundations by hand	B3	m³	1.88	4.89	63.54	0.00	9.53	73.08
Earthwork support to sides of trenches	B4	m²	16.64	6.66	86.53	30.78	17.60	134.91
Backfilling with excavated material	B5	m³	0.62	0.37	4.84	0.00	0.73	5.56
Hardcore 225mm thick	B6	m²	13.69	2.74	35.59	86.11	18.26	139.96
Hardcore filling to trench	B7	m³	0.14	0.07	0.91	6.29	1.08	8.28
Concrete grade (1:3:6) in foundations	B8	m³	1.54	2.08	27.03	131.85	23.83	182.71
Concrete grade (1:2:4) in bed 150mm thick	B9	m²	13.69	4.11	53.39	209.73	39.47	302.59
Concrete (1:2:4) in cavity wall filling	B10	m²	5.70	1.14	11.97	13.57	3.83	29.37
Carried forward				40.78	527.29	478.34	150.84	1,156.47

	Ref	Unit	Qty	Hours	Hours £	Mat'ls £	O & P £	Total £
Brought forward				40.78	527.29	478.34	150.84	1,156.47
Damp-proof membrane	B11	m²	13.13	0.53	6.83	30.46	5.59	42.88
Reinforcement ref A193 in foundation	B12	m²	5.70	0.68	8.89	11.51	3.06	23.47
Steel fabric reinforcement ref A193 in slab	B13	m²	13.69	2.05	26.70	27.38	8.11	62.19
Solid blockwork 140mm thick in cavity wall	B14	m²	7.41	9.63	163.76	123.60	43.10	330.46
Common bricks 112.5mm thick in cavity wall	B15	m²	5.70	9.69	164.73	93.77	38.77	297.27
Facing bricks in 112.5mm thick in skin of cavity wall	B16	m²	1.71	3.08	52.33	28.55	12.13	93.01
Form cavity 50mm wide in cavity wall	B17	m²	7.41	0.22	3.78	11.26	2.26	17.30
DPC 112mm wide	B18	m	11.40	0.57	9.69	11.63	3.20	24.52
DPC 140mm wide	B19	m	11.40	0.68	11.63	15.73	4.10	31.46
Bond in block wall	B20	m	1.30	0.57	9.72	3.78	2.03	15.53
Bond in half brick wall	B21	m	0.30	0.11	1.79	0.80	0.39	2.98
50mm thick insulation board	B22	m²	13.69	4.11	69.82	98.57	25.26	193.65
Carried to summary				72.70	1,056.95	935.38	298.85	2,291.18

PART C
EXTERNAL WALLS

	Ref	Unit	Qty	Hours	Hours £	Mat'ls £	O & P £	Total £
Solid blockwork 140mm thick in cavity wall	C1	m²	44.65	58.05	986.77	744.76	259.73	1,991.26
Facing brickwork 112.5mm thick in cavity wall	C2	m²	44.65	80.37	1,366.29	1,274.76	396.16	3,037.20
75mm thick insulation in cavity wall	C3	m²	44.65	9.82	166.99	238.43	60.81	466.24
Carried forward				148.24	2,520.05	2,257.95	716.70	5,494.70

	Ref	Unit	Qty	Hours	Hours £	Mat'ls £	O & P £	Total £
Brought forward				148.24	2,520.05	2,257.95	716.70	5,494.70
Steel lintel 2400mm long	C4	nr	2.00	0.50	8.50	296.58	45.76	350.84
Steel lintel 1500mm long	C5	nr	6.00	1.20	20.40	728.16	112.28	860.84
Steel lintel 1500mm long	C6	nr	1.00	0.15	2.55	98.72	15.19	116.46
Close cavity wall at jambs	C7	m	24.52	1.23	20.84	68.90	13.46	103.20
Close cavity wall at cills	C8	m	9.31	0.47	7.91	26.16	5.11	39.19
Close cavity wall at top	C9	m	11.40	0.57	9.69	32.03	6.26	47.98
DPC 112mm wide at jambs	C10	m	24.52	1.23	20.84	25.01	6.88	52.73
DPC 112mm wide at cills	C11	m	9.31	0.47	7.91	9.50	2.61	20.02
Carried to summary				154.04	2,618.70	3,543.01	924.26	7,085.97

PART D
FLAT ROOF

	Ref	Unit	Qty	Hours	Hours £	Mat'ls £	O & P £	Total £
200 x 50mm sawn softwood joists	D1	m	37.35	9.34	158.74	133.71	43.87	336.32
200 x 50mm sawn softwood sprocket pieces	D2	nr	16.00	2.24	38.08	28.64	10.01	76.73
18mm thick WPB grade decking	D3	m²	17.85	16.07	273.11	204.74	71.68	549.52
50 x 50mm (avg) wide sawn softwood firrings	D4	m	37.35	6.72	114.29	71.34	27.84	213.47
High density polyethylene vapour barrier 150mm thick	D5	m²	16.00	3.20	54.40	179.04	35.02	268.46
100 x 75mm sawn softwood wall plate	D6	m	12.00	3.60	61.20	32.88	14.11	108.19
100 x 75mm sawn softwood tilt fillet	D7	m	4.30	1.08	18.28	11.22	4.42	33.92
Build in ends of 200 x 50mm joists	D8	nr	9.00	3.15	53.55	1.98	8.33	63.86
Carried forward				45.39	771.64	663.55	215.28	1,650.47

	Ref	Unit	Qty	Hours	Hours £	Mat'ls £	O & P £	Total £
Brought forward				45.39	771.64	663.55	215.28	1,650.47
Rake out joint for flashing	D9	m	4.30	1.51	25.59	1.03	3.99	30.61
6mm thick soffit 150mm wide	D10	m	12.30	4.92	83.64	25.95	16.44	126.03
19mm wrought softwood fascia 200mm high	D11	m	12.30	6.15	104.55	33.33	20.68	158.57
Three layer fibre-based roofing felt	D12	m²	17.85	9.82	166.90	232.76	59.95	459.61
Felt turn-down 100mm girth	D13	m	12.30	1.23	20.91	17.22	5.72	43.85
Felt flashing 150mm girth	D14	m	4.30	0.43	7.31	7.40	2.21	16.91
112mm diameter PVC-U gutter	D15	m	4.30	1.12	19.01	13.93	4.94	37.88
Stop end	D16	nr	1.00	0.14	2.38	2.12	0.68	5.18
Stop end outlet	D17	nr	1.00	0.25	4.25	4.19	1.27	9.71
68mm diameter PVC-U down pipe	D18	m	4.80	1.20	20.40	27.36	7.16	54.92
Shoe	D19	nr	1.00	0.30	5.10	3.15	1.24	9.49
Paint fascia and soffit	D20	m²	3.69	0.74	12.55	5.90	2.77	21.22
Carried to summary				73.19	1,244.21	1,037.91	342.32	2,624.44
PART E **PITCHED ROOF**				N/A	N/A	N/A	N/A	N/A
PART F **WINDOWS AND** **EXTERNAL DOORS**								
PVC-U door size 840 x 1980mm complete (B)	F1	nr	2.00	5.00	85.00	278.67	54.55	418.22
PVC-U sliding patio door size 1700 x 2075mm (C)	F2	nr	2.00	14.00	238.00	616.94	128.24	983.18
Carried forward				19.00	323.00	895.61	182.79	1,401.40

	Ref	Unit	Qty	Hours	Hours £	Mat'ls £	O & P £	Total £
Brought forward				19.00	323.00	895.61	182.79	1,401.40
PVC-U window size 1200 x 1200mm complete (A)	F3	nr	6.00	12.00	204.00	1,092.84	194.53	1,491.37
25 x 225mm wrought softwood window board	F4	m	7.20	2.16	36.72	28.94	9.85	75.51
Paint window board	F5	m	7.20	0.14	2.38	6.62	1.35	10.35
Carried to summary				33.30	566.10	2,024.02	388.52	2,978.64
PART G INTERNAL PARTITIONS AND DOORS				N/A	N/A	N/A	N/A	N/A
PART H WALL FINISHES								
19 x 100mm wrought softwood skirting	H1	m	19.56	3.33	56.53	49.10	15.84	121.47
12mm plasterboard fixed to walls ith dabs	H2	m²	40.00	15.60	265.20	114.00	56.88	436.08
12mm plasterboard fixed to walls less than 300mm wide	H3	m	33.83	6.09	103.52	31.12	20.20	154.84
Two coats emulsion paint to walls	H4	m²	45.07	11.72	199.21	43.27	36.37	278.85
Paint skirting	H5	m	19.56	3.91	66.50	18.00	12.67	97.17
Carried to summary				40.64	690.96	255.48	141.97	1,088.41
PART J FLOOR FINISHES								
Cement and sand floor screed 40mm thick	J1	m²	12.58	3.15	53.47	59.88	17.00	130.35
Vinyl floor tiles, size 300 x 300mm	J2	m²	12.58	2.14	36.36	97.62	20.10	154.07
Carried forward				5.28	89.82	157.50	37.10	284.42

250 Two storey extension, size 4 x 4m, flat roof

	Ref	Unit	Qty	Hours	Hours £	Mat'ls £	O & P £	Total £
Brought forward				5.28	89.82	157.50	37.10	284.42
25mm thick tongued and grooved boarding	J3	m²	12.58	9.31	158.26	157.00	47.29	362.54
150 x 50mm softwood joists	J4	m	27.30	6.01	102.10	85.18	28.09	215.37
Cut and pin end of joists to existing brick wall	J5	nr	7.00	1.26	21.42	2.10	3.53	27.05
Build in ends of joists to blockwork	J6	nr	7.00	0.70	11.90	1.68	2.04	15.62
Carried to summary				22.56	383.50	403.46	118.04	905.00

PART K
CEILING FINISHES

	Ref	Unit	Qty	Hours	Hours £	Mat'ls £	O & P £	Total £
Plasterboard with taped butt joints fixed to joists	K1	m²	25.16	9.06	153.98	71.71	33.85	259.54
5mm skim coat to plasterboard ceilings	K2	m²	25.16	12.58	213.86	43.53	38.61	295.99
Two coats emulsion paint to ceilings	K3	m²	25.16	6.54	111.21	24.15	20.30	155.66
Carried to summary				28.18	479.05	139.39	92.76	711.20

PART L
ELECTRICAL WORK

	Ref	Unit	Qty	Hours	Hours £	Mat'ls £	O & P £	Total £
13 amp double switched socket outlet with neon	L1	nr	6.00	4.80	91.20	52.86	21.61	165.67
Lighting point	L2	nr	4.00	2.80	53.20	29.36	12.38	94.94
Lighting switch	L3	nr	5.00	3.50	66.50	22.10	13.29	101.89
Lighting wiring	L4	m	16.00	3.20	60.80	29.76	13.58	104.14
Power cable	L5	m	36.00	10.80	205.20	97.56	45.41	348.17
Carried to summary				25.10	476.90	231.64	106.28	814.82

	Ref	Unit	Qty	Hours	Hours £	Mat'ls £	O & P £	Total £
PART M **HEATING WORK**								
15mm copper pipe	M1	m	19.00	8.36	150.48	40.85	28.70	220.03
Elbow	M2	nr	8.00	4.48	80.64	19.52	15.02	115.18
Tee	M3	nr	2.00	1.12	20.16	4.88	3.76	28.80
Radiator, double convector size 1400 x 520mm	M4	nr	4.00	5.20	93.60	556.44	97.51	747.55
Break into existing pipe and insert tee	M5	nr	1.00	0.75	13.50	4.42	2.69	20.61
Carried to summary				19.91	358.38	626.11	147.67	1,132.16
PART N **ALTERATION WORK**								
Take out existing window size 1500 x 1000mm and lintel over, adapt opening to receive 1770 x 2000mm patio door and insert new lintel over (both measured separately) and make good	N1	nr	1.00	20.00	340.00	32.10	55.82	427.92
Take out existing window size 1500 x 1000mm, enlarge opening to receive new PVC–U door (measured separately)	N2	nr	1.00	24.00	408.00	58.16	69.92	536.08
Carried to summary				44.00	748.00	90.26	125.74	964.00

SUMMARY

	Hours	Hours £	Mat'ls £	O & P £	Total £
PART A **PRELIMINARIES**	0.00	0.00	0.00	0.00	3,476.00
PART B SUBSTRUCTURE TO **DPC LEVEL**	72.70	1,056.95	935.38	298.85	2,291.18
PART C **EXTERNAL WALLS**	154.04	2,618.70	3,543.01	924.26	7,085.97
PART D **FLAT ROOF**	73.19	1,244.21	1,037.91	342.32	2,624.44
PART E **PITCHED ROOF**	0.00	0.00	0.00	0.00	0.00
PART F WINDOWS AND **EXTERNAL DOORS**	33.30	566.10	2,024.02	388.52	2,978.64
PART G INTERNAL **PARTITIONS AND DOORS**	0.00	0.00	0.00	0.00	0.00
PART H **WALL FINISHES**	40.64	690.96	255.48	141.97	1,088.41
PART J **FLOOR FINISHES**	22.56	383.50	403.46	118.04	905.00
PART K **CEILING FINISHES**	28.18	479.05	139.39	92.76	711.20
PART L **ELECTRICAL WORK**	25.10	476.90	231.64	106.28	814.82
PART M **HEATING WORK**	19.91	358.38	626.11	147.67	1,132.16
PART N **ALTERATION WORK**	44.00	748.00	90.26	125.74	964.00
Final total	513.62	8,622.75	9,286.66	2,686.41	24,071.82

	Ref	Unit	Qty	Hours	Hours £	Mat'ls £	O & P £	Total £
PART A								
PRELIMINARIES								
Concrete mixer	A1	wks	12.00					600.00
Small tools	A2	wks	12.00					480.00
Scaffolding (m²/weeks)	A3		280					1,680.00
Skip	A4	wks	8.00					880.00
Clean up	A5	hrs	12.00					156.00
Carried to summary								3,796.00
PART B								
SUBSTRUCTURE TO								
DPC LEVEL								
Excavate topsoil 150mm thick by hand	B1	m²	22.00	6.60	85.80	0.00	12.87	98.67
Excavate to reduce levels	B2	m³	6.60	16.50	214.50	0.00	32.18	246.68
Excavate for trench foundations by hand	B3	m³	2.05	5.33	69.29	0.00	10.39	79.68
Earthwork support to sides of trenches	B4	m²	18.10	7.24	94.12	33.49	19.14	146.75
Backfilling with excavated material	B5	m³	0.70	0.42	5.46	0.00	0.82	6.28
Hardcore 225mm thick	B6	m²	16.28	3.26	42.33	102.40	21.71	166.44
Hardcore filling to trench	B7	m³	0.15	0.08	0.98	6.29	1.09	8.35
Concrete grade (1:3:6) in foundations	B8	m³	1.67	2.25	29.31	142.99	25.84	198.14
Concrete grade (1:2:4) in bed 150mm thick	B9	m²	16.28	4.88	63.49	249.41	46.94	359.84
Concrete (1:2:4) in cavity wall filling	B10	m²	6.20	1.24	11.97	14.76	4.01	30.73
Carried forward				47.80	617.24	549.33	174.99	1,341.56

	Ref	Unit	Qty	Hours	Hours £	Mat'ls £	O & P £	Total £
Brought forward				47.80	617.24	549.33	174.99	1,341.56
Damp-proof membrane	B11	m²	16.88	0.68	8.78	39.16	7.19	55.13
Reinforcement ref A193 in foundation	B12	m²	6.20	0.74	9.67	12.52	3.33	25.53
Steel fabric reinforcement ref A193 in slab	B13	m²	16.28	2.44	31.75	32.56	9.65	73.95
Solid blockwork 140mm thick in cavity wall	B14	m²	8.06	10.48	178.13	134.44	46.89	359.45
Common bricks 112.5mm thick in cavity wall	B15	m²	6.20	10.54	179.18	101.99	42.18	323.35
Facing bricks in 112.5mm thick in skin of cavity wall	B16	m²	1.86	3.35	56.92	28.55	12.82	98.29
Form cavity 50mm wide in cavity wall	B17	m²	8.06	0.24	4.11	12.25	2.45	18.82
DPC 112mm wide	B18	m	12.40	0.62	10.54	12.65	3.48	26.67
DPC 140mm wide	B19	m	12.40	0.74	12.65	17.11	4.46	34.22
Bond in block wall	B20	m	1.30	0.57	9.72	3.78	2.03	15.53
Bond in half brick wall	B21	m	0.30	0.11	1.79	0.80	0.39	2.98
50mm thick insulation board	B22	m²	16.28	4.88	83.03	117.22	30.04	230.28
Carried to summary				83.19	1,203.50	1,062.37	339.88	2,605.74

PART C
EXTERNAL WALLS

	Ref	Unit	Qty	Hours	Hours £	Mat'ls £	O & P £	Total £
Solid blockwork 140mm thick in cavity wall	C1	m²	46.82	60.87	1,034.72	780.96	272.35	2,088.03
Facing brickwork 112.5mm thick n cavity wall	C2	m²	46.82	84.28	1,432.69	1,336.71	415.41	3,184.81
75mm thick insulation in cavity wall	C3	m²	46.82	10.30	175.11	250.02	63.77	488.89
Carried forward				155.44	2,642.52	2,367.69	751.53	5,761.74

	Ref	Unit	Qty	Hours	Hours £	Mat'ls £	O & P £	Total £
Brought forward				155.44	2,642.52	2,367.69	751.53	5,761.74
Steel lintel 2400mm long	C4	nr	2.00	0.50	8.50	296.58	45.76	350.84
Steel lintel 1500mm long	C5	nr	8.00	1.60	27.20	970.88	149.71	1,147.79
Steel lintel 1500mm long	C6	nr	1.00	0.15	2.55	98.72	15.19	116.46
Close cavity wall at jambs	C7	m	29.32	1.47	24.92	82.39	16.10	123.41
Close cavity wall at cills	C8	m	11.71	0.59	9.95	32.91	6.43	49.29
Close cavity wall at top	C9	m	12.40	0.62	10.54	34.84	6.81	52.19
DPC 112mm wide at jambs	C10	m	29.32	1.47	24.92	29.91	8.22	63.05
DPC 112mm wide at cills	C11	m	11.71	0.59	9.95	11.94	3.28	25.18
Carried to summary				162.42	2,761.06	3,925.86	1,003.04	7,689.96

PART D
FLAT ROOF

	Ref	Unit	Qty	Hours	Hours £	Mat'ls £	O & P £	Total £
200 x 50mm sawn softwood joists	D1	m	45.65	11.41	194.01	163.43	53.62	411.06
200 x 50mm sawn softwood sprocket pieces	D2	nr	16.00	2.24	38.08	28.64	10.01	76.73
18mm thick WPB grade decking	D3	m²	22.00	19.80	336.60	252.34	88.34	677.28
50 x 50mm (avg) wide sawn softwood firrings	D4	m	45.65	8.22	139.69	87.19	34.03	260.91
High density polyethylene vapour barrier 150mm thick	D5	m²	20.00	4.00	68.00	223.80	43.77	335.57
100 x 75mm sawn softwood wall plate	D6	m	13.00	3.90	66.30	35.62	15.29	117.21
100 x 75mm sawn softwood tilt fillet	D7	m	5.30	1.33	22.53	13.83	5.45	41.81
Build in ends of 200 x 50mm joists	D8	nr	11.00	3.85	65.45	2.42	10.18	78.05
Carried forward				54.74	930.66	807.27	260.69	1,998.62

	Ref	Unit	Qty	Hours	Hours £	Mat'ls £	O & P £	Total £
Brought forward				54.74	930.66	807.27	260.69	1,998.62
Rake out joint for flashing	D9	m	5.30	1.86	31.54	1.27	4.92	37.73
6mm thick soffit 150mm wide	D10	m	13.30	5.32	90.44	28.06	17.78	136.28
19mm wrought softwood fascia 200mm high	D11	m	13.30	6.65	113.05	36.04	22.36	171.46
Three layer fibre-based roofing felt	D12	m²	22.00	12.10	205.70	286.88	73.89	566.47
Felt turn-down 100mm girth	D13	m	13.30	1.33	22.61	18.62	6.18	47.41
Felt flashing 150mm girth	D14	m	5.30	0.53	9.01	9.12	2.72	20.84
112mm diameter PVC-U gutter	D15	m	5.30	1.38	23.43	17.17	6.09	46.69
Stop end	D16	nr	1.00	0.14	2.38	2.12	0.68	5.18
Stop end outlet	D17	nr	1.00	0.25	4.25	4.19	1.27	9.71
68mm diameter PVC-U down pipe	D18	m	4.80	1.20	20.40	27.36	7.16	54.92
Shoe	D19	nr	1.00	0.30	5.10	3.15	1.24	9.49
Paint fascia and soffit	D20	m²	3.99	0.80	13.57	6.38	2.99	22.94
Carried to summary				86.60	1,472.12	1,247.64	407.96	3,127.73
PART E **PITCHED ROOF**				N/A	N/A	N/A	N/A	N/A
PART F **WINDOWS AND** **EXTERNAL DOORS**								
PVC-U door size 840 x 1980mm complete (B)	F1	nr	2.00	5.00	85.00	278.67	54.55	418.22
PVC-U sliding patio door size 1700 x 2075mm (C)	F2	nr	2.00	14.00	238.00	616.94	128.24	983.18
Carried forward				19.00	323.00	895.61	182.79	1,401.40

	Ref	Unit	Qty	Hours	Hours £	Mat'ls £	O & P £	Total £
Brought forward				19.00	323.00	895.61	182.79	1,401.40
PVC-U window size 1200 x 1200mm complete (A)	F3	nr	8.00	16.00	272.00	1,457.12	259.37	1,988.49
25 x 225mm wrought softwood window board	F4	m	9.60	2.88	48.96	38.59	13.13	100.68
Paint window board	F5	m	9.60	0.14	2.38	8.83	1.68	12.89
Carried to summary				38.02	646.34	2,400.15	456.97	3,503.47
PART G INTERNAL PARTITIONS AND DOORS				N/A	N/A	N/A	N/A	N/A
PART H WALL FINISHES								
19 x 100mm wrought softwood skirting	H1	m	21.50	3.66	62.14	53.97	17.42	133.52
12mm plasterboard fixed to walls with dabs	H2	m²	42.02	16.39	278.59	119.76	59.75	458.10
12mm plasterboard fixed to walls less than 300mm wide	H3	m	41.03	7.39	125.55	37.75	24.49	187.79
Two coats emulsion paint to walls	H4	m²	48.17	12.52	212.91	46.24	38.87	298.03
Paint skirting	H5	m	21.50	4.30	73.10	19.78	13.93	106.81
Carried to summary				44.25	752.29	277.49	154.47	1,184.25
PART J FLOOR FINISHES								
Cement and sand floor screed 40mm thick	J1	m²	16.28	4.07	69.19	77.49	22.00	168.69
Vinyl floor tiles, size 300 x 300mm	J2	m²	16.28	2.77	47.05	126.33	26.01	199.39
Carried forward				6.84	116.24	203.83	48.01	368.07

	Ref	Unit	Qty	Hours	Hours £	Mat'ls £	O & P £	Total £
Brought forward				6.84	116.24	203.83	48.01	368.07
25mm thick tongued and grooved boarding	J3	m²	16.28	12.05	204.80	203.17	61.20	469.17
150 x 50mm softwood joists	J4	m	31.20	6.86	116.69	97.34	32.10	246.14
Cut and pin end of joists to existing brick wall	J5	nr	8.00	1.44	24.48	2.40	4.03	30.91
Build in ends of joists to blockwork	J6	nr	8.00	0.80	13.60	1.92	2.33	17.85
Carried to summary				27.99	475.81	508.66	147.67	1,132.14

PART K
CEILING FINISHES

	Ref	Unit	Qty	Hours	Hours £	Mat'ls £	O & P £	Total £
Plasterboard with taped butt joints fixed to joists	K1	m²	33.56	12.08	205.39	95.65	45.15	346.19
5mm skim coat to plasterboard ceilings	K2	m²	33.56	16.78	285.26	58.06	51.50	394.82
Two coats emulsion paint to ceilings	K3	m²	33.56	8.73	148.34	32.22	27.08	207.64
Carried to summary				37.59	638.98	185.92	123.74	948.64

PART L
ELECTRICAL WORK

	Ref	Unit	Qty	Hours	Hours £	Mat'ls £	O & P £	Total £
13 amp double switched socket outlet with neon	L1	nr	8.00	6.40	121.60	70.48	28.81	220.89
Lighting point	L2	nr	6.00	4.20	79.80	44.04	18.58	142.42
Lighting switch	L3	nr	5.00	3.50	66.50	22.10	13.29	101.89
Lighting wiring	L4	m	18.00	3.60	68.40	33.48	15.28	117.16
Power cable	L5	m	40.00	12.00	228.00	108.40	50.46	386.86
Carried to summary				29.70	564.30	278.50	126.42	969.22

	Ref	Unit	Qty	Hours	Hours £	Mat'ls £	O & P £	Total £
PART M **HEATING WORK**								
15mm copper pipe	M1	m	21.00	9.24	166.32	45.15	31.72	243.19
Elbow	M2	nr	8.00	4.48	80.64	19.52	15.02	115.18
Tee	M3	nr	2.00	1.12	20.16	4.88	3.76	28.80
Radiator, double convector size 1400 x 520mm	M4	nr	6.00	7.80	140.40	834.66	146.26	1,121.32
Break into existing pipe and insert tee	M5	nr	1.00	0.75	13.50	4.42	2.69	20.61
Carried to summary				23.39	421.02	908.63	199.45	1,529.10
PART N **ALTERATION WORK**								
Take out existing window size 1500 x 1000mm and lintel over, adapt opening to receive 1770 x 2000mm patio door and insert new lintel over (both measured separately) and make good	N1	nr	1.00	20.00	340.00	32.10	55.82	427.92
Take out existing window size 1500 x 1000mm, enlarge opening to receive new PVC–U door (measured separately)	N2	nr	1.00	24.00	408.00	58.16	69.92	536.08
Carried to summary				44.00	748.00	90.26	125.74	964.00

260 Two storey extension, size 4 x 5m, flat roof

SUMMARY

	Hours	Hours £	Mat'ls £	O & P £	Total £
PART A **PRELIMINARIES**	0.00	0.00	0.00	0.00	3,796.00
PART B SUBSTRUCTURE TO **DPC LEVEL**	83.19	1,203.50	1,062.37	339.88	2,605.74
PART C **EXTERNAL WALLS**	164.42	2,761.06	3,925.86	1,003.04	7,689.96
PART D **FLAT ROOF**	86.60	1,472.12	1,247.64	407.96	3,127.73
PART E **PITCHED ROOF**	0.00	0.00	0.00	0.00	0.00
PART F WINDOWS AND **EXTERNAL DOORS**	38.02	646.34	2,400.15	456.97	3,503.47
PART G INTERNAL **PARTITIONS AND DOORS**	0.00	0.00	0.00	0.00	0.00
PART H **WALL FINISHES**	44.25	752.29	277.49	154.47	1,184.25
PART J **FLOOR FINISHES**	27.99	475.81	508.66	147.67	1,132.14
PART K **CEILING FINISHES**	37.59	638.98	185.92	123.74	948.64
PART L **ELECTRICAL WORK**	29.70	564.30	278.50	126.42	969.22
PART M **HEATING WORK**	23.39	421.02	908.63	199.45	1,529.10
PART N **ALTERATION WORK**	44.00	748.00	90.26	125.74	964.00
Final total	579.15	9,683.42	10,885.48	3,085.34	27,450.25

	Ref	Unit	Qty	Hours	Hours £	Mat'ls £	O & P £	Total £
PART A **PRELIMINARIES**								
Concrete mixer	A1	wks	12.00					600.00
Small tools	A2	wks	12.00					480.00
Scaffolding (m²/weeks)	A3		300					1,800.00
Skip	A4	wks	9.00					990.00
Clean up	A5	hrs	12.00					156.00
Carried to summary								4,026.00
PART B **SUBSTRUCTURE TO** **DPC LEVEL**								
Excavate topsoil 150mm thick by hand	B1	m²	26.15	7.85	101.99	0.00	15.30	117.28
Excavate to reduce levels	B2	m³	7.85	19.63	255.13	0.00	38.27	293.39
Excavate for trench foundations by hand	B3	m³	2.21	5.75	74.70	0.00	11.20	85.90
Earthwork support to sides of trenches	B4	m²	19.56	7.82	101.71	36.19	20.68	158.58
Backfilling with excavated material	B5	m³	0.77	0.46	6.01	0.00	0.90	6.91
Hardcore 225mm thick	B6	m²	19.98	4.00	51.95	125.67	26.64	204.27
Hardcore filling to trench	B7	m³	0.17	0.09	1.11	6.29	1.11	8.50
Concrete grade (1:3:6) in foundations	B8	m³	1.81	2.44	31.77	154.97	28.01	214.75
Concrete grade (1:2:4) in bed 150mm thick	B9	m²	19.98	5.99	77.92	306.09	57.60	441.62
Concrete (1:2:4) in cavity wall filling	B10	m²	6.70	1.34	11.97	15.95	4.19	32.10
Carried forward				55.36	714.24	645.16	203.91	1,563.31

	Ref	Unit	Qty	Hours	Hours £	Mat'ls £	O & P £	Total £
Brought forward				55.36	714.24	645.16	203.91	1,563.31
Damp-proof membrane	B11	m²	20.63	0.83	10.73	47.86	8.79	67.38
Reinforcement ref A193 in foundation	B12	m²	6.70	0.80	10.45	13.53	3.60	27.58
Steel fabric reinforcement ref A193 in slab	B13	m²	19.98	3.00	38.96	39.96	11.84	90.76
Solid blockwork 140mm thick in cavity wall	B14	m²	8.71	11.32	192.49	145.28	50.67	388.44
Common bricks 112.5mm thick in cavity wall	B15	m²	6.70	11.39	193.63	110.22	45.58	349.42
Facing bricks in 112.5mm thick in skin of cavity wall	B16	m²	2.01	3.62	61.51	28.55	13.51	103.56
Form cavity 50mm wide in cavity wall	B17	m²	8.71	0.26	4.44	13.24	2.65	20.33
DPC 112mm wide	B18	m	13.40	0.67	11.39	13.67	3.76	28.82
DPC 140mm wide	B19	m	13.40	0.80	13.67	18.49	4.82	36.98
Bond in block wall	B20	m	1.30	0.57	9.72	3.78	2.03	15.53
Bond in half brick wall	B21	m	0.30	0.11	1.79	0.80	0.39	2.98
50mm thick insulation board	B22	m²	19.98	5.99	101.90	143.86	36.86	282.62
Carried to summary				94.72	1,364.91	1,224.41	388.40	2,977.72

PART C
EXTERNAL WALLS

	Ref	Unit	Qty	Hours	Hours £	Mat'ls £	O & P £	Total £
Solid blockwork 140mm thick in cavity wall	C1	m²	51.87	67.43	1,146.33	865.19	301.73	2,313.25
Facing brickwork 112.5mm thick in cavity wall	C2	m²	51.87	93.37	1,587.22	1,480.89	460.22	3,528.33
75mm thick insulation in cavity wall	C3	m²	51.87	11.41	193.99	276.99	70.65	541.63
Carried forward				172.21	2,927.54	2,623.07	832.60	6,383.20

	Ref	Unit	Qty	Hours	Hours £	Mat'ls £	O & P £	Total £
Brought forward				172.21	2,927.54	2,623.07	832.60	6,383.20
Steel lintel 2400mm long	C4	nr	2.00	0.50	8.50	296.58	45.76	350.84
Steel lintel 1500mm long	C5	nr	8.00	1.60	27.20	970.88	149.71	1,147.79
Steel lintel 1500mm long	C6	nr	1.00	0.15	2.55	98.72	15.19	116.46
Close cavity wall at jambs	C7	m	29.32	1.47	24.92	82.39	16.10	123.41
Close cavity wall at cills	C8	m	11.71	0.59	9.95	32.91	6.43	49.29
Close cavity wall at top	C9	m	13.40	0.67	11.39	37.65	7.36	56.40
DPC 112mm wide at jambs	C10	m	29.32	1.47	24.92	29.91	8.22	63.05
DPC 112mm wide at cills	C11	m	11.71	0.59	9.95	11.94	3.28	25.18
Carried to summary				179.23	3,046.93	4,184.04	1,084.65	8,315.63

PART D
FLAT ROOF

	Ref	Unit	Qty	Hours	Hours £	Mat'ls £	O & P £	Total £
200 x 50mm sawn softwood joists	D1	m	53.95	13.49	229.29	193.14	63.36	485.79
200 x 50mm sawn softwood sprocket pieces	D2	nr	16.00	2.24	38.08	28.64	10.01	76.73
18mm thick WPB grade decking	D3	m²	26.15	23.54	400.10	299.94	105.01	805.04
50 x 50mm (avg) wide sawn softwood firrings	D4	m	53.95	9.71	165.09	103.04	40.22	308.35
High density polyethylene vapour barrier 150mm thick	D5	m²	24.00	4.80	81.60	268.56	52.52	402.68
100 x 75mm sawn softwood wall plate	D6	m	14.00	4.20	71.40	38.36	16.46	126.22
100 x 75mm sawn softwood tilt fillet	D7	m	6.30	1.58	26.78	16.44	6.48	49.70
Build in ends of 200 x 50mm joists	D8	nr	13.00	4.55	77.35	2.86	12.03	92.24
Carried forward				64.10	1,089.67	950.99	306.10	2,346.76

	Ref	Unit	Qty	Hours	Hours £	Mat'ls £	O & P £	Total £
Brought forward				64.10	1,089.67	950.99	306.10	2,346.76
Rake out joint for flashing	D9	m	6.30	2.21	37.49	1.51	5.85	44.85
6mm thick soffit 150mm wide	D10	m	14.30	5.72	97.24	30.17	19.11	146.52
19mm wrought softwood fascia 200mm high	D11	m	14.30	7.15	121.55	38.75	24.05	184.35
Three layer fibre-based roofing felt	D12	m²	26.15	14.38	244.50	341.00	87.82	673.32
Felt turn-down 100mm girth	D13	m	14.30	1.43	24.31	20.02	6.65	50.98
Felt flashing 150mm girth	D14	m	6.30	0.63	10.71	10.84	3.23	24.78
112mm diameter PVC-U gutter	D15	m	6.30	1.64	27.85	20.41	7.24	55.50
Stop end	D16	nr	1.00	0.14	2.38	2.12	0.68	5.18
Stop end outlet	D17	nr	1.00	0.25	4.25	4.19	1.27	9.71
68mm diameter PVC-U down pipe	D18	m	4.80	1.20	20.40	27.36	7.16	54.92
Shoe	D19	nr	1.00	0.30	5.10	3.15	1.24	9.49
Paint fascia and soffit	D20	m²	4.29	0.86	14.59	6.86	3.22	24.67
Carried to summary				100.00	1,700.03	1,457.38	473.61	3,631.02
PART E **PITCHED ROOF**				N/A	N/A	N/A	N/A	N/A
PART F **WINDOWS AND** **EXTERNAL DOORS**								
PVC-U door size 840 x 1980mm complete (B)	F1	nr	2.00	5.00	85.00	557.34	96.35	738.69
PVC-U sliding patio door size 1700 x 2075mm (C)	F2	nr	2.00	14.00	238.00	724.20	144.33	1,106.53
Carried forward				19.00	323.00	1,281.54	240.68	1,845.22

	Ref	Unit	Qty	Hours	Hours £	Mat'ls £	O & P £	Total £
Brought forward				19.00	323.00	1,281.54	240.68	1,845.22
PVC-U window size 1200 x 1200mm complete (A)	F3	nr	8.00	16.00	272.00	1,457.12	259.37	1,988.49
25 x 225mm wrought softwood window board	F4	m	9.60	2.88	48.96	38.59	13.13	100.68
Paint window board	F5	m	9.60	0.14	2.38	8.83	1.68	12.89
Carried to summary				38.02	646.34	2,786.08	514.86	3,947.29

PART G INTERNAL PARTITIONS AND DOORS

	Ref	Unit	Qty	Hours	Hours £	Mat'ls £	O & P £	Total £
50 x 75mm softwood sole plate	G1	m	3.70	0.81	13.84	4.96	2.82	21.62
50 x 75mm sawn softwood head	G2	m	3.70	0.81	13.84	4.96	2.82	21.62
50 x 75mm sawn softwood studs	G3	m	17.15	4.80	81.63	22.98	15.69	120.31
50 x 75mm sawn softwood noggings	G4	m	3.70	1.04	17.61	4.96	3.39	25.96
Plasterboard	G5	m²	14.80	5.33	90.58	39.52	19.51	149.61
35mm thick veneered internal door size 762 x 1981mm	G6	nr	1.00	1.25	21.25	69.91	13.67	104.83
35 x 150mm wrought softwood lining	G7	m	4.87	1.07	18.21	22.50	6.11	46.82
13 x 38mm wrought softwood stop	G8	m	4.87	0.97	16.56	3.31	2.98	22.85
19 x 50mm wrought softwood architrave	G9	m	9.74	1.46	24.84	13.05	5.68	43.57
19 x 100mm wrought softwood skirting	G10	m	5.88	1.00	16.99	14.76	4.76	36.51
100mm rising steel butts	G11	pr	1.00	0.30	5.10	1.92	1.05	8.07
SAA mortice latch with lever furniture	G12	nr	1.00	0.30	5.10	1.92	1.05	8.07
Carried forward				19.15	325.55	204.74	79.54	609.84

	Ref	Unit	Qty	Hours	Hours £	Mat'ls £	O & P £	Total £
Brought forward				19.15	325.55	204.74	79.54	609.84
Two coats emulsion on plaster–board walls	G13	m²	14.80	3.85	65.42	14.21	11.94	91.57
Paint general surfaces	G14	m²	3.29	0.66	11.19	5.26	2.47	18.92
Carried to summary				23.66	402.15	224.21	93.95	720.33
PART H **WALL FINISHES**								
19 x 100mm softwood skirting	H1	m	28.50	4.85	82.37	71.54	23.09	176.99
12mm plasterboard fixed to walls with dabs	H2	m²	46.22	18.03	306.44	131.73	65.72	503.89
12mm plasterboard fixed to walls less than 300mm wide	H3	m	41.03	7.39	125.55	37.75	24.49	187.79
Two coats emulsion paint to walls	H4	m²	48.17	12.52	212.91	46.24	38.87	298.03
Paint skirting	H5	m	21.50	4.30	73.10	19.78	13.93	106.81
Carried to summary				47.08	800.37	307.03	166.11	1,273.51
PART J **FLOOR FINISHES**								
Cement and sand floor screed 40mm thick	J1	m²	19.88	4.97	84.49	94.63	26.87	205.99
Vinyl floor tiles, size 300 x 300mm	J2	m²	19.88	3.38	57.45	154.27	31.76	243.48
25mm thick tongued and grooved boarding	J3	m²	19.88	14.71	250.09	248.10	74.73	572.92
150 x 50mm softwood joists	J4	m	39.00	8.58	145.86	121.68	40.13	307.67
Cut and pin end of joists to excisting brick wall	J5	nr	8.00	1.44	24.48	2.40	4.03	30.91
Build in ends of joists to blockwork	J6	nr	8.00	0.80	13.60	1.92	2.33	17.85
Carried to summary				33.88	575.97	623.00	179.85	1,378.82

	Ref	Unit	Qty	Hours	Hours £	Mat'ls £	O & P £	Total £

PART K
CEILING FINISHES

	Ref	Unit	Qty	Hours	Hours £	Mat'ls £	O & P £	Total £
Plasterboard with taped butt joints fixed to joists	K1	m²	39.96	14.39	244.56	113.89	53.77	412.21
5mm skim coat to plasterboard ceilings	K2	m²	39.96	19.98	339.66	69.13	61.32	470.11
Two coats emulsion paint to ceilings	K3	m²	39.96	10.39	176.62	38.36	32.25	247.23
Carried to summary				44.76	760.84	221.38	147.33	1,129.55

PART L
ELECTRICAL WORK

	Ref	Unit	Qty	Hours	Hours £	Mat'ls £	O & P £	Total £
13 amp double switched socket outlet with neon	L1	nr	10.00	8.00	152.00	88.10	36.02	276.12
Lighting point	L2	nr	6.00	4.20	79.80	44.04	18.58	142.42
Lighting switch	L3	nr	5.00	3.50	66.50	22.10	13.29	101.89
Lighting wiring	L4	m	22.00	4.40	83.60	40.92	18.68	143.20
Power cable	L5	m	44.00	13.20	250.80	119.24	55.51	425.55
Carried to summary				33.30	632.70	314.40	142.07	1,089.17

PART M
HEATING WORK

	Ref	Unit	Qty	Hours	Hours £	Mat'ls £	O & P £	Total £
15mm copper pipe	M1	m	23.00	10.12	182.16	49.45	34.74	266.35
Elbow	M2	nr	8.00	4.48	80.64	19.52	15.02	115.18
Tee	M3	nr	2.00	1.12	20.16	4.88	3.76	28.80
Radiator, double convector size 1400 x 520mm	M4	nr	6.00	7.80	140.40	834.66	146.26	1,121.32
Break into existing pipe and insert tee	M5	nr	1.00	0.75	13.50	4.42	2.69	20.61
Carried to summary				24.27	436.86	912.93	202.47	1,552.26

	Ref	Unit	Qty	Hours	Hours £	Mat'ls £	O & P £	Total £

PART N
ALTERATION WORK

	Ref	Unit	Qty	Hours	Hours £	Mat'ls £	O & P £	Total £
Take out existing window size 1500 x 1000mm and lintel over, adapt opening to receive 1770 x 2000mm patio door and insert new lintel over (both measured separately) and make good	N1	nr	1.00	20.00	340.00	32.10	55.82	427.92
Take out existing window size 1500 x 1000mm, enlarge opening to receive new PVC–U door (measured separately)	N2	nr	1.00	24.00	408.00	58.16	69.92	536.08
Carried to summary				44.00	748.00	90.26	125.74	964.00

SUMMARY

	Hours	Hours £	Mat'ls £	O & P £	Total £
PART A **PRELIMINARIES**	0.00	0.00	0.00	0.00	4,026.00
PART B SUBSTRUCTURE TO **DPC LEVEL**	94.72	1,364.91	1,224.41	388.40	2,977.72
PART C **EXTERNAL WALLS**	179.23	3,046.93	4,184.04	1,084.65	8,315.63
PART D **FLAT ROOF**	100.00	1,700.03	1,457.38	473.61	3,631.02
PART E **PITCHED ROOF**	0.00	0.00	0.00	0.00	0.00
PART F WINDOWS AND **EXTERNAL DOORS**	38.02	646.34	2,786.08	514.86	3,947.29
PART G INTERNAL **PARTITIONS AND DOORS**	23.66	402.15	224.21	93.95	720.33
PART H **WALL FINISHES**	47.08	800.37	307.03	166.11	1,273.51
PART J **FLOOR FINISHES**	33.88	575.97	623.00	179.85	1,378.82
PART K **CEILING FINISHES**	44.76	760.84	221.38	147.33	1,129.55
PART L **ELECTRICAL WORK**	33.30	632.70	314.40	142.07	1,089.17
PART M **HEATING WORK**	24.27	436.86	912.93	202.47	1,552.26
PART N **ALTERATION WORK**	44.00	748.00	90.26	125.74	964.00
Final total	662.92	11,115.10	12,345.12	3,519.04	31,005.30

	Ref	Unit	Qty	Hours	Hours £	Mat'ls £	O & P £	Total £
PART A **PRELIMINARIES**								
Concrete mixer	A1	wks	7.00					350.00
Small tools	A2	wks	8.00					320.00
Scaffolding (m²/weeks)	A3		140					840.00
Skip	A4	wks	4.00					440.00
Clean up	A5	hrs	8.00					104.00
Carried to summary								2,054.00
PART B **SUBSTRUCTURE TO** **DPC LEVEL**								
Excavate topsoil 150mm thick by hand	B1	m²	7.10	2.13	27.69	0.00	4.15	31.84
Excavate to reduce levels	B2	m³	2.13	5.33	69.23	0.00	10.38	79.61
Excavate for trench foundations by hand	B3	m³	1.06	2.76	35.83	0.00	5.37	41.20
Earthwork support to sides of trenches	B4	m²	9.34	3.74	48.57	17.28	9.88	75.72
Backfilling with excavated material	B5	m³	0.40	0.24	3.12	0.00	0.47	3.59
Hardcore 225mm thick	B6	m²	4.08	0.82	10.61	25.66	5.44	41.71
Hardcore filling to trench	B7	m³	0.08	0.04	0.52	6.29	1.02	7.83
Concrete grade (1:3:6) in foundations	B8	m³	0.86	1.16	15.09	73.63	13.31	102.04
Concrete grade (1:2:4) in bed 150mm thick	B9	m²	4.08	1.22	15.91	62.51	11.76	90.18
Concrete (1:2:4) in cavity wall filling	B10	m²	3.20	0.64	11.97	7.62	2.94	22.52
Carried forward				18.07	238.53	192.99	64.73	496.25

	Ref	Unit	Qty	Hours	Hours £	Mat'ls £	O & P £	Total £
Brought forward				18.07	238.53	192.99	64.73	496.25
Damp-proof membrane	B11	m²	4.38	0.18	2.28	10.16	1.87	14.31
Reinforcement ref A193 in foundation	B12	m²	3.22	0.39	5.02	6.50	1.73	13.26
Steel fabric reinforcement ref A193 in slab	B13	m²	4.08	0.61	7.96	8.16	2.42	18.53
Solid blockwork 140mm thick in cavity wall	B14	m²	4.16	5.41	91.94	69.39	24.20	185.52
Common bricks 112.5mm thick in cavity wall	B15	m²	3.20	5.44	92.48	52.64	21.77	166.89
Facing bricks in 112.5mm thick in skin of cavity wall	B16	m²	0.96	1.73	29.38	28.55	8.69	66.61
Form cavity 50mm wide in cavity wall	B17	m²	4.16	0.12	2.12	6.32	1.27	9.71
DPC 112mm wide	B18	m	6.40	0.32	5.44	6.53	1.80	13.76
DPC 140mm wide	B19	m	6.40	0.38	6.53	8.83	2.30	17.66
Bond in block wall	B20	m	1.30	0.57	9.72	3.78	2.03	15.53
Bond in half brick wall	B21	m	0.30	0.11	1.79	0.80	0.39	2.98
50mm thick insulation board	B22	m²	4.08	1.22	20.81	29.38	7.53	57.71
Carried to summary				34.55	513.99	424.04	140.70	1,078.73

PART C
EXTERNAL WALLS

	Ref	Unit	Qty	Hours	Hours £	Mat'ls £	O & P £	Total £
Solid blockwork 140mm thick in cavity wall	C1	m²	23.94	31.12	529.07	399.32	139.26	1,067.65
Facing brickwork 112.5mm thick in cavity wall	C2	m²	23.94	43.09	732.56	683.49	212.41	1,628.46
75mm thick insulation in cavity wall	C3	m²	23.94	5.27	89.54	127.84	32.61	249.98
Carried forward				79.48	1,351.17	1,210.65	384.27	2,946.09

	Ref	Unit	Qty	Hours	Hours £	Mat'ls £	O & P £	Total £
Brought forward				79.48	1,351.17	1,210.65	384.27	2,946.09
Steel lintel 2400mm long	C4	nr	2.00	0.50	8.50	296.58	45.76	350.84
Steel lintel 1500mm long	C5	nr	4.00	0.80	13.60	485.44	74.86	573.90
Close cavity wall at jambs	C7	m	13.76	0.69	11.70	38.67	7.55	57.92
Close cavity wall at cills	C8	m	5.07	0.25	4.31	14.25	2.78	21.34
Close cavity wall at top	C9	m	6.40	0.32	5.44	17.98	3.51	26.94
DPC 112mm wide at jambs	C10	m	13.76	0.69	11.70	14.04	3.86	29.59
DPC 112mm wide at cills	C11	m	5.07	0.25	4.31	5.17	1.42	10.90
Carried to summary				82.98	1,410.72	2,082.77	524.02	4,017.52
PART D **FLAT ROOF**				N/A	N/A	N/A	N/A	N/A
PART E **PITCHED ROOF**								
100 x 75mm sawn softwood wall plate	E1	m	3.00	0.90	15.30	7.83	3.47	26.60
200 x 50mm sawn softwood pole plate	E2	nr	3.30	0.99	16.83	10.96	4.17	31.95
100 x 50mm sawn softwood rafters	E3	m	17.50	3.50	59.50	43.40	15.44	118.34
100 x 50mm sawn softwood purlin	E4	m	3.30	0.66	11.22	9.57	3.12	23.91
150 x 50mm softwood joists	E5	m	12.50	2.50	42.50	39.00	12.23	93.73
150 x 50mm sawn softwood sprockets	E6	nr	14.00	1.68	28.56	21.84	7.56	57.96
100mm layer of insulation quilt laid over and between joists	E7	m²	6.00	2.88	48.96	69.60	17.78	136.34
Carried forward				13.11	222.87	202.20	63.76	488.83

	Ref	Unit	Qty	Hours	Hours £	Mat'ls £	O & P £	Total £
Brought forward				13.11	222.87	202.20	63.76	488.83
6mm softwood soffit 150mm wide	E8	m	8.30	3.32	56.44	17.51	12.43	95.29
19mm wrought softwood fascia/ barge board 200mm high	E9	m	3.30	1.65	28.05	8.94	5.55	42.54
Marley Plain roof tiles on felt and battens	E10	m²	11.55	21.95	373.07	418.57	118.75	910.38
Double eaves course	E11	m	3.30	1.16	19.64	13.89	5.03	38.56
Verge with tile undercloak	E12	m	7.00	1.75	29.75	46.27	11.40	87.42
Lead flashing code 5, 200mm girth	E13	m	3.30	1.98	33.66	28.45	9.32	71.42
Rake out joint for flashing	E14	m	3.30	1.16	19.64	0.79	3.06	23.49
112mm diameter PVC-U gutter	E15	m	3.30	0.86	14.59	10.69	3.79	29.07
Stop end	E16	nr	1.00	0.14	2.38	2.12	0.68	5.18
Stop end outlet	E17	nr	1.00	0.25	4.25	4.19	1.27	9.71
68mm diameter PVC-U down pipe	E18	m	2.50	0.63	10.63	14.25	3.73	28.61
Shoe	E19	nr	1.00	0.30	5.10	3.15	1.24	9.49
Paint fascia and soffit	E20	m²	2.59	0.52	8.81	2.75	1.73	13.28
Carried to summary				48.76	828.85	773.77	241.73	1,853.26

PART F
WINDOWS AND
EXTERNAL DOORS

	Ref	Unit	Qty	Hours	Hours £	Mat'ls £	O & P £	Total £
PVC-U door size 840 x 1980mm complete (B)	F1	nr	1.00	2.50	42.50	278.67	48.18	369.35
PVC-U sliding patio door size 1700 x 2075mm (C)	F2	nr	2.00	14.00	238.00	616.94	128.24	983.18
Carried forward				16.50	280.50	895.61	176.42	1,352.53

	Ref	Unit	Qty	Hours	Hours £	Mat'ls £	O & P £	Total £
Brought forward				16.50	280.50	895.61	176.42	1,352.53
PVC-U window size 1200 x 1200mm complete (A)	F3	nr	4.00	8.00	136.00	728.56	129.68	994.24
25 x 225mm wrought softwood window board	F4	m	4.80	1.44	24.48	19.30	6.57	50.34
Paint window board	F5	m	4.80	0.14	2.38	4.42	1.02	7.82
Carried to summary				26.08	443.36	1,647.88	313.69	2,404.93
PART G INTERNAL PARTITIONS AND DOORS				N/A	N/A	N/A	N/A	N/A
PART H WALL FINISHES								
19 x 100mm wrought softwood skirting	H1	m	10.34	1.76	29.88	25.95	8.38	64.21
12mm plasterboard fixed to walls with dabs	H2	m²	20.04	7.82	132.87	57.11	28.50	218.48
12mm plasterboard fixed to walls less than 300mm wide	H3	m	18.33	3.30	56.09	16.86	10.94	83.90
Two coats emulsion paint to walls	H4	m²	22.86	5.94	101.04	21.95	18.45	141.43
Paint skirting	H5	m	10.34	2.07	35.16	9.51	6.70	51.37
Carried to summary				20.88	355.03	131.39	72.96	559.39
PART J FLOOR FINISHES								
Cement and sand floor screed 40mm thick	J1	m²	4.08	1.02	17.34	19.42	5.51	42.27
Vinyl floor tiles, size 300 x 300mm	J2	m²	4.08	0.69	11.79	31.66	6.52	49.97
Carried forward				1.71	29.13	51.08	12.03	92.24

	Ref	Unit	Qty	Hours	Hours £	Mat'ls £	O & P £	Total £
Brought forward				1.71	29.13	51.08	12.03	92.24
25mm thick tongued and grooved boarding	J3	m²	4.08	3.02	51.33	50.92	15.34	117.58
150 x 50mm softwood joists	J4	m	9.50	2.09	35.53	29.64	9.78	74.95
Cut and pin end of joists to existing brick wall	J5	nr	5.00	0.90	15.30	1.50	2.52	19.32
Build in ends of joists to blockwork	J6	nr	5.00	0.50	8.50	1.20	1.46	11.16
Carried to summary				8.22	139.79	134.34	41.12	315.25

PART K
CEILING FINISHES

	Ref	Unit	Qty	Hours	Hours £	Mat'ls £	O & P £	Total £
Plasterboard with taped butt joints fixed to joists	K1	m²	8.16	2.94	49.94	23.26	10.98	84.17
5mm skim coat to plasterboard ceilings	K2	m²	8.16	4.08	69.36	14.12	12.52	96.00
Two coats emulsion paint to ceilings	K3	m²	8.16	2.12	36.07	7.83	6.59	50.49
Carried to summary				9.14	155.37	45.21	30.09	230.66

PART L
ELECTRICAL WORK

	Ref	Unit	Qty	Hours	Hours £	Mat'ls £	O & P £	Total £
13 amp double switched socket outlet with neon	L1	nr	4.00	3.20	60.80	35.24	14.41	110.45
Lighting point	L2	nr	2.00	1.40	26.60	14.68	6.19	47.47
Lighting switch	L3	nr	4.00	2.80	53.20	17.68	10.63	81.51
Lighting wiring	L4	m	10.00	2.00	38.00	18.60	8.49	65.09
Power cable	L5	m	24.00	7.20	136.80	65.04	30.28	232.12
Carried to summary				16.60	315.40	151.24	70.00	536.64

	Ref	Unit	Qty	Hours	Hours £	Mat'ls £	O & P £	Total £
PART M **HEATING WORK**								
15mm copper pipe	M1	m	9.00	3.96	71.28	19.35	13.59	104.22
Elbow	M2	nr	8.00	4.48	80.64	19.52	15.02	115.18
Tee	M3	nr	2.00	1.12	20.16	4.88	3.76	28.80
Radiator, double convector size 1400 x 520mm	M4	nr	4.00	5.20	93.60	556.44	97.51	747.55
Break into existing pipe and insert tee	M5	nr	1.00	0.75	13.50	4.42	2.69	20.61
Carried to summary				15.51	279.18	604.61	132.57	1,016.36
PART N **ALTERATION WORK**								
Take out existing window size 1500 x 1000mm and lintel over, adapt opening to receive 1770 x 2000mm patio door and insert new lintel over (both measured separately) and make good	N1	nr	1.00	20.00	340.00	32.10	55.82	427.92
Take out existing window size 1500 x 1000mm, enlarge opening to receive new PVC–U door (measured separately)	N2	nr	1.00	24.00	408.00	58.16	69.92	536.08
Carried to summary				44.00	748.00	90.26	125.74	964.00

SUMMARY

	Hours	Hours £	Mat'ls £	O & P £	Total £
PART A **PRELIMINARIES**	0.00	0.00	0.00	0.00	2,054.00
PART B SUBSTRUCTURE TO **DPC LEVEL**	34.55	513.99	424.04	140.70	1,078.73
PART C **EXTERNAL WALLS**	82.98	1,410.72	2,082.77	524.02	4,017.51
PART D **FLAT ROOF**	0.00	0.00	0.00	0.00	0.00
PART E **PITCHED ROOF**	48.76	828.85	773.77	241.73	1,853.26
PART F WINDOWS AND **EXTERNAL DOORS**	26.08	443.36	1,647.88	313.69	2,404.93
PART G INTERNAL **PARTITIONS AND DOORS**	0.00	0.00	0.00	0.00	0.00
PART H **WALL FINISHES**	20.88	355.03	131.39	72.96	559.39
PART J **FLOOR FINISHES**	8.22	139.79	134.34	41.12	315.25
PART K **CEILING FINISHES**	9.14	155.37	45.21	30.09	230.66
PART L **ELECTRICAL WORK**	16.60	315.40	151.24	70.00	536.64
PART M **HEATING WORK**	15.51	279.18	604.61	132.57	1,016.36
PART N **ALTERATION WORK**	44.00	748.00	90.26	125.74	964.00
Final total	306.72	5,189.69	6,085.51	1,692.62	15,030.73

	Ref	Unit	Qty	Hours	Hours £	Mat'ls £	O & P £	Total £
PART A **PRELIMINARIES**								
Concrete mixer	A1	wks	8.00					400.00
Small tools	A2	wks	9.00					360.00
Scaffolding (m²/weeks)	A3		160					960.00
Skip	A4	wks	5.00					550.00
Clean up	A5	hrs	10.00					130.00
Carried to summary								2,400.00
PART B **SUBSTRUCTURE TO** **DPC LEVEL**								
Excavate topsoil 150mm thick by hand	B1	m²	9.25	2.78	36.08	0.00	5.41	41.49
Excavate to reduce levels	B2	m³	2.77	6.93	90.03	0.00	13.50	103.53
Excavate for trench foundations by hand	B3	m³	1.22	3.17	41.24	0.00	6.19	47.42
Earthwork support to sides of trenches	B4	m²	10.80	4.32	56.16	19.98	11.42	87.56
Backfilling with excavated material	B5	m³	0.47	0.28	3.67	0.00	0.55	4.22
Hardcore 225mm thick	B6	m²	5.78	1.16	15.03	36.36	7.71	59.09
Hardcore filling to trench	B7	m³	0.09	0.05	0.59	6.29	1.03	7.91
Concrete grade (1:3:6) in foundations	B8	m³	1.00	1.35	17.55	85.62	15.48	118.65
Concrete grade (1:2:4) in bed 150mm thick	B9	m²	5.78	1.73	22.54	88.55	16.66	127.76
Concrete (1:2:4) in cavity wall filling	B10	m²	3.70	0.74	11.97	8.81	3.12	23.89
Carried forward				22.50	294.84	245.60	81.07	621.50

	Ref	Unit	Qty	Hours	Hours £	Mat'ls £	O & P £	Total £
Brought forward				22.50	294.84	245.60	81.07	621.50
Damp-proof membrane	B11	m²	6.13	0.25	3.19	14.22	2.61	20.02
Reinforcement ref A193 in foundation	B12	m²	3.70	0.44	5.77	7.47	1.99	15.23
Steel fabric reinforcement ref A193 in slab	B13	m²	5.78	0.87	11.27	11.56	3.42	26.26
Solid blockwork 140mm thick in cavity wall	B14	m²	4.81	6.25	106.30	80.23	27.98	214.51
Common bricks 112.5mm thick in cavity wall	B15	m²	3.70	6.29	106.93	60.87	25.17	192.96
Facing bricks in 112.5mm thick in skin of cavity wall	B16	m²	1.11	2.00	33.97	28.55	9.38	71.89
Form cavity 50mm wide in cavity wall	B17	m²	4.81	0.14	2.45	7.31	1.46	11.23
DPC 112mm wide	B18	m	7.40	0.37	6.29	7.55	2.08	15.91
DPC 140mm wide	B19	m	7.40	0.44	7.55	10.21	2.66	20.42
Bond in block wall	B20	m	1.30	0.57	9.72	3.78	2.03	15.53
Bond in half brick wall	B21	m	0.30	0.11	1.79	0.80	0.39	2.98
50mm thick insulation board	B22	m²	5.78	1.73	29.48	41.62	10.66	81.76
Carried to summary				41.97	619.54	519.78	170.90	1,310.22

**PART C
EXTERNAL WALLS**

	Ref	Unit	Qty	Hours	Hours £	Mat'ls £	O & P £	Total £
Solid blockwork 140mm thick in cavity wall	C1	m²	28.99	37.69	640.68	483.55	168.63	1,292.87
Facing brickwork 112.5mm thick in cavity wall	C2	m²	28.99	52.18	887.09	827.66	257.21	1,971.97
75mm thick insulation in cavity wall	C3	m²	28.99	6.38	108.42	154.81	39.48	302.71
Carried forward				96.25	1,636.20	1,466.02	465.33	3,567.55

	Ref	Unit	Qty	Hours	Hours £	Mat'ls £	O & P £	Total £
Brought forward				96.25	1,636.20	1,466.02	465.33	3,567.55
Steel lintel 2400mm long	C4	nr	2.00	0.50	8.50	296.58	45.76	350.84
Steel lintel 1500mm long	C5	nr	4.00	0.80	13.60	485.44	74.86	573.90
Close cavity wall at jambs	C7	m	13.76	0.69	11.70	38.67	7.55	57.92
Close cavity wall at cills	C8	m	5.07	0.25	4.31	14.25	2.78	21.34
Close cavity wall at top	C9	m	7.60	0.38	6.46	21.36	4.17	31.99
DPC 112mm wide at jambs	C10	m	13.76	0.69	11.70	14.04	3.86	29.59
DPC 112mm wide at cills	C11	m	5.07	0.25	4.31	5.17	1.42	10.90
Carried to summary				99.81	1,696.77	2,341.52	605.74	4,644.03
PART D **FLAT ROOF**				N/A	N/A	N/A	N/A	N/A

PART E
PITCHED ROOF

	Ref	Unit	Qty	Hours	Hours £	Mat'ls £	O & P £	Total £
100 x 75mm sawn softwood wall plate	E1	m	4.00	1.20	20.40	10.44	4.63	35.47
200 x 50mm sawn softwood pole plate	E2	nr	4.30	1.29	21.93	14.28	5.43	41.64
100 x 50mm sawn softwood rafters	E3	m	17.50	3.50	59.50	43.40	15.44	118.34
100 x 50mm sawn softwood purlin	E4	m	4.30	0.86	14.62	12.47	4.06	31.15
150 x 50mm softwood joists	E5	m	12.50	2.50	42.50	39.00	12.23	93.73
150 x 50mm sawn softwood sprockets	E6	nr	14.00	1.68	28.56	21.84	7.56	57.96
100mm layer of insulation quilt laid over and between joists	E7	m²	8.00	3.84	65.28	92.80	23.71	181.79
Carried forward				14.87	252.79	234.23	73.05	560.07

	Ref	Unit	Qty	Hours	Hours £	Mat'ls £	O & P £	Total £
Brought forward				14.87	252.79	234.23	73.05	560.07
6mm softwood soffit 150mm wide	E8	m	9.30	3.72	63.24	19.62	12.43	95.29
19mm wrought softwood fascia/ barge board 200mm high	E9	m	4.30	2.15	36.55	11.65	7.23	55.43
Marley Plain roof tiles on felt and battens	E10	m²	15.05	28.60	486.12	545.41	154.73	1,186.26
Double eaves course	E11	m	4.30	1.51	25.59	18.10	6.55	50.24
Verge with tile undercloak	E12	m	7.00	1.75	29.75	46.27	11.40	87.42
Lead flashing code 5, 200mm girth	E13	m	4.30	2.58	43.86	37.07	12.14	93.06
Rake out joint for flashing	E14	m	4.30	1.51	25.59	9.63	5.28	40.50
112mm diameter PVC-U gutter	E15	m	4.30	1.12	19.01	13.93	4.94	37.88
Stop end	E16	nr	1.00	0.14	2.38	2.12	0.68	5.18
Stop end outlet	E17	nr	1.00	0.25	4.25	4.19	1.27	9.71
68mm diameter PVC-U down pipe	E18	m	2.50	0.63	10.63	14.25	3.73	28.61
Shoe	E19	nr	1.00	0.30	5.10	3.15	1.24	9.49
Paint fascia and soffit	E20	m²	2.79	0.56	9.49	2.96	1.87	14.31
Carried to summary				59.67	1,014.32	962.58	296.54	2,273.44

**PART F
WINDOWS AND
EXTERNAL DOORS**

	Ref	Unit	Qty	Hours	Hours £	Mat'ls £	O & P £	Total £
PVC-U door size 840 x 1980mm complete (B)	F1	nr	1.00	2.50	42.50	278.67	48.18	369.35
PVC-U sliding patio door size 1700 x 2075mm (C)	F2	nr	2.00	14.00	238.00	616.94	128.24	983.18
Carried forward				16.50	280.50	895.61	176.42	1,352.53

	Ref	Unit	Qty	Hours	Hours £	Mat'ls £	O & P £	Total £
Brought forward				16.50	280.50	895.61	176.42	1,352.53
PVC-U window size 1200 x 1200mm complete (A)	F3	nr	4.00	8.00	136.00	728.56	129.68	994.24
25 x 225mm wrought softwood window board	F4	m	4.80	1.44	24.48	19.30	6.57	50.34
Paint window board	F5	m	4.80	0.14	2.38	4.42	1.02	7.82
Carried to summary				26.08	443.36	1,647.88	313.69	2,404.93
PART G INTERNAL PARTITIONS AND DOORS				N/A	N/A	N/A	N/A	N/A
PART H WALL FINISHES								
19 x 100mm wrought softwood skirting	H1	m	12.34	2.10	35.66	30.97	10.00	76.63
12mm plasterboard fixed to walls with dabs	H2	m²	24.94	9.73	165.35	71.08	35.46	271.90
12mm plasterboard fixed to walls less than 300mm wide	H3	m	18.83	3.39	57.62	17.32	11.24	86.18
Two coats emulsion paint to walls	H4	m²	26.76	6.96	118.28	25.69	21.60	165.56
Paint skirting	H5	m	12.34	2.47	41.96	11.35	8.00	61.31
Carried to summary				24.64	418.87	156.42	86.29	661.58
PART J FLOOR FINISHES								
Cement and sand floor screed 40mm thick	J1	m²	5.78	1.45	24.57	27.51	7.81	59.89
Vinyl floor tiles, size 300 x 300mm	J2	m²	5.78	0.98	16.70	44.85	9.23	70.79
Carried forward				2.43	41.27	72.37	17.05	130.68

	Ref	Unit	Qty	Hours	Hours £	Mat'ls £	O & P £	Total £
Brought forward				2.43	41.27	72.37	17.05	130.68
25mm thick tongued and grooved boarding	J3	m²	5.78	4.28	72.71	72.13	21.73	166.57
150 x 50mm softwood joists	J4	m	13.30	2.93	49.74	41.50	13.69	104.92
Cut and pin end of joists to existing brick wall	J5	nr	7.00	1.26	21.42	2.10	3.53	27.05
Build in ends of joists to blockwork	J6	nr	7.00	0.70	11.90	1.68	2.04	15.62
Carried to summary				11.59	197.04	189.78	58.02	444.84

**PART K
CEILING FINISHES**

	Ref	Unit	Qty	Hours	Hours £	Mat'ls £	O & P £	Total £
Plasterboard with taped butt joints fixed to joists	K1	m²	11.56	4.16	70.75	32.95	15.55	119.25
5mm skim coat to plasterboard ceilings	K2	m²	11.56	5.78	98.26	20.00	17.74	136.00
Two coats emulsion paint to ceilings	K3	m²	11.56	3.01	51.10	11.10	9.33	71.52
Carried to summary				12.95	220.10	64.04	42.62	326.77

**PART L
ELECTRICAL WORK**

	Ref	Unit	Qty	Hours	Hours £	Mat'ls £	O & P £	Total £
13 amp double switched socket outlet with neon	L1	nr	4.00	3.20	60.80	35.24	14.41	110.45
Lighting point	L2	nr	4.00	2.80	53.20	29.36	12.38	94.94
Lighting switch	L3	nr	4.00	2.80	53.20	17.68	10.63	81.51
Lighting wiring	L4	m	12.00	2.40	45.60	22.32	10.19	78.11
Power cable	L5	m	28.00	8.40	159.60	75.88	35.32	270.80
Carried to summary				19.60	372.40	180.48	82.93	635.81

	Ref	Unit	Qty	Hours	Hours £	Mat'ls £	O & P £	Total £
PART M **HEATING WORK**								
15mm copper pipe	M1	m	11.00	4.84	87.12	23.65	16.62	127.39
Elbow	M2	nr	8.00	4.48	80.64	19.52	15.02	115.18
Tee	M3	nr	2.00	1.12	20.16	4.88	3.76	28.80
Radiator, double convector size 1400 x 520mm	M4	nr	4.00	5.20	93.60	556.44	97.51	747.55
Break into existing pipe and insert tee	M5	nr	1.00	0.75	13.50	4.42	2.69	20.61
Carried to summary				16.39	295.02	608.91	135.59	1,039.52
PART N **ALTERATION WORK**								
Take out existing window size 1500 x 1000mm and lintel over, adapt opening to receive 1770 x 2000mm patio door and insert new lintel over (both measured separately) and make good	N1	nr	1.00	20.00	340.00	32.10	55.82	427.92
Take out existing window size 1500 x 1000mm, enlarge opening to receive new PVC–U door (measured separately)	N2	nr	1.00	24.00	408.00	58.16	69.92	536.08
Carried to summary				44.00	748.00	90.26	125.74	964.00

SUMMARY

	Hours	Hours £	Mat'ls £	O & P £	Total £
PART A **PRELIMINARIES**	0.00	0.00	0.00	0.00	2,400.00
PART B SUBSTRUCTURE TO **DPC LEVEL**	41.97	619.54	519.78	170.90	1,310.22
PART C **EXTERNAL WALLS**	99.81	1,696.77	2,341.52	605.74	4,644.03
PART D **FLAT ROOF**	0.00	0.00	0.00	0.00	0.00
PART E **PITCHED ROOF**	59.67	1,014.32	962.58	296.54	2,273.44
PART F WINDOWS AND **EXTERNAL DOORS**	26.08	443.36	1,647.88	313.69	2,404.93
PART G INTERNAL **PARTITIONS AND DOORS**	0.00	0.00	0.00	0.00	0.00
PART H **WALL FINISHES**	24.64	418.87	156.42	86.29	661.58
PART J **FLOOR FINISHES**	11.59	197.04	189.78	58.02	444.84
PART K **CEILING FINISHES**	12.95	220.10	64.04	42.62	326.77
PART L **ELECTRICAL WORK**	19.60	372.40	180.48	82.93	635.81
PART M **HEATING WORK**	16.39	295.02	608.91	135.59	1,039.52
PART N **ALTERATION WORK**	44.00	748.00	90.26	125.74	964.00
Final total	356.70	6,025.42	6,761.65	1,918.07	17,105.15

	Ref	Unit	Qty	Hours	Hours £	Mat'ls £	O & P £	Total £
PART A **PRELIMINARIES**								
Concrete mixer	A1	wks	9.00					450.00
Small tools	A2	wks	10.00					400.00
Scaffolding (m²/weeks)	A3		180					1,080.00
Skip	A4	wks	6.00					660.00
Clean up	A5	hrs	12.00					156.00
Carried to summary								2,746.00
PART B **SUBSTRUCTURE TO** **DPC LEVEL**								
Excavate topsoil 150mm thick by hand	B1	m²	11.40	3.42	44.46	0.00	6.67	51.13
Excavate to reduce levels	B2	m³	3.42	8.55	111.15	0.00	16.67	127.82
Excavate for trench foundations by hand	B3	m³	1.39	3.61	46.98	0.00	7.05	54.03
Earthwork support to sides of trenches	B4	m²	12.26	4.90	63.75	22.68	12.96	99.40
Backfilling with excavated material	B5	m³	0.55	0.33	4.29	0.00	0.64	4.93
Hardcore 225mm thick	B6	m²	7.48	1.50	19.45	47.05	9.97	76.47
Hardcore filling to trench	B7	m³	0.12	0.06	0.78	6.29	1.06	8.13
Concrete grade (1:3:6) in foundations	B8	m³	1.13	1.53	19.83	96.75	17.49	134.07
Concrete grade (1:2:4) in bed 150mm thick	B9	m²	7.48	2.24	29.17	114.59	21.56	165.33
Concrete (1:2:4) in cavity wall filling	B10	m²	4.20	0.84	11.97	10.00	3.29	25.26
Carried forward				26.98	351.84	297.36	97.38	746.58

	Ref	Unit	Qty	Hours	Hours £	Mat'ls £	O & P £	Total £
Brought forward				26.98	351.84	297.36	97.38	746.58
Damp-proof membrane	B11	m²	7.88	0.32	4.10	18.28	3.36	25.74
Reinforcement ref A193 in foundation	B12	m²	4.20	0.50	6.55	8.48	2.26	17.29
Steel fabric reinforcement ref A193 in slab	B13	m²	7.48	1.12	14.59	14.96	4.43	33.98
Solid blockwork 140mm thick in cavity wall	B14	m²	5.46	7.10	120.67	91.07	31.76	243.50
Common bricks 112.5mm thick in cavity wall	B15	m²	4.20	7.14	121.38	69.09	28.57	219.04
Facing bricks in 112.5mm thick in skin of cavity wall	B16	m²	1.26	2.27	38.56	28.55	10.07	77.17
Form cavity 50mm wide in cavity wall	B17	m²	5.46	0.16	2.78	8.30	1.66	12.75
DPC 112mm wide	B18	m	8.40	0.42	7.14	8.57	2.36	18.06
DPC 140mm wide	B19	m	8.40	0.50	8.57	11.59	3.02	23.18
Bond in block wall	B20	m	1.30	0.57	9.72	3.78	2.03	15.53
Bond in half brick wall	B21	m	0.30	0.11	1.79	0.80	0.39	2.98
50mm thick insulation board	B22	m²	7.48	2.24	38.15	53.86	13.80	105.80
Carried to summary				49.44	725.82	614.70	201.08	1,541.60

PART C
EXTERNAL WALLS

	Ref	Unit	Qty	Hours	Hours £	Mat'ls £	O & P £	Total £
Solid blockwork 140mm thick in cavity wall	C1	m²	34.04	44.25	752.28	567.79	198.01	1,518.08
Facing brickwork 112.5mm thick in cavity wall	C2	m²	34.04	61.27	1,041.62	971.84	302.02	2,315.49
75mm thick insulation in cavity wall	C3	m²	34.04	7.49	127.31	181.77	46.36	355.45
Carried forward				113.01	1,921.22	1,721.40	546.39	4,189.01

	Ref	Unit	Qty	Hours	Hours £	Mat'ls £	O & P £	Total £
Brought forward				113.01	1,921.22	1,721.40	546.39	4,189.01
Steel lintel 2400mm long	C4	nr	2.00	0.50	8.50	296.58	45.76	350.84
Steel lintel 1500mm long	C5	nr	4.00	0.80	13.60	485.44	74.86	573.90
Close cavity wall at jambs	C7	m	13.76	0.69	11.70	38.67	7.55	57.92
Close cavity wall at cills	C8	m	5.07	0.25	4.31	14.25	2.78	21.34
Close cavity wall at top	C9	m	8.40	0.42	7.14	23.60	4.61	35.36
DPC 112mm wide at jambs	C10	m	13.76	0.69	11.70	14.04	3.86	29.59
DPC 112mm wide at cills	C11	m	5.07	0.25	4.31	5.17	1.42	10.90
Carried to summary				116.62	1,982.47	2,599.15	687.24	5,268.86
PART D **FLAT ROOF**				N/A	N/A	N/A	N/A	N/A
PART E **PITCHED ROOF**								
100 x 75mm sawn softwood wall plate	E1	m	5.00	1.50	25.50	13.05	5.78	44.33
200 x 50mm sawn softwood pole plate	E2	nr	5.30	1.59	27.03	17.60	6.69	51.32
100 x 50mm sawn softwood rafters	E3	m	17.50	3.50	59.50	43.40	15.44	118.34
100 x 50mm sawn softwood purlin	E4	m	5.30	1.06	18.02	15.37	5.01	38.40
150 x 50mm softwood joists	E5	m	12.50	2.50	42.50	39.00	12.23	93.73
150 x 50mm sawn softwood sprockets	E6	nr	14.00	1.68	28.56	21.84	7.56	57.96
100mm layer of insulation quilt laid over and between joists	E7	m²	10.00	4.80	81.60	116.00	29.64	227.24
Carried forward				16.63	282.71	266.26	82.34	631.31

	Ref	Unit	Qty	Hours	Hours £	Mat'ls £	O & P £	Total £
Brought forward				16.63	282.71	266.26	82.34	631.31
6mm softwood soffit 150mm wide	E8	m	10.30	4.12	70.04	21.73	13.77	105.54
19mm wrought softwood fascia/ barge board 200mm high	E9	m	5.30	2.65	45.05	14.36	8.91	68.32
Marley Plain roof tiles on felt and battens	E10	m²	18.55	35.25	599.17	672.25	190.71	1,462.13
Double eaves course	E11	m	5.30	1.86	31.54	22.31	8.08	61.93
Verge with tile undercloak	E12	m	7.00	1.75	29.75	46.27	11.40	87.42
Lead flashing code 5, 200mm girth	E13	m	5.30	3.18	54.06	45.69	14.96	114.71
Rake out joint for flashing	E14	m	5.30	1.86	31.54	1.27	4.92	37.73
112mm diameter PVC-U gutter	E15	m	5.30	1.38	23.43	17.17	6.09	46.69
Stop end	E16	nr	1.00	0.14	2.38	2.12	0.68	5.18
Stop end outlet	E17	nr	1.00	0.25	4.25	4.19	1.27	9.71
68mm diameter PVC-U down pipe	E18	m	4.80	1.20	20.40	27.36	7.16	54.92
Shoe	E19	nr	1.00	0.30	5.10	3.15	1.24	9.49
Paint fascia and soffit	E20	m²	3.09	0.62	0.03	3.28	0.50	3.80
Carried to summary				69.32	1,167.90	1,125.10	343.95	2,636.95

**PART F
WINDOWS AND
EXTERNAL DOORS**

	Ref	Unit	Qty	Hours	Hours £	Mat'ls £	O & P £	Total £
PVC-U door size 840 x 1980mm complete (B)	F1	nr	1.00	2.50	42.50	278.67	48.18	369.35
PVC-U sliding patio door size 1700 x 2075mm (C)	F2	nr	2.00	14.00	238.00	616.94	128.24	983.18
Carried forward				16.50	280.50	895.61	176.42	1,352.53

	Ref	Unit	Qty	Hours	Hours £	Mat'ls £	O & P £	Total £
Brought forward				16.50	280.50	895.61	176.42	1,352.53
PVC-U window size 1200 x 1200mm complete (A)	F3	nr	4.00	8.00	136.00	728.56	129.68	994.24
25 x 225mm wrought softwood window board	F4	m	4.80	1.44	24.48	19.30	6.57	50.34
Paint window board	F5	m	4.80	0.14	2.38	4.42	1.02	7.82
Carried to summary				26.08	443.36	1,647.88	313.69	2,404.93
PART G INTERNAL PARTITIONS AND DOORS				N/A	N/A	N/A	N/A	N/A
PART H WALL FINISHES								
19 x 100mm wrought softwood skirting	H1	m	14.34	2.44	41.44	35.99	11.62	89.05
12mm plasterboard fixed to walls with dabs	H2	m²	29.84	11.64	197.84	85.04	42.43	325.32
12mm plasterboard fixed to walls less than 300mm wide	H3	m	18.33	3.30	56.09	16.86	10.94	83.90
Two coats emulsion paint to walls	H4	m²	32.66	8.49	144.36	31.35	26.36	202.07
Paint skirting	H5	m	14.34	2.87	48.76	13.19	9.29	71.24
Carried to summary				28.73	488.48	182.45	100.64	771.57
PART J FLOOR FINISHES								
Cement and sand floor screed 40mm thick	J1	m²	7.48	1.87	31.79	35.60	10.11	77.50
Vinyl floor tiles, size 300 x 300mm	J2	m²	7.48	1.27	21.62	58.04	11.95	91.61
Carried forward				3.14	53.41	93.65	22.06	169.12

	Ref	Unit	Qty	Hours	Hours £	Mat'ls £	O & P £	Total £
Brought forward				3.14	53.41	93.65	22.06	169.12
25mm thick tongued and grooved boarding	J3	m²	7.48	5.54	94.10	93.35	28.12	215.57
150 x 50mm softwood joists	J4	m	15.20	3.34	56.85	47.42	15.64	119.91
Cut and pin end of joists to exising brick wall	J5	nr	8.00	1.44	24.48	2.40	4.03	30.91
Build in ends of joists to blockwork	J6	nr	8.00	0.80	13.60	1.92	2.33	17.85
Carried to summary				14.26	242.43	238.74	72.18	553.35

PART K
CEILING FINISHES

	Ref	Unit	Qty	Hours	Hours £	Mat'ls £	O & P £	Total £
Plasterboard with taped butt joints fixed to joists	K1	m²	14.96	5.39	91.56	42.64	20.13	154.32
5mm skim coat to plasterboard ceilings	K2	m²	14.96	7.48	127.16	25.88	22.96	176.00
Two coats emulsion paint to ceilings	K3	m²	14.96	3.89	66.12	14.36	12.07	92.56
Carried to summary				16.76	284.84	82.88	55.16	422.88

PART L
ELECTRICAL WORK

	Ref	Unit	Qty	Hours	Hours £	Mat'ls £	O & P £	Total £
13 amp double switched socket outlet with neon	L1	nr	6.00	4.80	91.20	52.86	21.61	165.67
Lighting point	L2	nr	4.00	2.80	53.20	29.36	12.38	94.94
Lighting switch	L3	nr	4.00	2.80	53.20	17.68	10.63	81.51
Lighting wiring	L4	m	14.00	2.80	53.20	26.04	11.89	91.13
Power cable	L5	m	32.00	9.60	182.40	86.72	40.37	309.49
Carried to summary				22.80	433.20	212.66	96.88	742.74

	Ref	Unit	Qty	Hours	Hours £	Mat'ls £	O & P £	Total £
PART M **HEATING WORK**								
15mm copper pipe	M1	m	13.00	5.72	102.96	27.95	19.64	150.55
Elbow	M2	nr	8.00	4.48	80.64	19.52	15.02	115.18
Tee	M3	nr	2.00	1.12	20.16	4.88	3.76	28.80
Radiator, double convector size 1400 x 520mm	M4	nr	4.00	5.20	93.60	556.44	97.51	747.55
Break into existing pipe and insert tee	M5	nr	1.00	0.75	13.50	4.42	2.69	20.61
Carried to summary				17.27	310.86	613.21	138.61	1,062.68
PART N **ALTERATION WORK**								
Take out existing window size 1500 x 1000mm and lintel over, adapt opening to receive 1770 x 2000mm patio door and insert new lintel over (both measured separately) and make good	N1	nr	1.00	20.00	340.00	32.10	55.82	427.92
Take out existing window size 1500 x 1000mm, enlarge opening to receive new PVC–U door (measured separately)	N2	nr	1.00	24.00	408.00	58.16	69.92	536.08
Carried to summary				44.00	748.00	90.26	125.74	964.00

SUMMARY

	Hours	Hours £	Mat'ls £	O & P £	Total £
PART A **PRELIMINARIES**	0.00	0.00	0.00	0.00	2,746.00
PART B SUBSTRUCTURE TO **DPC LEVEL**	49.44	725.82	614.70	201.08	1,541.60
PART C **EXTERNAL WALLS**	116.62	1,982.47	2,599.15	687.24	5,268.86
PART D **FLAT ROOF**	0.00	0.00	0.00	0.00	0.00
PART E **PITCHED ROOF**	69.32	1,167.90	1,125.10	343.95	2,636.95
PART F WINDOWS AND **EXTERNAL DOORS**	26.08	443.36	1,647.88	313.69	2,404.93
PART G INTERNAL **PARTITIONS AND DOORS**	0.00	0.00	0.00	0.00	0.00
PART H **WALL FINISHES**	28.73	488.48	182.45	100.64	771.57
PART J **FLOOR FINISHES**	14.26	242.43	238.74	72.18	553.35
PART K **CEILING FINISHES**	16.76	284.84	82.88	55.16	422.88
PART L **ELECTRICAL WORK**	22.80	433.20	212.66	96.88	742.74
PART M **HEATING WORK**	17.27	310.86	613.21	138.61	1,062.68
PART N **ALTERATION WORK**	44.00	748.00	90.26	125.74	964.00
Final total	405.28	6,827.36	7,407.03	2,135.17	19,115.56

	Ref	Unit	Qty	Hours	Hours £	Mat'ls £	O & P £	Total £
PART A **PRELIMINARIES**								
Concrete mixer	A1	wks	9.00					450.00
Small tools	A2	wks	9.00					360.00
Scaffolding (m²/weeks)	A3		180					1,080.00
Skip	A4	wks	5.00					550.00
Clean up	A5	hrs	10.00					130.00
Carried to summary								2,570.00
PART B **SUBSTRUCTURE TO** **DPC LEVEL**								
Excavate topsoil 150mm thick by hand	B1	m²	10.40	3.12	40.56	0.00	6.08	46.64
Excavate to reduce levels	B2	m³	3.12	7.80	101.40	0.00	15.21	116.61
Excavate for trench foundations by hand	B3	m³	1.39	3.61	46.98	0.00	7.05	54.03
Earthwork support to sides of trenches	B4	m²	12.26	4.90	63.75	22.68	12.96	99.40
Backfilling with excavated material	B5	m³	0.47	0.28	3.67	0.00	0.55	4.22
Hardcore 225mm thick	B6	m²	6.48	1.30	16.85	40.76	8.64	66.25
Hardcore filling to trench	B7	m³	0.09	0.05	0.59	6.29	1.03	7.91
Concrete grade (1:3:6) in foundations	B8	m³	1.13	1.53	19.83	96.75	17.49	134.07
Concrete grade (1:2:4) in bed 150mm thick	B9	m²	6.48	1.94	25.27	99.27	18.68	143.23
Concrete (1:2:4) in cavity wall filling	B10	m²	4.20	0.84	11.97	10.00	3.29	25.26
Carried forward				25.37	330.87	275.75	90.99	697.61

	Ref	Unit	Qty	Hours	Hours £	Mat'ls £	O & P £	Total £
Brought forward				25.37	330.87	275.75	90.99	697.61
Damp-proof membrane	B11	m²	6.88	0.28	3.58	15.96	2.93	22.47
Reinforcement ref A193 in foundation	B12	m²	4.20	0.50	6.55	8.48	2.26	17.29
Steel fabric reinforcement ref A193 in slab	B13	m²	6.48	0.97	12.64	12.96	3.84	29.44
Solid blockwork 140mm thick in cavity wall	B14	m²	5.46	7.10	120.67	91.07	31.76	243.50
Common bricks 112.5mm thick in cavity wall	B15	m²	4.20	7.14	121.38	69.09	28.57	219.04
Facing bricks in 112.5mm thick in skin of cavity wall	B16	m²	1.26	2.27	38.56	28.55	10.07	77.17
Form cavity 50mm wide in cavity wall	B17	m²	5.46	0.16	2.78	8.30	1.66	12.75
DPC 112mm wide	B18	m	8.40	0.42	7.14	8.57	2.36	18.06
DPC 140mm wide	B19	m	8.40	0.50	8.57	11.59	3.02	23.18
Bond in block wall	B20	m	1.30	0.57	9.72	3.78	2.03	15.53
Bond in half brick wall	B21	m	0.30	0.11	1.79	0.80	0.39	2.98
50mm thick insulation board	B22	m²	6.48	1.94	33.05	46.66	11.96	91.66
Carried to summary				47.34	697.28	581.57	191.83	1,470.68
PART C **EXTERNAL WALLS**								
Solid blockwork 140mm thick in cavity wall	C1	m²	34.04	44.25	752.28	567.79	198.01	1,518.08
Facing brickwork 112.5mm thick in cavity wall	C2	m²	34.04	61.27	1,041.62	971.84	302.02	2,315.49
75mm thick insulation in cavity wall	C3	m²	34.04	7.49	127.31	181.77	46.36	355.45
Carried forward				113.01	1,921.22	1,721.40	546.39	4,189.01

	Ref	Unit	Qty	Hours	Hours £	Mat'ls £	O & P £	Total £
Brought forward				113.01	1,921.22	1,721.40	546.39	4,189.01
Steel lintel 2400mm long	C4	nr	2.00	0.50	8.50	296.58	45.76	350.84
Steel lintel 1500mm long	C5	nr	4.00	0.80	13.60	485.44	74.86	573.90
Close cavity wall at jambs	C7	m	13.76	0.69	11.70	38.67	7.55	57.92
Close cavity wall at cills	C8	m	5.07	0.25	4.31	14.25	2.78	21.34
Close cavity wall at top	C9	m	8.40	0.42	7.14	23.60	4.61	35.36
DPC 112mm wide at jambs	C10	m	13.76	0.69	11.70	14.04	3.86	29.59
DPC 112mm wide at cills	C11	m	5.07	0.25	4.31	5.17	1.42	10.90
Carried to summary				116.62	1,982.47	2,599.15	687.24	5,268.86
PART D **FLAT ROOF**				N/A	N/A	N/A	N/A	N/A
PART E **PITCHED ROOF**								
100 x 75mm sawn softwood wall plate	E1	m	3.00	0.90	15.30	7.83	3.47	26.60
200 x 50mm sawn softwood pole plate	E2	nr	3.30	0.99	16.83	10.96	4.17	31.95
100 x 50mm sawn softwood rafters	E3	m	31.50	6.30	107.10	78.12	27.78	213.00
100 x 50mm sawn softwood purlin	E4	m	3.30	0.66	11.22	9.57	3.12	23.91
150 x 50mm softwood joists	E5	m	15.50	3.10	52.70	48.36	15.16	116.22
150 x 50mm sawn softwood sprockets	E6	nr	18.00	2.16	36.72	28.08	9.72	74.52
100mm layer of insulation quilt laid over and between joists	E7	m²	9.00	4.32	73.44	104.40	26.68	204.52
Carried forward				18.43	313.31	287.32	90.09	690.72

	Ref	Unit	Qty	Hours	Hours £	Mat'ls £	O & P £	Total £
Brought forward				18.43	313.31	287.32	90.09	690.72
6mm softwood soffit 150mm wide	E8	m	10.30	4.12	70.04	21.73	13.77	105.54
19mm wrought softwood fascia/ barge board 200mm high	E9	m	3.30	1.65	28.05	8.94	5.55	42.54
Marley Plain roof tiles on felt and battens	E10	m²	14.85	28.22	479.66	538.16	152.67	1,170.49
Double eaves course	E11	m	3.30	1.16	19.64	13.89	5.03	38.56
Verge with tile undercloak	E12	m	9.00	2.25	38.25	59.49	14.66	112.40
Lead flashing code 5, 200mm girth	E13	m	3.30	1.98	33.66	28.45	9.32	71.42
Rake out joint for flashing	E14	m	3.30	1.16	19.64	0.79	3.06	23.49
112mm diameter PVC-U gutter	E15	m	3.30	0.86	14.59	10.69	3.79	29.07
Stop end	E16	nr	1.00	0.14	2.38	2.12	0.68	5.18
Stop end outlet	E17	nr	1.00	0.25	4.25	4.19	1.27	9.71
68mm diameter PVC-U down pipe	E18	m	2.50	0.63	10.63	14.25	3.73	28.61
Shoe	E19	nr	1.00	0.30	5.10	3.15	1.24	9.49
Paint fascia and soffit	E20	m²	3.09	0.62	0.03	3.28	0.50	3.80
Carried to summary				60.59	1,019.57	982.56	300.32	2,302.45

**PART F
WINDOWS AND
EXTERNAL DOORS**

	Ref	Unit	Qty	Hours	Hours £	Mat'ls £	O & P £	Total £
PVC-U door size 840 x 1980mm complete (B)	F1	nr	1.00	2.50	42.50	278.67	48.18	369.35
PVC-U sliding patio door size 1700 x 2075mm (C)	F2	nr	2.00	14.00	238.00	616.94	128.24	983.18
Carried forward				16.50	280.50	895.61	176.42	1,352.53

	Ref	Unit	Qty	Hours	Hours £	Mat'ls £	O & P £	Total £
Brought forward				16.50	280.50	895.61	176.42	1,352.53
PVC-U window size 1200 x 1200mm complete (A)	F3	nr	4.00	8.00	136.00	728.56	129.68	994.24
25 x 225mm wrought softwood window board	F4	m	4.80	1.44	24.48	19.30	6.57	50.34
Paint window board	F5	m	4.80	0.14	2.38	4.42	1.02	7.82
Carried to summary				26.08	443.36	1,647.88	313.69	2,404.93
PART G INTERNAL PARTITIONS AND DOORS				N/A	N/A	N/A	N/A	N/A
PART H WALL FINISHES								
19 x 100mm wrought softwood skirting	H1	m	14.34	2.44	41.44	35.99	11.62	89.05
12mm plasterboard fixed to walls with dabs	H2	m²	29.84	11.64	197.84	85.04	42.43	325.32
12mm plasterboard fixed to walls less than 300mm wide	H3	m	18.33	3.30	56.09	16.86	10.94	83.90
Two coats emulsion paint to walls	H4	m²	32.66	8.49	144.36	31.35	26.36	202.07
Paint skirting	H5	m	14.34	2.87	48.76	13.19	9.29	71.24
Carried to summary				28.73	488.48	182.45	100.64	771.57
PART J FLOOR FINISHES								
Cement and sand floor screed 40mm thick	J1	m²	6.48	1.62	27.54	30.84	8.76	67.14
Vinyl floor tiles, size 300 x 300mm	J2	m²	6.48	1.10	18.73	50.28	10.35	79.36
Carried forward				2.72	46.27	81.13	19.11	146.51

	Ref	Unit	Qty	Hours	Hours £	Mat'ls £	O & P £	Total £
Brought forward				2.72	46.27	81.13	19.11	146.51
25mm thick tongued and grooved boarding	J3	m²	6.48	4.80	81.52	80.87	24.36	186.75
150 x 50mm softwood joists	J4	m	14.50	3.19	54.23	45.24	14.92	114.39
Cut and pin end of joists to existing brick wall	J5	nr	5.00	0.90	15.30	1.50	2.52	19.32
Build in ends of joists to blockwork	J6	nr	5.00	0.50	8.50	1.20	1.46	11.16
Carried to summary				12.11	205.82	209.94	62.36	478.12

PART K
CEILING FINISHES

	Ref	Unit	Qty	Hours	Hours £	Mat'ls £	O & P £	Total £
Plasterboard with taped butt joints fixed to joists	K1	m²	12.96	4.67	79.32	36.94	17.44	133.69
5mm skim coat to plasterboard ceilings	K2	m²	12.96	6.48	110.16	22.42	19.89	152.47
Two coats emulsion paint to ceilings	K3	m²	12.96	3.37	57.28	12.44	10.46	80.18
Carried to summary				14.52	246.76	71.80	47.78	366.34

PART L
ELECTRICAL WORK

	Ref	Unit	Qty	Hours	Hours £	Mat'ls £	O & P £	Total £
13 amp double switched socket outlet with neon	L1	nr	4.00	3.20	60.80	35.24	14.41	110.45
Lighting point	L2	nr	4.00	2.80	53.20	29.36	12.38	94.94
Lighting switch	L3	nr	4.00	2.80	53.20	17.68	10.63	81.51
Lighting wiring	L4	m	12.00	2.40	45.60	22.32	10.19	78.11
Power cable	L5	m	28.00	8.40	159.60	75.88	35.32	270.80
Carried to summary				19.60	372.40	180.48	82.93	635.81

	Ref	Unit	Qty	Hours	Hours £	Mat'ls £	O & P £	Total £
PART M **HEATING WORK**								
15mm copper pipe	M1	m	11.00	4.84	87.12	23.65	16.62	127.39
Elbow	M2	nr	8.00	4.48	80.64	19.52	15.02	115.18
Tee	M3	nr	2.00	1.12	20.16	4.88	3.76	28.80
Radiator, double convector size 1400 x 520mm	M4	nr	4.00	5.20	93.60	556.44	97.51	747.55
Break into existing pipe and insert tee	M5	nr	1.00	0.75	13.50	4.42	2.69	20.61
Carried to summary				16.39	295.02	608.91	135.59	1,039.52
PART N **ALTERATION WORK**								
Take out existing window size 1500 x 1000mm and lintel over, adapt opening to receive 1770 x 2000mm patio door and insert new lintel over (both measured separately) and make good	N1	nr	1.00	20.00	340.00	32.10	55.82	427.92
Take out existing window size 1500 x 1000mm, enlarge opening to receive new PVC–U door (measured separately)	N2	nr	1.00	24.00	408.00	58.16	69.92	536.08
Carried to summary				44.00	748.00	90.26	125.74	964.00

SUMMARY

	Hours	Hours £	Mat'ls £	O & P £	Total £
PART A **PRELIMINARIES**	0.00	0.00	0.00	0.00	2,570.00
PART B SUBSTRUCTURE TO **DPC LEVEL**	47.34	697.28	581.57	191.83	1,470.68
PART C **EXTERNAL WALLS**	116.62	1,982.47	2,599.15	687.24	5,268.86
PART D **FLAT ROOF**	0.00	0.00	0.00	0.00	0.00
PART E **PITCHED ROOF**	60.59	1,019.57	982.56	300.32	2,302.45
PART F WINDOWS AND **EXTERNAL DOORS**	26.08	443.36	1,647.88	313.69	2,404.93
PART G INTERNAL **PARTITIONS AND DOORS**	0.00	0.00	0.00	0.00	0.00
PART H **WALL FINISHES**	28.73	488.48	182.45	100.64	771.57
PART J **FLOOR FINISHES**	12.11	205.82	209.94	62.36	478.12
PART K **CEILING FINISHES**	14.52	246.76	71.80	47.78	366.34
PART L **ELECTRICAL WORK**	19.60	372.40	180.48	82.93	635.81
PART M **HEATING WORK**	16.39	295.02	608.91	135.59	1,039.52
PART N **ALTERATION WORK**	44.00	748.00	90.26	125.74	964.00
Final total	385.98	6,499.16	7,155.00	2,048.13	18,272.29

	Ref	Unit	Qty	Hours	Hours £	Mat'ls £	O & P £	Total £
PART A **PRELIMINARIES**								
Concrete mixer	A1	wks	10.00					500.00
Small tools	A2	wks	10.00					400.00
Scaffolding (m²/weeks)	A3		200					1,200.00
Skip	A4	wks	6.00					660.00
Clean up	A5	hrs	12.00					156.00
Carried to summary								2,916.00
PART B **SUBSTRUCTURE TO** **DPC LEVEL**								
Excavate topsoil 150mm thick by hand	B1	m²	14.19	4.26	55.34	0.00	8.30	63.64
Excavate to reduce levels	B2	m³	4.06	10.15	131.95	0.00	19.79	151.74
Excavate for trench foundations by hand	B3	m³	1.39	3.61	46.98	0.00	7.05	54.03
Earthwork support to sides of trenches	B4	m²	13.72	5.49	71.34	25.38	14.51	111.23
Backfilling with excavated material	B5	m³	0.55	0.33	4.29	0.00	0.64	4.93
Hardcore 225mm thick	B6	m²	9.18	1.84	23.87	57.74	12.24	93.85
Hardcore filling to trench	B7	m³	0.12	0.06	0.78	6.29	1.06	8.13
Concrete grade (1:3:6) in foundations	B8	m³	1.27	1.71	22.29	108.74	19.65	150.68
Concrete grade (1:2:4) in bed 150mm thick	B9	m²	9.18	2.75	35.80	140.64	26.47	202.91
Concrete (1:2:4) in cavity wall filling	B10	m²	4.70	0.94	11.97	11.19	3.47	26.63
Carried forward				31.14	404.62	349.98	113.19	867.78

	Ref	Unit	Qty	Hours	Hours £	Mat'ls £	O & P £	Total £
Brought forward				31.14	404.62	349.98	113.19	867.78
Damp-proof membrane	B11	m²	9.63	0.39	5.01	22.34	4.10	31.45
Reinforcement ref A193 in foundation	B12	m²	4.70	0.56	7.33	9.49	2.52	19.35
Steel fabric reinforcement ref A193 in slab	B13	m²	9.18	1.38	17.90	18.36	5.44	41.70
Solid blockwork 140mm thick in cavity wall	B14	m²	6.11	7.94	135.03	101.91	35.54	272.49
Common bricks 112.5mm thick in cavity wall	B15	m²	4.70	7.99	135.83	77.32	31.97	245.12
Facing bricks in 112.5mm thick in skin of cavity wall	B16	m²	1.41	2.54	43.15	28.55	10.75	82.45
Form cavity 50mm wide in cavity wall	B17	m²	5.46	0.16	2.78	8.30	1.66	12.75
DPC 112mm wide	B18	m	9.40	0.47	7.99	9.59	2.64	20.21
DPC 140mm wide	B19	m	9.40	0.56	9.59	12.97	3.38	25.94
Bond in block wall	B20	m	1.30	0.57	9.72	3.78	2.03	15.53
Bond in half brick wall	B21	m	0.30	0.11	1.79	0.80	0.39	2.98
50mm thick insulation board	B22	m²	9.18	2.75	46.82	66.10	16.94	129.85
Carried to summary				56.57	827.55	709.49	230.56	1,767.60

PART C
EXTERNAL WALLS

	Ref	Unit	Qty	Hours	Hours £	Mat'ls £	O & P £	Total £
Solid blockwork 140mm thick in cavity wall	C1	m²	39.09	50.82	863.89	652.02	227.39	1,743.30
Facing brickwork 112.5mm thick in cavity wall	C2	m²	39.09	70.36	1,196.15	1,116.02	346.83	2,659.00
75mm thick insulation in cavity wall	C3	m²	39.09	8.60	146.20	208.74	53.24	408.18
Carried forward				129.78	2,206.24	1,976.78	627.45	4,810.47

	Ref	Unit	Qty	Hours	Hours £	Mat'ls £	O & P £	Total £
Brought forward				129.78	2,206.24	1,976.78	627.45	4,810.47
Steel lintel 2400mm long	C4	nr	2.00	0.50	8.50	296.58	45.76	350.84
Steel lintel 1500mm long	C5	nr	4.00	0.80	13.60	485.44	74.86	573.90
Close cavity wall at jambs	C7	m	13.76	0.69	11.70	38.67	7.55	57.92
Close cavity wall at cills	C8	m	5.07	0.25	4.31	14.25	2.78	21.34
Close cavity wall at top	C9	m	9.40	0.47	7.99	26.41	5.16	39.56
DPC 112mm wide at jambs	C10	m	13.76	0.69	11.70	14.04	3.86	29.59
DPC 112mm wide at cills	C11	m	5.07	0.25	4.31	5.17	1.42	10.90
Carried to summary				133.43	2,268.34	2,857.33	768.85	5,894.53
PART D **FLAT ROOF**				N/A	N/A	N/A	N/A	N/A
PART E **PITCHED ROOF**								
100 x 75mm sawn softwood wall plate	E1	m	4.00	1.20	20.40	10.44	4.63	35.47
200 x 50mm sawn softwood pole plate	E2	nr	4.30	1.29	21.93	14.28	5.43	41.64
100 x 50mm sawn softwood rafters	E3	m	31.50	6.30	107.10	78.12	27.78	213.00
100 x 50mm sawn softwood purlin	E4	m	4.30	0.86	14.62	12.47	4.06	31.15
150 x 50mm softwood joists	E5	m	24.50	4.90	83.30	76.44	23.96	183.70
150 x 50mm sawn softwood sprockets	E6	nr	18.00	2.16	36.72	28.08	9.72	74.52
100mm layer of insulation quilt laid over and between joists	E7	m²	12.00	5.76	97.92	139.20	35.57	272.69
Carried forward				22.47	381.99	359.03	111.15	852.17

	Ref	Unit	Qty	Hours	Hours £	Mat'ls £	O & P £	Total £
Brought forward				22.47	381.99	359.03	111.15	852.17
6mm softwood soffit 150mm wide	E8	m	11.30	4.52	76.84	23.84	15.10	115.79
19mm wrought softwood fascia/ barge board 200mm high	E9	m	4.30	2.15	36.55	11.65	7.23	55.43
Marley Plain roof tiles on felt and battens	E10	m²	19.55	37.15	631.47	708.49	200.99	1,540.95
Double eaves course	E11	m	4.30	1.51	25.59	18.10	6.55	50.24
Verge with tile undercloak	E12	m	9.00	2.25	38.25	59.49	14.66	112.40
Lead flashing code 5, 200mm girth	E13	m	4.30	2.58	43.86	37.07	12.14	93.06
Rake out joint for flashing	E14	m	4.30	1.51	25.59	1.03	3.99	30.61
112mm diameter PVC-U gutter	E15	m	4.30	1.12	19.01	13.93	4.94	37.88
Stop end	E16	nr	1.00	0.14	2.38	2.12	0.68	5.18
Stop end outlet	E17	nr	1.00	0.25	4.25	4.19	1.27	9.71
68mm diameter PVC-U down pipe	E18	m	2.50	0.63	10.63	14.25	3.73	28.61
Shoe	E19	nr	1.00	0.30	5.10	3.15	1.24	9.49
Paint fascia and soffit	E20	m²	3.09	0.62	0.03	3.28	0.50	3.80
Carried to summary				75.67	1,275.93	1,241.52	377.62	2,895.07

**PART F
WINDOWS AND
EXTERNAL DOORS**

	Ref	Unit	Qty	Hours	Hours £	Mat'ls £	O & P £	Total £
PVC-U door size 840 x 1980mm complete (B)	F1	nr	1.00	2.50	42.50	278.67	48.18	369.35
PVC-U sliding patio door size 1700 x 2075mm (C)	F2	nr	2.00	14.00	238.00	616.94	128.24	983.18
Carried forward				16.50	280.50	895.61	176.42	1,352.53

	Ref	Unit	Qty	Hours	Hours £	Mat'ls £	O & P £	Total £
Brought forward				16.50	280.50	895.61	176.42	1,352.53
PVC-U window size 1200 x 1200mm complete (A)	F3	nr	4.00	8.00	136.00	728.56	129.68	994.24
25 x 225mm wrought softwood window board	F4	m	4.80	1.44	24.48	19.30	6.57	50.34
Paint window board	F5	m	4.80	0.14	2.38	4.42	1.02	7.82
Carried to summary				26.08	443.36	1,647.88	313.69	2,404.93
PART G INTERNAL PARTITIONS AND DOORS				N/A	N/A	N/A	N/A	N/A
PART H WALL FINISHES								
19 x 100mm wrought softwood skirting	H1	m	16.34	2.78	47.22	41.01	13.24	101.47
12mm plasterboard fixed to walls with dabs	H2	m²	34.74	13.55	230.33	99.01	49.40	378.74
12mm plasterboard fixed to walls less than 300mm wide	H3	m	18.33	3.30	56.09	16.86	10.94	83.90
Two coats emulsion paint to walls	H4	m²	38.56	10.03	170.44	37.02	31.12	238.57
Paint skirting	H5	m	16.34	3.27	55.56	15.03	10.59	81.18
Carried to summary				32.92	559.63	208.94	115.28	883.85
PART J FLOOR FINISHES								
Cement and sand floor screed 40mm thick	J1	m²	9.18	2.30	39.02	43.70	12.41	95.12
Vinyl floor tiles, size 300 x 300mm	J2	m²	9.18	1.56	26.53	71.24	14.67	112.43
Carried forward				3.86	65.55	114.93	27.07	207.55

	Ref	Unit	Qty	Hours	Hours £	Mat'ls £	O & P £	Total £
Brought forward				3.86	65.55	114.93	27.07	207.55
25mm thick tongued and grooved boarding	J3	m²	9.18	6.79	115.48	114.57	34.51	264.56
150 x 50mm softwood joists	J4	m	20.30	4.47	75.92	63.34	20.89	160.15
Cut and pin end of joists to existing brick wall	J5	nr	7.00	1.26	21.42	2.10	3.53	27.05
Build in ends of joists to blockwork	J6	nr	7.00	0.70	11.90	1.68	2.04	15.62
Carried to summary				17.07	290.27	296.62	88.03	674.92

PART K
CEILING FINISHES

	Ref	Unit	Qty	Hours	Hours £	Mat'ls £	O & P £	Total £
Plasterboard with taped butt joints fixed to joists	K1	m²	18.36	6.61	112.36	52.33	24.70	189.39
5mm skim coat to plasterboard ceilings	K2	m²	18.36	9.18	156.06	31.76	28.17	216.00
Two coats emulsion paint to ceilings	K3	m²	18.36	4.77	81.15	17.63	14.82	113.59
Carried to summary				20.56	349.57	101.71	67.69	518.98

PART L
ELECTRICAL WORK

	Ref	Unit	Qty	Hours	Hours £	Mat'ls £	O & P £	Total £
13 amp double switched socket outlet with neon	L1	nr	6.00	4.80	91.20	52.86	21.61	165.67
Lighting point	L2	nr	4.00	2.80	53.20	29.36	12.38	94.94
Lighting switch	L3	nr	4.00	2.80	53.20	17.68	10.63	81.51
Lighting wiring	L4	m	12.00	2.40	45.60	22.32	10.19	78.11
Power cable	L5	m	32.00	9.60	182.40	86.72	40.37	309.49
Carried to summary				22.40	425.60	208.94	95.18	729.72

	Ref	Unit	Qty	Hours	Hours £	Mat'ls £	O & P £	Total £
PART M **HEATING WORK**								
15mm copper pipe	M1	m	13.00	5.72	102.96	27.95	19.64	150.55
Elbow	M2	nr	8.00	4.48	80.64	19.52	15.02	115.18
Tee	M3	nr	2.00	1.12	20.16	4.88	3.76	28.80
Radiator, double convector size 1400 x 520mm	M4	nr	4.00	5.20	93.60	556.44	97.51	747.55
Break into existing pipe and insert tee	M5	nr	1.00	0.75	13.50	4.42	2.69	20.61
Carried to summary				17.27	310.86	613.21	138.61	1,062.68
PART N **ALTERATION WORK**								
Take out existing window size 1500 x 1000mm and lintel over, adapt opening to receive 1770 x 2000mm patio door and insert new lintel over (both measured separately) and make good	N1	nr	1.00	20.00	340.00	32.10	55.82	427.92
Take out existing window size 1500 x 1000mm, enlarge opening to receive new PVC–U door (measured separately)	N2	nr	1.00	24.00	408.00	58.16	69.92	536.08
Carried to summary				44.00	748.00	90.26	125.74	964.00

310   Two storey extension, size 3 x 4m, pitched roof

SUMMARY

	Hours	Hours £	Mat'ls £	O & P £	Total £
PART A **PRELIMINARIES**	0.00	0.00	0.00	0.00	2,916.00
PART B SUBSTRUCTURE TO **DPC LEVEL**	56.57	827.55	709.49	230.56	1,767.60
PART C **EXTERNAL WALLS**	133.43	2,268.34	2,857.33	768.85	5,894.52
PART D **FLAT ROOF**	0.00	0.00	0.00	0.00	0.00
PART E **PITCHED ROOF**	75.67	1,275.93	1,241.52	377.62	2,895.07
PART F WINDOWS AND **EXTERNAL DOORS**	26.08	443.36	1,647.88	313.69	2,404.93
PART G INTERNAL **PARTITIONS AND DOORS**	0.00	0.00	0.00	0.00	0.00
PART H **WALL FINISHES**	32.92	559.63	208.94	115.28	883.85
PART J **FLOOR FINISHES**	17.07	290.27	296.62	88.03	674.92
PART K **CEILING FINISHES**	20.56	349.57	101.71	67.69	518.98
PART L **ELECTRICAL WORK**	22.40	425.60	208.94	95.18	729.72
PART M **HEATING WORK**	17.27	310.86	613.21	138.61	1,062.68
PART N **ALTERATION WORK**	44.00	748.00	90.26	125.74	964.00
Final total	445.97	7,499.11	7,975.90	2,321.25	20,712.27

	Ref	Unit	Qty	Hours	Hours £	Mat'ls £	O & P £	Total £
PART A **PRELIMINARIES**								
Concrete mixer	A1	wks	11.00					550.00
Small tools	A2	wks	11.00					440.00
Scaffolding (m²/weeks)	A3		220					1,320.00
Skip	A4	wks	7.00					770.00
Clean up	A5	hrs	12.00					156.00
Carried to summary								3,236.00
PART B **SUBSTRUCTURE TO DPC LEVEL**								
Excavate topsoil 150mm thick by hand	B1	m²	17.50	5.25	68.25	0.00	10.24	78.49
Excavate to reduce levels	B2	m³	5.00	12.50	162.50	0.00	24.38	186.88
Excavate for trench foundations by hand	B3	m³	1.72	4.47	58.14	0.00	8.72	66.86
Earthwork support to sides of trenches	B4	m²	15.18	6.07	78.94	28.08	16.05	123.07
Backfilling with excavated material	B5	m³	0.62	0.37	4.84	0.00	0.73	5.56
Hardcore 225mm thick	B6	m²	11.88	2.38	30.89	74.73	15.84	121.46
Hardcore filling to trench	B7	m³	0.14	0.07	0.91	6.29	1.08	8.28
Concrete grade (1:3:6) in foundations	B8	m³	1.40	1.89	24.57	119.87	21.67	166.10
Concrete grade (1:2:4) in bed 150mm thick	B9	m²	11.88	3.56	46.33	182.00	34.25	262.58
Concrete (1:2:4) in cavity wall filling	B10	m²	5.20	1.04	11.97	12.38	3.65	28.00
Carried forward				37.61	487.33	423.34	136.60	1,047.27

	Ref	Unit	Qty	Hours	Hours £	Mat'ls £	O & P £	Total £
Brought forward				37.61	487.33	423.34	136.60	1,047.27
Damp-proof membrane	B11	m²	12.38	0.50	6.44	28.72	5.27	40.43
Reinforcement ref A193 in foundation	B12	m²	5.20	0.62	8.11	10.50	2.79	21.41
Steel fabric reinforcement ref A193 in slab	B13	m²	11.58	1.74	22.58	23.16	6.86	52.60
Solid blockwork 140mm thick in cavity wall	B14	m²	6.76	8.79	149.40	112.76	39.32	301.48
Common bricks 112.5mm thick in cavity wall	B15	m²	5.20	8.84	150.28	85.54	35.37	271.19
Facing bricks in 112.5mm thick in skin of cavity wall	B16	m²	1.56	2.81	47.74	28.55	11.44	87.73
Form cavity 50mm wide in cavity wall	B17	m²	6.76	0.20	3.45	10.28	2.06	15.78
DPC 112mm wide	B18	m	10.40	0.52	8.84	10.61	2.92	22.37
DPC 140mm wide	B19	m	10.40	0.62	10.61	14.35	3.74	28.70
Bond in block wall	B20	m	1.30	0.57	9.72	3.78	2.03	15.53
Bond in half brick wall	B21	m	0.30	0.11	1.79	0.80	0.39	2.98
50mm thick insulation board	B22	m²	11.88	3.56	60.59	85.54	21.92	168.04
Carried to summary				66.49	966.86	837.93	270.72	2,075.52

**PART C
EXTERNAL WALLS**

	Ref	Unit	Qty	Hours	Hours £	Mat'ls £	O & P £	Total £
Solid blockwork 140mm thick in cavity wall	C1	m²	44.14	57.38	975.49	736.26	256.76	1,968.51
Facing brickwork 112.5mm thick n cavity wall	C2	m²	41.14	74.05	1,258.88	1,174.55	365.01	2,798.45
75mm thick insulation in cavity wall	C3	m²	41.14	9.05	153.86	219.69	56.03	429.58
Carried forward				140.48	2,388.24	2,130.49	677.81	5,196.54

	Ref	Unit	Qty	Hours	Hours £	Mat'ls £	O & P £	Total £
Brought forward				140.48	2,388.24	2,130.49	677.81	5,196.54
Steel lintel 2400mm long	C4	nr	2.00	0.50	8.50	296.58	45.76	350.84
Steel lintel 1500mm long	C5	nr	4.00	0.80	13.60	485.44	74.86	573.90
Close cavity wall at jambs	C7	m	13.76	0.69	11.70	38.67	7.55	57.92
Close cavity wall at cills	C8	m	5.07	0.25	4.31	14.25	2.78	21.34
Close cavity wall at top	C9	m	10.40	0.52	8.84	29.22	5.71	43.77
DPC 112mm wide at jambs	C10	m	13.60	0.68	11.56	13.87	3.81	29.25
DPC 112mm wide at cills	C11	m	5.07	0.25	4.31	5.17	1.42	10.90
Carried to summary				144.18	2,451.06	3,013.69	819.71	6,284.46
PART D **FLAT ROOF**				N/A	N/A	N/A	N/A	N/A
PART E **PITCHED ROOF**								
100 x 75mm sawn softwood wall plate	E1	m	5.00	1.50	25.50	13.05	5.78	44.33
200 x 50mm sawn softwood pole plate	E2	nr	5.30	1.59	27.03	17.60	6.69	51.32
100 x 50mm sawn softwood rafters	E3	m	31.50	6.30	107.10	78.12	27.78	213.00
100 x 50mm sawn softwood purlin	E4	m	5.30	1.06	18.02	15.37	5.01	38.40
150 x 50mm softwood joists	E5	m	24.50	4.90	83.30	76.44	23.96	183.70
150 x 50mm sawn softwood sprockets	E6	nr	18.00	2.16	36.72	28.08	9.72	74.52
100mm layer of insulation quilt laid over and between joists	E7	m²	15.00	7.20	122.40	174.00	44.46	340.86
Carried forward				24.71	420.07	402.66	123.41	946.13

	Ref	Unit	Qty	Hours	Hours £	Mat'ls £	O & P £	Total £
Brought forward				24.71	420.07	402.66	123.41	946.13
6mm softwood soffit 150mm wide	E8	m	12.30	4.92	83.64	25.95	16.44	126.03
19mm wrought softwood fascia/ barge board 200mm high	E9	m	5.30	2.65	45.05	14.36	8.91	68.32
Marley Plain roof tiles on felt and battens	E10	m²	23.85	45.32	770.36	864.32	245.20	1,879.88
Double eaves course	E11	m	5.30	1.86	31.54	22.31	8.08	61.93
Verge with tile undercloak	E12	m	9.00	2.25	38.25	59.49	14.66	112.40
Lead flashing code 5, 200mm girth	E13	m	5.30	3.18	54.06	45.69	14.96	114.71
Rake out joint for flashing	E14	m	5.30	1.86	31.54	1.27	4.92	37.73
112mm diameter PVC-U gutter	E15	m	5.30	1.38	23.43	17.17	6.09	46.69
Stop end	E16	nr	1.00	0.14	2.38	2.12	0.68	5.18
Stop end outlet	E17	nr	1.00	0.25	4.25	4.19	1.27	9.71
68mm diameter PVC-U down pipe	E18	m	2.50	0.63	10.63	14.25	3.73	28.61
Shoe	E19	nr	1.00	0.30	5.10	3.15	1.24	9.49
Paint fascia and soffit	E20	m²	3.69	0.74	0.04	3.91	0.59	4.54
Carried to summary				88.31	1,488.78	1,458.54	442.10	3,389.41

**PART F
WINDOWS AND
EXTERNAL DOORS**

	Ref	Unit	Qty	Hours	Hours £	Mat'ls £	O & P £	Total £
PVC-U door size 840 x 1980mm complete (B)	F1	nr	1.00	2.50	42.50	278.67	48.18	369.35
PVC-U sliding patio door size 1700 x 2075mm (C)	F2	nr	2.00	14.00	238.00	616.94	128.24	983.18
Carried forward				16.50	280.50	895.61	176.42	1,352.53

	Ref	Unit	Qty	Hours	Hours £	Mat'ls £	O & P £	Total £
Brought forward				16.50	280.50	895.61	176.42	1,352.53
PVC-U window size 1200 x 1200mm complete (A)	F3	nr	4.00	8.00	136.00	728.56	129.68	994.24
25 x 225mm wrought softwood window board	F4	m	4.80	1.44	24.48	19.30	6.57	50.34
Paint window board	F5	m	4.80	0.14	2.38	4.42	1.02	7.82
Carried to summary				26.08	443.36	1,647.88	313.69	2,404.93
PART G INTERNAL PARTITIONS AND DOORS				N/A	N/A	N/A	N/A	N/A
PART H WALL FINISHES								
19 x 100mm wrought softwood skirting	H1	m	18.34	3.12	53.00	46.03	14.86	113.89
12mm plasterboard fixed to walls with dabs	H2	m²	39.64	15.46	262.81	112.97	56.37	432.16
12mm plasterboard fixed to walls less than 300mm wide	H3	m	18.33	3.30	56.09	16.86	10.94	83.90
Two coats emulsion paint to walls	H4	m²	42.46	11.04	187.67	40.76	34.27	262.70
Paint skirting	H5	m	18.34	3.67	62.36	16.87	11.88	91.11
Carried to summary				36.58	621.93	233.51	128.32	983.76
PART J FLOOR FINISHES								
Cement and sand floor screed 40mm thick	J1	m²	11.88	2.97	50.49	56.55	16.06	123.09
Vinyl floor tiles, size 300 x 300mm	J2	m²	11.88	2.02	34.33	92.19	18.98	145.50
Carried forward				4.99	84.82	148.74	35.03	268.59

	Ref	Unit	Qty	Hours	Hours £	Mat'ls £	O & P £	Total £
Brought forward				4.99	84.82	148.74	35.03	268.59
25mm thick tongued and grooved boarding	J3	m²	11.88	8.79	149.45	148.26	44.66	342.37
150 x 50mm softwood joists	J4	m	23.20	5.10	86.77	72.38	23.87	183.02
Cut and pin end of joists to existing brick wall	J5	nr	8.00	1.44	24.48	2.40	4.03	30.91
Build in ends of joists to blockwork	J6	nr	8.00	0.80	13.60	1.92	2.33	17.85
Carried to summary				21.12	359.12	373.70	109.92	842.75

PART K
CEILING FINISHES

	Ref	Unit	Qty	Hours	Hours £	Mat'ls £	O & P £	Total £
Plasterboard with taped butt joints fixed to joists	K1	m²	23.76	8.55	145.41	67.72	31.97	245.10
5mm skim coat to plasterboard ceilings	K2	m²	23.76	11.88	201.96	41.10	36.46	279.52
Two coats emulsion paint to ceilings	K3	m²	23.76	6.18	105.02	22.81	19.17	147.00
Carried to summary				26.61	452.39	131.63	87.60	671.62

PART L
ELECTRICAL WORK

	Ref	Unit	Qty	Hours	Hours £	Mat'ls £	O & P £	Total £
13 amp double switched socket outlet with neon	L1	nr	6.00	4.80	91.20	52.86	21.61	165.67
Lighting point	L2	nr	4.00	2.80	53.20	29.36	12.38	94.94
Lighting switch	L3	nr	4.00	2.80	53.20	17.68	10.63	81.51
Lighting wiring	L4	m	14.00	2.80	53.20	26.04	11.89	91.13
Power cable	L5	m	36.00	10.80	205.20	97.56	45.41	348.17
Carried to summary				24.00	456.00	223.50	101.93	781.43

	Ref	Unit	Qty	Hours	Hours £	Mat'ls £	O & P £	Total £
PART M **HEATING WORK**								
15mm copper pipe	M1	m	15.00	6.60	118.80	32.25	22.66	173.71
Elbow	M2	nr	8.00	4.48	80.64	19.52	15.02	115.18
Tee	M3	nr	2.00	1.12	20.16	4.88	3.76	28.80
Radiator, double convector size 1400 x 520mm	M4	nr	4.00	5.20	93.60	556.44	97.51	747.55
Break into existing pipe and insert tee	M5	nr	1.00	0.75	13.50	4.42	2.69	20.61
Carried to summary				18.15	326.70	617.51	141.63	1,085.84
PART N **ALTERATION WORK**								
Take out existing window size 1500 x 1000mm and lintel over, adapt opening to receive 1770 x 2000mm patio door and insert new lintel over (both measured separately) and make good	N1	nr	1.00	20.00	340.00	32.10	55.82	427.92
Take out existing window size 1500 x 1000mm, enlarge opening to receive new PVC–U door (measured separately)	N2	nr	1.00	24.00	408.00	58.16	69.92	536.08
Carried to summary				44.00	748.00	90.26	125.74	964.00

SUMMARY

	Hours	Hours £	Mat'ls £	O & P £	Total £
PART A **PRELIMINARIES**	0.00	0.00	0.00	0.00	3,236.00
PART B SUBSTRUCTURE TO **DPC LEVEL**	66.49	966.86	837.93	270.72	2,075.51
PART C **EXTERNAL WALLS**	144.18	2,451.06	3,013.69	819.71	6,284.46
PART D **FLAT ROOF**	0.00	0.00	0.00	0.00	0.00
PART E **PITCHED ROOF**	88.31	1,488.78	1,458.54	442.10	3,389.41
PART F WINDOWS AND **EXTERNAL DOORS**	26.08	443.36	1,647.88	313.69	2,404.93
PART G INTERNAL **PARTITIONS AND DOORS**	0.00	0.00	0.00	0.00	0.00
PART H **WALL FINISHES**	36.58	621.93	233.51	128.32	983.76
PART J **FLOOR FINISHES**	21.12	359.12	373.70	109.92	842.74
PART K **CEILING FINISHES**	26.61	452.39	131.63	87.60	671.62
PART L **ELECTRICAL WORK**	24.00	456.00	223.50	101.93	781.43
PART M **HEATING WORK**	18.15	326.70	617.51	141.63	1,085.84
PART N **ALTERATION WORK**	44.00	748.00	90.26	125.74	964.00
Final total	495.52	8,314.20	8,628.15	2,541.36	22,719.70

	Ref	Unit	Qty	Hours	Hours £	Mat'ls £	O & P £	Total £
PART A **PRELIMINARIES**								
Concrete mixer	A1	wks	12.00					600.00
Small tools	A2	wks	12.00					480.00
Scaffolding (m²/weeks)	A3		260					1,560.00
Skip	A4	wks	8.00					880.00
Clean up	A5	hrs	12.00					156.00
Carried to summary								3,676.00
PART B **SUBSTRUCTURE TO** **DPC LEVEL**								
Excavate topsoil 150mm thick by hand	B1	m²	19.85	5.96	77.42	0.00	11.61	89.03
Excavate to reduce levels	B2	m³	5.95	14.88	193.38	0.00	29.01	222.38
Excavate for trench foundations by hand	B3	m³	1.88	4.89	63.54	0.00	9.53	73.08
Earthwork support to sides of trenches	B4	m²	16.64	6.66	86.53	30.78	17.60	134.91
Backfilling with excavated material	B5	m³	0.70	0.42	5.46	0.00	0.82	6.28
Hardcore 225mm thick	B6	m²	14.58	2.92	37.91	91.71	19.44	149.06
Hardcore filling to trench	B7	m³	0.15	0.08	0.98	6.29	1.09	8.35
Concrete grade (1:3:6) in foundations	B8	m³	1.54	2.08	27.03	131.85	23.83	182.71
Concrete grade (1:2:4) in bed 150mm thick	B9	m²	14.58	4.37	56.86	223.37	42.03	322.26
Concrete (1:2:4) in cavity wall filling	B10	m²	5.70	1.14	11.97	13.57	3.83	29.37
Carried forward				43.38	561.06	497.57	158.79	1,217.43

	Ref	Unit	Qty	Hours	Hours £	Mat'ls £	O & P £	Total £
Brought forward				43.38	561.06	497.57	158.79	1,217.43
Damp-proof membrane	B11	m²	15.13	0.61	7.87	35.10	6.45	49.41
Reinforcement ref A193 in foundation	B12	m²	5.70	0.68	8.89	11.51	3.06	23.47
Steel fabric reinforcement ref A193 in slab	B13	m²	14.58	2.19	28.43	29.16	8.64	66.23
Solid blockwork 140mm thick in cavity wall	B14	m²	7.41	9.63	163.76	123.60	43.10	330.46
Common bricks 112.5mm thick in cavity wall	B15	m²	5.70	9.69	164.73	93.77	38.77	297.27
Facing bricks in 112.5mm thick in skin of cavity wall	B16	m²	1.71	3.08	52.33	28.55	12.13	93.01
Form cavity 50mm wide in cavity wall	B17	m²	7.41	0.22	3.78	11.26	2.26	17.30
DPC 112mm wide	B18	m	11.40	0.57	9.69	11.63	3.20	24.52
DPC 140mm wide	B19	m	11.40	0.68	11.63	15.73	4.10	31.46
Bond in block wall	B20	m	1.30	0.57	9.72	3.78	2.03	15.53
Bond in half brick wall	B21	m	0.30	0.11	1.79	0.80	0.39	2.98
50mm thick insulation board	B22	m²	14.58	4.37	74.36	104.98	26.90	206.23
Carried to summary				75.78	1,098.04	967.44	309.82	2,375.30

PART C
EXTERNAL WALLS

	Ref	Unit	Qty	Hours	Hours £	Mat'ls £	O & P £	Total £
Solid blockwork 140mm thick in cavity wall	C1	m²	46.65	60.65	1,030.97	778.12	271.36	2,080.45
Facing brickwork 112.5mm thick in cavity wall	C2	m²	46.65	83.97	1,427.49	1,331.86	413.90	3,173.25
75mm thick insulation in cavity wall	C3	m²	46.65	10.26	174.47	249.11	63.54	487.12
Carried forward				154.88	2,632.93	2,359.09	748.80	5,740.82

	Ref	Unit	Qty	Hours	Hours £	Mat'ls £	O & P £	Total £
Brought forward				154.88	2,632.93	2,359.09	748.80	5,740.82
Steel lintel 2400mm long	C4	nr	2.00	0.50	8.50	296.58	45.76	350.84
Steel lintel 1500mm long	C5	nr	6.00	1.20	20.40	728.16	112.28	860.84
Steel lintel 1150mm long	C6	nr	1.00	0.15	2.55	98.72	15.19	116.46
Close cavity wall at jambs	C7	m	13.76	0.69	11.70	38.67	7.55	57.92
Close cavity wall at cills	C8	m	5.07	0.25	4.31	14.25	2.78	21.34
Close cavity wall at top	C9	m	10.40	0.52	8.84	29.22	5.71	43.77
DPC 112mm wide at jambs	C10	m	13.60	0.68	11.56	13.87	3.81	29.25
DPC 112mm wide at cills	C11	m	5.07	0.25	4.31	5.17	1.42	10.90
Carried to summary				278.44	2,813.40	3,665.06	1,055.77	7,294.23
PART D **FLAT ROOF**				N/A	N/A	N/A	N/A	N/A
PART E **PITCHED ROOF**								
100 x 75mm sawn softwood wall plate	E1	m	6.00	1.80	30.60	15.66	6.94	53.20
200 x 50mm sawn softwood pole plate	E2	nr	6.30	1.89	32.13	20.92	7.96	61.00
100 x 50mm sawn softwood rafters	E3	m	31.50	6.30	107.10	78.12	27.78	213.00
100 x 50mm sawn softwood purlin	E4	m	6.30	1.26	21.42	18.27	5.95	45.64
150 x 50mm softwood joists	E5	m	24.10	4.82	81.94	75.19	23.57	180.70
150 x 50mm sawn softwood sprockets	E6	nr	18.00	2.16	36.72	28.08	9.72	74.52
100mm layer of insulation quilt laid over and between joists	E7	m²	18.00	8.64	146.88	208.80	53.35	409.03
Carried forward				26.87	456.79	445.04	135.27	1,037.10

	Ref	Unit	Qty	Hours	Hours £	Mat'ls £	O & P £	Total £
Brought forward				26.87	456.79	445.04	135.27	1,037.10
6mm softwood soffit 150mm wide	E8	m	13.30	5.32	90.44	28.06	17.78	136.28
19mm wrought softwood fascia/ barge board 200mm high	E9	m	6.30	3.15	53.55	17.07	10.59	81.22
Marley Plain roof tiles on felt and battens	E10	m²	28.35	53.87	915.71	1,027.40	291.47	2,234.58
Double eaves course	E11	m	6.30	2.21	37.49	26.52	9.60	73.61
Verge with tile undercloak	E12	m	9.00	2.25	38.25	59.49	14.66	112.40
Lead flashing code 5, 200mm girth	E13	m	6.30	3.78	64.26	54.31	17.78	136.35
Rake out joint for flashing	E14	m	6.30	2.21	37.49	1.51	5.85	44.85
112mm diameter PVC-U gutter	E15	m	6.30	1.64	27.85	20.41	7.24	55.50
Stop end	E16	nr	1.00	0.14	2.38	2.12	0.68	5.18
Stop end outlet	E17	nr	1.00	0.25	4.25	4.19	1.27	9.71
68mm diameter PVC-U down pipe	E18	m	2.50	0.63	10.63	14.25	3.73	28.61
Shoe	E19	nr	1.00	0.30	5.10	3.15	1.24	9.49
Paint fascia and soffit	E20	m²	3.69	0.74	0.04	3.91	0.59	4.54
Carried to summary				101.13	1,706.72	1,680.92	508.15	3,895.78

**PART F
WINDOWS AND
EXTERNAL DOORS**

	Ref	Unit	Qty	Hours	Hours £	Mat'ls £	O & P £	Total £
PVC-U door size 840 x 1980mm complete (B)	F1	nr	2.00	5.00	85.00	278.67	54.55	418.22
PVC-U sliding patio door size 1700 x 2075mm (C)	F2	nr	2.00	14.00	238.00	616.94	128.24	983.18
Carried forward				19.00	323.00	895.61	182.79	1,401.40

	Ref	Unit	Qty	Hours	Hours £	Mat'ls £	O & P £	Total £
Brought forward				19.00	323.00	895.61	182.79	1,401.40
PVC-U window size 1200 x 1200mm complete (A)	F3	nr	6.00	12.00	204.00	1,092.84	194.53	1,491.37
25 x 225mm wrought softwood window board	F4	m	7.20	2.16	36.72	28.94	9.85	75.51
Paint window board	F5	m	7.20	0.14	2.38	6.62	1.35	10.35
Carried to summary				33.30	566.10	2,024.02	388.52	2,978.64
PART G INTERNAL PARTITIONS AND DOORS				N/A	N/A	N/A	N/A	N/A
PART H WALL FINISHES								
19 x 100mm wrought softwood skirting	H1	m	19.56	3.33	56.53	49.10	15.84	121.47
12mm plasterboard fixed to walls with dabs	H2	m²	40.00	15.60	265.20	114.00	56.88	436.08
12mm plasterboard fixed to walls less than 300mm wide	H3	m	33.83	6.09	103.52	31.12	20.20	154.84
Two coats emulsion paint to walls	H4	m²	45.07	11.72	199.21	43.27	36.37	278.85
Paint skirting	H5	m	19.56	3.91	66.50	18.00	12.67	97.17
Carried to summary				40.64	690.96	255.48	141.97	1,088.41
PART J FLOOR FINISHES								
Cement and sand floor screed 40mm thick	J1	m²	14.58	3.65	61.97	69.40	19.70	151.07
Vinyl floor tiles, size 300 x 300mm	J2	m²	14.58	2.48	42.14	113.14	23.29	178.57
Carried forward				6.12	104.10	182.54	43.00	329.64

324   *Two storey extension, size 3 x 6m, pitched roof*

	Ref	Unit	Qty	Hours	Hours £	Mat'ls £	O & P £	Total £
Brought forward				6.12	104.10	182.54	43.00	329.64
25mm thick tongued and grooved boarding	J3	m²	14.58	10.79	183.42	181.96	54.81	420.18
150 x 50mm softwood joists	J4	m	29.00	6.38	108.46	90.48	29.84	228.78
Cut and pin end of joists to existing brick wall	J5	nr	10.00	1.80	30.60	3.00	5.04	38.64
Build in ends of joists to blockwork	J6	nr	10.00	1.00	17.00	2.40	2.91	22.31
Carried to summary				26.09	443.58	460.38	135.59	1,039.55

PART K
CEILING FINISHES

	Ref	Unit	Qty	Hours	Hours £	Mat'ls £	O & P £	Total £
Plasterboard with taped butt joints fixed to joists	K1	m²	29.18	10.50	178.58	83.16	39.26	301.01
5mm skim coat to plasterboard ceilings	K2	m²	29.18	14.59	248.03	50.48	44.78	343.29
Two coats emulsion paint to ceilings	K3	m²	29.18	7.59	128.98	28.01	23.55	180.54
Carried to summary				32.68	555.59	161.66	107.59	824.84

PART L
ELECTRICAL WORK

	Ref	Unit	Qty	Hours	Hours £	Mat'ls £	O & P £	Total £
13 amp double switched socket outlet with neon	L1	nr	8.00	6.40	121.60	70.48	28.81	220.89
Lighting point	L2	nr	4.00	2.80	53.20	29.36	12.38	94.94
Lighting switch	L3	nr	5.00	3.50	66.50	22.10	13.29	101.89
Lighting wiring	L4	m	16.00	3.20	60.80	29.76	13.58	104.14
Power cable	L5	m	40.00	12.00	228.00	108.40	50.46	386.86
Carried to summary				27.90	530.10	260.10	118.53	908.73

	Ref	Unit	Qty	Hours	Hours £	Mat'ls £	O & P £	Total £
PART M **HEATING WORK**								
15mm copper pipe	M1	m	18.00	7.92	142.56	38.70	27.19	208.45
Elbow	M2	nr	8.00	4.48	80.64	19.52	15.02	115.18
Tee	M3	nr	2.00	1.12	20.16	4.88	3.76	28.80
Radiator, double convector size 1400 x 520mm	M4	nr	4.00	5.20	93.60	556.44	97.51	747.55
Break into existing pipe and insert tee	M5	nr	1.00	0.75	13.50	4.42	2.69	20.61
Carried to summary				19.47	350.46	623.96	146.16	1,120.58
PART N **ALTERATION WORK**								
Take out existing window size 1500 x 1000mm and lintel over, adapt opening to receive 1770 x 2000mm patio door and insert new lintel over (both measured separately) and make good	N1	nr	1.00	20.00	340.00	32.10	55.82	427.92
Take out existing window size 1500 x 1000mm, enlarge opening to receive new PVC–U door (measured separately)	N2	nr	1.00	24.00	408.00	58.16	69.92	536.08
Carried to summary				44.00	748.00	90.26	125.74	964.00

SUMMARY

	Hours	Hours £	Mat'ls £	O & P £	Total £
PART A **PRELIMINARIES**	0.00	0.00	0.00	0.00	3,676.00
PART B SUBSTRUCTURE TO **DPC LEVEL**	75.78	1,098.04	967.44	309.82	2,375.30
PART C **EXTERNAL WALLS**	278.44	2,813.40	3,665.06	1,055.77	7,294.23
PART D **FLAT ROOF**	0.00	0.00	0.00	0.00	0.00
PART E **PITCHED ROOF**	101.13	1,706.72	1,680.92	508.15	3,895.78
PART F WINDOWS AND **EXTERNAL DOORS**	33.30	566.10	2,024.02	388.52	2,978.64
PART G INTERNAL **PARTITIONS AND DOORS**	0.00	0.00	0.00	0.00	0.00
PART H **WALL FINISHES**	40.64	690.96	255.48	141.97	1,088.41
PART J **FLOOR FINISHES**	26.09	443.58	460.38	135.59	1,039.55
PART K **CEILING FINISHES**	32.68	555.59	161.66	107.59	824.84
PART L **ELECTRICAL WORK**	27.90	530.10	260.10	118.53	908.73
PART M **HEATING WORK**	19.47	350.46	623.96	146.16	1,120.58
PART N **ALTERATION WORK**	44.00	748.00	90.26	125.74	964.00
Final total	679.43	9,502.95	10,189.28	3,037.84	26,166.06

	Ref	Unit	Qty	Hours	Hours £	Mat'ls £	O & P £	Total £
PART A **PRELIMINARIES**								
Concrete mixer	A1	wks	11.00					550.00
Small tools	A2	wks	11.00					440.00
Scaffolding (m²/weeks)	A3		260					1,560.00
Skip	A4	wks	7.00					770.00
Clean up	A5	hrs	12.00					156.00
Carried to summary								3,476.00
PART B **SUBSTRUCTURE TO** **DPC LEVEL**								
Excavate topsoil 150mm thick by hand	B1	m²	17.85	5.36	69.62	0.00	10.44	80.06
Excavate to reduce levels	B2	m³	5.35	13.38	173.88	0.00	26.08	199.96
Excavate for trench foundations by hand	B3	m³	1.88	4.89	63.54	0.00	9.53	73.08
Earthwork support to sides of trenches	B4	m²	16.64	6.66	86.53	30.78	17.60	134.91
Backfilling with excavated material	B5	m³	0.62	0.37	4.84	0.00	0.73	5.56
Hardcore 225mm thick	B6	m²	13.69	2.74	35.59	86.11	18.26	139.96
Hardcore filling to trench	B7	m³	0.14	0.07	0.91	6.29	1.08	8.28
Concrete grade (1:3:6) in foundations	B8	m³	1.54	2.08	27.03	131.85	23.83	182.71
Concrete grade (1:2:4) in bed 150mm thick	B9	m²	14.58	4.37	56.86	223.37	42.03	322.26
Concrete (1:2:4) in cavity wall filling	B10	m²	5.70	1.14	11.97	13.57	3.83	29.37
Carried forward				41.05	530.76	491.97	153.41	1,176.14

	Ref	Unit	Qty	Hours	Hours £	Mat'ls £	O & P £	Total £
Brought forward				41.05	530.76	491.97	153.41	1,176.14
Damp-proof membrane	B11	m²	13.13	0.53	6.83	30.46	5.59	42.88
Reinforcement ref A193 in foundation	B12	m²	5.70	0.68	8.89	11.51	3.06	23.47
Steel fabric reinforcement ref A193 in slab	B13	m²	13.69	2.05	26.70	27.38	8.11	62.19
Solid blockwork 140mm thick in cavity wall	B14	m²	7.41	9.63	163.76	123.60	43.10	330.46
Common bricks 112.5mm thick in cavity wall	B15	m²	5.70	9.69	164.73	93.77	38.77	297.27
Facing bricks in 112.5mm thick in skin of cavity wall	B16	m²	1.71	3.08	52.33	28.55	12.13	93.01
Form cavity 50mm wide in cavity wall	B17	m²	7.41	0.22	3.78	11.26	2.26	17.30
DPC 112mm wide	B18	m	11.40	0.57	9.69	11.63	3.20	24.52
DPC 140mm wide	B19	m	11.40	0.68	11.63	15.73	4.10	31.46
Bond in block wall	B20	m	1.30	0.57	9.72	3.78	2.03	15.53
Bond in half brick wall	B21	m	0.30	0.11	1.79	0.80	0.39	2.98
50mm thick insulation board	B22	m²	13.69	4.11	69.82	98.57	25.26	193.65
Carried to summary				72.97	1,060.42	949.02	301.42	2,310.85

PART C
EXTERNAL WALLS

	Ref	Unit	Qty	Hours	Hours £	Mat'ls £	O & P £	Total £
Solid blockwork 140mm thick in cavity wall	C1	m²	46.65	60.65	1,030.97	778.12	271.36	2,080.45
Facing brickwork 112.5mm thick in cavity wall	C2	m²	46.65	83.97	1,427.49	1,331.86	413.90	3,173.25
75mm thick insulation in cavity wall	C3	m²	46.65	10.26	174.47	249.11	63.54	487.12
Carried forward				154.88	2,632.93	2,359.09	748.80	5,740.82

	Ref	Unit	Qty	Hours	Hours £	Mat'ls £	O & P £	Total £
Brought forward				154.88	2,632.93	2,359.09	748.80	5,740.82
Steel lintel 2400mm long	C4	nr	2.00	0.50	8.50	296.58	45.76	350.84
Steel lintel 1500mm long	C5	nr	6.00	1.20	20.40	728.16	112.28	860.84
Steel lintel 1150mm long	C6	nr	1.00	0.15	2.55	98.72	15.19	116.46
Close cavity wall at jambs	C7	m	24.52	1.23	20.84	68.90	13.46	103.20
Close cavity wall at cills	C8	m	9.31	0.47	7.91	26.16	5.11	39.19
Close cavity wall at top	C9	m	11.40	0.57	9.69	32.03	6.26	47.98
DPC 112mm wide at jambs	C10	m	24.52	1.23	20.84	25.01	6.88	52.73
DPC 112mm wide at cills	C11	m	9.31	0.47	7.91	9.50	2.61	20.02
Carried to summary				279.46	2,830.74	3,695.25	1,062.90	7,348.89
PART D **FLAT ROOF**				N/A	N/A	N/A	N/A	N/A
PART E **PITCHED ROOF**								
100 x 75mm sawn softwood wall plate	E1	m	4.00	1.20	20.40	10.44	4.63	35.47
200 x 50mm sawn softwood pole plate	E2	nr	4.30	1.29	21.93	14.28	5.43	41.64
100 x 50mm sawn softwood rafters	E3	m	49.50	9.90	168.30	122.76	43.66	334.72
100 x 50mm sawn softwood purlin	E4	m	4.30	0.86	14.62	12.47	4.06	31.15
150 x 50mm softwood joists	E5	m	40.50	8.10	137.70	126.36	39.61	303.67
150 x 50mm sawn softwood sprockets	E6	nr	20.00	2.40	40.80	31.20	10.80	82.80
100mm layer of insulation quilt laid over and between joists	E7	m²	16.00	7.68	130.56	185.60	47.42	363.58
Carried forward				31.43	534.31	503.11	155.61	1,193.03

	Ref	Unit	Qty	Hours	Hours £	Mat'ls £	O & P £	Total £
Brought forward				31.43	534.31	503.11	155.61	1,193.03
6mm softwood soffit 150mm wide	E8	m	13.30	5.32	90.44	28.06	17.78	136.28
19mm wrought softwood fascia/ barge board 200mm high	E9	m	4.30	2.15	36.55	11.65	7.23	55.43
Marley Plain roof tiles on felt and battens	E10	m²	23.65	44.94	763.90	857.08	243.15	1,864.12
Double eaves course	E11	m	4.30	1.51	25.59	18.10	6.55	50.24
Verge with tile undercloak	E12	m	11.00	2.75	46.75	72.71	17.92	137.38
Lead flashing code 5, 200mm girth	E13	m	4.30	2.58	43.86	37.07	12.14	93.06
Rake out joint for flashing	E14	m	4.30	1.51	25.59	1.03	3.99	30.61
112mm diameter PVC-U gutter	E15	m	4.30	1.12	19.01	13.93	4.94	37.88
Stop end	E16	nr	1.00	0.14	2.38	2.12	0.68	5.18
Stop end outlet	E17	nr	1.00	0.25	4.25	4.19	1.27	9.71
68mm diameter PVC-U down pipe	E18	m	2.50	0.63	10.63	14.25	3.73	28.61
Shoe	E19	nr	1.00	0.30	5.10	3.15	1.24	9.49
Paint fascia and soffit	E20	m²	3.99	0.80	0.04	4.23	0.64	4.91
Carried to summary				93.90	1,582.79	1,552.58	470.31	3,605.67

**PART F
WINDOWS AND
EXTERNAL DOORS**

	Ref	Unit	Qty	Hours	Hours £	Mat'ls £	O & P £	Total £
PVC-U door size 840 x 1980mm complete (B)	F1	nr	2.00	5.00	85.00	278.67	54.55	418.22
PVC-U sliding patio door size 1700 x 2075mm (C)	F2	nr	2.00	14.00	238.00	616.94	128.24	983.18
Carried forward				19.00	323.00	895.61	182.79	1,401.40

	Ref	Unit	Qty	Hours	Hours £	Mat'ls £	O & P £	Total £
Brought forward				19.00	323.00	895.61	182.79	1,401.40
PVC-U window size 1200 x 1200mm complete (A)	F3	nr	6.00	12.00	204.00	1,092.84	194.53	1,491.37
25 x 225mm wrought softwood window board	F4	m	7.20	2.16	36.72	28.94	9.85	75.51
Paint window board	F5	m	7.20	0.14	2.38	6.62	1.35	10.35
Carried to summary				33.30	566.10	2,024.02	388.52	2,978.64
PART G INTERNAL PARTITIONS AND DOORS				N/A	N/A	N/A	N/A	N/A
PART H WALL FINISHES								
19 x 100mm wrought softwood skirting	H1	m	19.56	3.33	56.53	49.10	15.84	121.47
12mm plasterboard fixed to walls with dabs	H2	m²	40.00	15.60	265.20	114.00	56.88	436.08
12mm plasterboard fixed to walls less than 300mm wide	H3	m	33.83	6.09	103.52	31.12	20.20	154.84
Two coats emulsion paint to walls	H4	m²	45.07	11.72	199.21	43.27	36.37	278.85
Paint skirting	H5	m	19.56	3.91	66.50	18.00	12.67	97.17
Carried to summary				40.64	690.96	255.48	141.97	1,088.41
PART J FLOOR FINISHES								
Cement and sand floor screed 40mm thick	J1	m²	12.58	3.15	53.47	59.88	17.00	130.35
Vinyl floor tiles, size 300 x 300mm	J2	m²	12.58	2.14	36.36	97.62	20.10	154.07
Carried forward				5.28	89.82	157.50	37.10	284.42

	Ref	Unit	Qty	Hours	Hours £	Mat'ls £	O & P £	Total £
Brought forward				5.28	89.82	157.50	37.10	284.42
25mm thick tongued and grooved boarding	J3	m²	12.58	9.31	158.26	157.00	47.29	362.54
150 x 50mm softwood joists	J4	m	27.30	6.01	102.10	85.18	28.09	215.37
Cut and pin end of joists to existing brick wall	J5	nr	7.00	1.26	21.42	2.10	3.53	27.05
Build in ends of joists to blockwork	J6	nr	7.00	0.70	11.90	1.68	2.04	15.62
Carried to summary				22.56	383.50	403.46	118.04	905.00

**PART K
CEILING FINISHES**

	Ref	Unit	Qty	Hours	Hours £	Mat'ls £	O & P £	Total £
Plasterboard with taped butt joints fixed to joists	K1	m²	25.16	9.06	153.98	71.71	33.85	259.54
5mm skim coat to plasterboard ceilings	K2	m²	25.16	12.58	213.86	43.53	38.61	295.99
Two coats emulsion paint to ceilings	K3	m²	25.16	6.54	111.21	24.15	20.30	155.66
Carried to summary				28.18	479.05	139.39	92.76	711.20

**PART L
ELECTRICAL WORK**

	Ref	Unit	Qty	Hours	Hours £	Mat'ls £	O & P £	Total £
13 amp double switched socket outlet with neon	L1	nr	6.00	4.80	91.20	52.86	21.61	165.67
Lighting point	L2	nr	4.00	2.80	53.20	29.36	12.38	94.94
Lighting switch	L3	nr	5.00	3.50	66.50	22.10	13.29	101.89
Lighting wiring	L4	m	16.00	3.20	60.80	29.76	13.58	104.14
Power cable	L5	m	36.00	10.80	205.20	97.56	45.41	348.17
Carried to summary				25.10	476.90	231.64	106.28	814.82

	Ref	Unit	Qty	Hours	Hours £	Mat'ls £	O & P £	Total £

PART M
HEATING WORK

	Ref	Unit	Qty	Hours	Hours £	Mat'ls £	O & P £	Total £
15mm copper pipe	M1	m	19.00	8.36	150.48	40.85	28.70	220.03
Elbow	M2	nr	8.00	4.48	80.64	19.52	15.02	115.18
Tee	M3	nr	2.00	1.12	20.16	4.88	3.76	28.80
Radiator, double convector size 1400 x 520mm	M4	nr	4.00	5.20	93.60	556.44	97.51	747.55
Break into existing pipe and insert tee	M5	nr	1.00	0.75	13.50	4.42	2.69	20.61
Carried to summary				19.91	358.38	626.11	147.67	1,132.16

PART N
ALTERATION WORK

	Ref	Unit	Qty	Hours	Hours £	Mat'ls £	O & P £	Total £
Take out existing window size 1500 x 1000mm and lintel over, adapt opening to receive 1770 x 2000mm patio door and insert new lintel over (both measured separately) and make good	N1	nr	1.00	20.00	340.00	32.10	55.82	427.92
Take out existing window size 1500 x 1000mm, enlarge opening to receive new PVC–U door (measured separately)	N2	nr	1.00	24.00	408.00	58.16	69.92	536.08
Carried to summary				44.00	748.00	90.26	125.74	964.00

SUMMARY

	Hours	Hours £	Mat'ls £	O & P £	Total £
PART A PRELIMINARIES	0.00	0.00	0.00	0.00	3,476.00
PART B SUBSTRUCTURE TO DPC LEVEL	72.97	1,060.42	949.02	301.42	2,310.86
PART C EXTERNAL WALLS	279.46	2,830.74	3,695.25	1,062.90	7,348.89
PART D FLAT ROOF	0.00	0.00	0.00	0.00	0.00
PART E PITCHED ROOF	93.90	1,582.79	1,552.58	470.31	3,605.67
PART F WINDOWS AND EXTERNAL DOORS	33.30	566.10	2,024.02	388.52	2,978.64
PART G INTERNAL PARTITIONS AND DOORS	0.00	0.00	0.00	0.00	0.00
PART H WALL FINISHES	40.64	690.96	255.48	141.97	1,088.41
PART J FLOOR FINISHES	22.56	383.50	403.46	118.04	905.00
PART K CEILING FINISHES	28.18	479.05	139.39	92.76	711.20
PART L ELECTRICAL WORK	25.10	476.90	231.64	106.28	814.82
PART M HEATING WORK	19.91	358.38	626.11	147.67	1,132.16
PART N ALTERATION WORK	44.00	748.00	90.26	125.74	964.00
Final total	660.02	9,176.84	9,967.21	2,955.61	25,335.65

	Ref	Unit	Qty	Hours	Hours £	Mat'ls £	O & P £	Total £
PART A **PRELIMINARIES**								
Concrete mixer	A1	wks	12.00					600.00
Small tools	A2	wks	12.00					480.00
Scaffolding (m²/weeks)	A3		280					1,680.00
Skip	A4	wks	8.00					880.00
Clean up	A5	hrs	12.00					156.00
Carried to summary								3,796.00
PART B **SUBSTRUCTURE TO** **DPC LEVEL**								
Excavate topsoil 150mm thick by hand	B1	m²	22.00	6.60	85.80	0.00	12.87	98.67
Excavate to reduce levels	B2	m³	6.60	16.50	214.50	0.00	32.18	246.68
Excavate for trench foundations by hand	B3	m³	2.05	5.33	69.29	0.00	10.39	79.68
Earthwork support to sides of trenches	B4	m²	18.10	7.24	94.12	33.49	19.14	146.75
Backfilling with excavated material	B5	m³	0.70	0.42	5.46	0.00	0.82	6.28
Hardcore 225mm thick	B6	m²	16.28	3.26	42.33	102.40	21.71	166.44
Hardcore filling to trench	B7	m³	0.15	0.08	0.98	6.29	1.09	8.35
Concrete grade (1:3:6) in foundations	B8	m³	1.67	2.25	29.31	142.99	25.84	198.14
Concrete grade (1:2:4) in bed 150mm thick	B9	m²	16.28	4.88	63.49	249.41	46.94	359.84
Concrete (1:2:4) in cavity wall filling	B10	m²	6.20	1.24	11.97	14.76	4.01	30.73
Carried forward				47.80	617.24	549.33	174.99	1,341.56

	Ref	Unit	Qty	Hours	Hours £	Mat'ls £	O & P £	Total £
Brought forward				47.80	617.24	549.33	174.99	1,341.56
Damp-proof membrane	B11	m²	16.88	0.68	8.78	39.16	7.19	55.13
Reinforcement ref A193 in foundation	B12	m²	6.20	0.74	9.67	12.52	3.33	25.53
Steel fabric reinforcement ref A193 in slab	B13	m²	16.28	2.44	31.75	32.56	9.65	73.95
Solid blockwork 140mm thick in cavity wall	B14	m²	8.06	10.48	178.13	134.44	46.89	359.45
Common bricks 112.5mm thick in cavity wall	B15	m²	6.20	10.54	179.18	101.99	42.18	323.35
Facing bricks in 112.5mm thick in skin of cavity wall	B16	m²	1.86	3.35	56.92	28.55	12.82	98.29
Form cavity 50mm wide in cavity wall	B17	m²	8.06	0.24	4.11	12.25	2.45	18.82
DPC 112mm wide	B18	m	12.40	0.62	10.54	12.65	3.48	26.67
DPC 140mm wide	B19	m	12.40	0.74	12.65	17.11	4.46	34.22
Bond in block wall	B20	m	1.30	0.57	9.72	3.78	2.03	15.53
Bond in half brick wall	B21	m	0.30	0.11	1.79	0.80	0.39	2.98
50mm thick insulation board	B22	m²	16.28	4.88	83.03	117.22	30.04	230.28
Carried to summary				83.19	1,203.50	1,062.37	339.88	2,605.74

**PART C
EXTERNAL WALLS**

	Ref	Unit	Qty	Hours	Hours £	Mat'ls £	O & P £	Total £
Solid blockwork 140mm thick in cavity wall	C1	m²	46.82	60.87	1,034.72	780.96	272.35	2,088.03
Facing brickwork 112.5mm thick in cavity wall	C2	m²	46.82	84.28	1,432.69	1,336.71	415.41	3,184.81
75mm thick insulation in cavity wall	C3	m²	46.82	10.30	175.11	250.02	63.77	488.89
Carried forward				155.44	2,642.52	2,367.69	751.53	5,761.74

	Ref	Unit	Qty	Hours	Hours £	Mat'ls £	O & P £	Total £
Brought forward				155.44	2,642.52	2,367.69	751.53	5,761.74
Steel lintel 2400mm long	C4	nr	2.00	0.50	8.50	296.58	45.76	350.84
Steel lintel 1500mm long	C5	nr	8.00	1.60	27.20	970.88	149.71	1,147.79
Steel lintel 1150mm long	C6	nr	1.00	0.15	2.55	98.72	15.19	116.46
Close cavity wall at jambs	C7	m	29.32	1.47	24.92	82.39	16.10	123.41
Close cavity wall at cills	C8	m	11.71	0.59	9.95	32.91	6.43	49.29
Close cavity wall at top	C9	m	12.40	0.62	10.54	34.84	6.81	52.19
DPC 112mm wide at jambs	C10	m	29.32	1.47	24.92	29.91	8.22	63.05
DPC 112mm wide at cills	C11	m	11.71	0.59	9.95	11.94	3.28	25.18
Carried to summary				280.95	2,856.14	3,963.47	1,106.94	7,686.55
PART D **FLAT ROOF**				N/A	N/A	N/A	N/A	N/A
PART E **PITCHED ROOF**								
100 x 75mm sawn softwood wall plate	E1	m	5.00	1.50	25.50	13.05	5.78	44.33
200 x 50mm sawn softwood pole plate	E2	nr	5.30	1.59	27.03	17.60	6.69	51.32
100 x 50mm sawn softwood rafters	E3	m	49.50	9.90	168.30	122.76	43.66	334.72
100 x 50mm sawn softwood purlin	E4	m	5.30	1.06	18.02	15.37	5.01	38.40
150 x 50mm softwood joists	E5	m	40.50	8.10	137.70	126.36	39.61	303.67
150 x 50mm sawn softwood sprockets	E6	nr	20.00	2.40	40.80	31.20	10.80	82.80
100mm layer of insulation quilt laid over and between joists	E7	m²	20.00	9.60	163.20	232.00	59.28	454.48
Carried forward				34.15	580.55	558.34	170.83	1,309.72

	Ref	Unit	Qty	Hours	Hours £	Mat'ls £	O & P £	Total £
Brought forward				34.15	580.55	558.34	170.83	1,309.72
6mm softwood soffit 150mm wide	E8	m	16.30	6.52	110.84	34.39	21.78	167.02
19mm wrought softwood fascia/ barge board 200mm high	E9	m	5.30	2.65	45.05	14.36	8.91	68.32
Marley Plain roof tiles on felt and battens	E10	m²	29.15	55.39	941.55	1,056.40	299.69	2,297.63
Double eaves course	E11	m	5.30	1.86	31.54	22.31	8.08	61.93
Verge with tile undercloak	E12	m	11.00	2.75	46.75	72.71	17.92	137.38
Lead flashing code 5, 200mm girth	E13	m	5.30	3.18	54.06	45.69	14.96	114.71
Rake out joint for flashing	E14	m	5.60	1.96	33.32	1.34	5.20	39.86
112mm diameter PVC-U gutter	E15	m	5.30	1.38	23.43	17.17	6.09	46.69
Stop end	E16	nr	1.00	0.14	2.38	2.12	0.68	5.18
Stop end outlet	E17	nr	1.00	0.25	4.25	4.19	1.27	9.71
68mm diameter PVC-U down pipe	E18	m	2.50	0.63	10.63	14.25	3.73	28.61
Shoe	E19	nr	1.00	0.30	5.10	3.15	1.24	9.49
Paint fascia and soffit	E20	m²	4.29	0.86	0.04	4.55	0.69	5.28
Carried to summary				110.15	1,857.94	1,828.66	552.99	4,239.59

PART F
WINDOWS AND
EXTERNAL DOORS

	Ref	Unit	Qty	Hours	Hours £	Mat'ls £	O & P £	Total £
PVC-U door size 840 x 1980mm complete (B)	F1	nr	2.00	5.00	85.00	278.67	54.55	418.22
PVC-U sliding patio door size 1700 x 2075mm (C)	F2	nr	2.00	14.00	238.00	616.94	128.24	983.18
Carried forward				19.00	323.00	895.61	182.79	1,401.40

	Ref	Unit	Qty	Hours	Hours £	Mat'ls £	O & P £	Total £
Brought forward				19.00	323.00	895.61	182.79	1,401.40
PVC-U window size 1200 x 1200mm complete (A)	F3	nr	8.00	16.00	272.00	1,457.12	259.37	1,988.49
25 x 225mm wrought softwood window board	F4	m	9.60	2.88	48.96	38.59	13.13	100.68
Paint window board	F5	m	9.60	0.14	2.38	8.83	1.68	12.89
Carried to summary				38.02	646.34	2,400.15	456.97	3,503.47
PART G **INTERNAL** **PARTITIONS AND** **DOORS**				N/A	N/A	N/A	N/A	N/A
PART H **WALL FINISHES**								
19 x 100mm wrought softwood skirting	H1	m	21.50	3.66	62.14	53.97	17.42	133.52
12mm plasterboard fixed to walls with dabs	H2	m²	42.02	16.39	278.59	119.76	59.75	458.10
12mm plasterboard fixed to walls less than 300mm wide	H3	m	41.03	7.39	125.55	37.75	24.49	187.79
Two coats emulsion paint to walls	H4	m²	48.17	12.52	212.91	46.24	38.87	298.03
Paint skirting	H5	m	21.50	4.30	73.10	19.78	13.93	106.81
Carried to summary				44.25	752.29	277.49	154.47	1,184.25
PART J **FLOOR FINISHES**								
Cement and sand floor screed 40mm thick	J1	m²	16.28	4.07	69.19	77.49	22.00	168.69
Vinyl floor tiles, size 300 x 300mm	J2	m²	16.28	2.77	47.05	126.33	26.01	199.39
Carried forward				6.84	116.24	203.83	48.01	368.07

	Ref	Unit	Qty	Hours	Hours £	Mat'ls £	O & P £	Total £
Brought forward				6.84	116.24	203.83	48.01	368.07
25mm thick tongued and grooved boarding	J3	m²	16.28	12.05	204.80	203.17	61.20	469.17
150 x 50mm softwood joists	J4	m	31.20	6.86	116.69	97.34	32.10	246.14
Cut and pin end of joists to existing brick wall	J5	nr	8.00	1.44	24.48	2.40	4.03	30.91
Build in ends of joists to blockwork	J6	nr	8.00	0.80	13.60	1.92	2.33	17.85
Carried to summary				27.99	475.81	508.66	147.67	1,132.14

PART K
CEILING FINISHES

	Ref	Unit	Qty	Hours	Hours £	Mat'ls £	O & P £	Total £
Plasterboard with taped butt joints fixed to joists	K1	m²	33.56	12.08	205.39	95.65	45.15	346.19
5mm skim coat to plasterboard ceilings	K2	m²	33.56	16.78	285.26	58.06	51.50	394.82
Two coats emulsion paint to ceilings	K3	m²	33.56	8.73	148.34	32.22	27.08	207.64
Carried to summary				37.59	638.98	185.92	123.74	948.64

PART L
ELECTRICAL WORK

	Ref	Unit	Qty	Hours	Hours £	Mat'ls £	O & P £	Total £
13 amp double switched socket outlet with neon	L1	nr	8.00	6.40	121.60	70.48	28.81	220.89
Lighting point	L2	nr	6.00	4.20	79.80	44.04	18.58	142.42
Lighting switch	L3	nr	5.00	3.50	66.50	22.10	13.29	101.89
Lighting wiring	L4	m	18.00	3.60	68.40	33.48	15.28	117.16
Power cable	L5	m	40.00	12.00	228.00	108.40	50.46	386.86
Carried to summary				29.70	564.30	278.50	126.42	969.22

	Ref	Unit	Qty	Hours	Hours £	Mat'ls £	O & P £	Total £

PART M
HEATING WORK

	Ref	Unit	Qty	Hours	Hours £	Mat'ls £	O & P £	Total £
15mm copper pipe	M1	m	21.00	9.24	166.32	45.15	31.72	243.19
Elbow	M2	nr	8.00	4.48	80.64	19.52	15.02	115.18
Tee	M3	nr	2.00	1.12	20.16	4.88	3.76	28.80
Radiator, double convector size 1400 x 520mm	M4	nr	6.00	7.80	140.40	834.66	146.26	1,121.32
Break into existing pipe and insert tee	M5	nr	1.00	0.75	13.50	4.42	2.69	20.61
Carried to summary				23.39	421.02	908.63	199.45	1,529.10

PART N
ALTERATION WORK

	Ref	Unit	Qty	Hours	Hours £	Mat'ls £	O & P £	Total £
Take out existing window size 1500 x 1000mm and lintel over, adapt opening to receive 1770 x 2000mm patio door and insert new lintel over (both measured separately) and make good	N1	nr	1.00	20.00	340.00	32.10	55.82	427.92
Take out existing window size 1500 x 1000mm, enlarge opening to receive new PVC–U door (measured separately)	N2	nr	1.00	24.00	408.00	58.16	69.92	536.08
Carried to summary				44.00	748.00	90.26	125.74	964.00

SUMMARY

	Hours	Hours £	Mat'ls £	O & P £	Total £
PART A **PRELIMINARIES**	0.00	0.00	0.00	0.00	3,796.00
PART B SUBSTRUCTURE TO **DPC LEVEL**	83.19	1,203.50	1,062.37	339.88	2,605.75
PART C **EXTERNAL WALLS**	280.95	2,856.14	3,963.47	1,106.94	7,686.55
PART D **FLAT ROOF**	0.00	0.00	0.00	0.00	0.00
PART E **PITCHED ROOF**	110.15	1,857.94	1,828.66	552.99	4,239.59
PART F WINDOWS AND **EXTERNAL DOORS**	38.02	646.34	2,400.15	456.97	3,503.47
PART G INTERNAL **PARTITIONS AND DOORS**	0.00	0.00	0.00	0.00	0.00
PART H **WALL FINISHES**	44.25	752.29	277.49	154.47	1,184.25
PART J **FLOOR FINISHES**	27.99	475.81	508.66	147.67	1,132.14
PART K **CEILING FINISHES**	37.59	638.98	185.92	123.74	948.64
PART L **ELECTRICAL WORK**	29.70	564.30	278.50	126.42	969.22
PART M **HEATING WORK**	23.39	421.02	908.63	199.45	1,529.10
PART N **ALTERATION WORK**	44.00	748.00	90.26	125.74	964.00
Final total	719.23	10,164.32	11,504.11	3,334.27	28,558.71

	Ref	Unit	Qty	Hours	Hours £	Mat'ls £	O & P £	Total £
PART A **PRELIMINARIES**								
Concrete mixer	A1	wks	12.00					600.00
Small tools	A2	wks	12.00					480.00
Scaffolding (m²/weeks)	A3		300					1,800.00
Skip	A4	wks	9.00					990.00
Clean up	A5	hrs	12.00					156.00
Carried to summary								4,026.00
PART B **SUBSTRUCTURE TO DPC LEVEL**								
Excavate topsoil 150mm thick by hand	B1	m²	26.15	7.85	101.99	0.00	15.30	117.28
Excavate to reduce levels	B2	m³	7.85	19.63	255.13	0.00	38.27	293.39
Excavate for trench foundations by hand	B3	m³	2.21	5.75	74.70	0.00	11.20	85.90
Earthwork support to sides of trenches	B4	m²	19.56	7.82	101.71	36.19	20.68	158.58
Backfilling with excavated material	B5	m³	0.77	0.46	6.01	0.00	0.90	6.91
Hardcore 225mm thick	B6	m²	19.98	4.00	51.95	125.67	26.64	204.27
Hardcore filling to trench	B7	m³	0.17	0.09	1.11	6.29	1.11	8.50
Concrete grade (1:3:6) in foundations	B8	m³	1.81	2.44	31.77	154.97	28.01	214.75
Concrete grade (1:2:4) in bed 150mm thick	B9	m²	19.98	5.99	77.92	306.09	57.60	441.62
Concrete (1:2:4) in cavity wall filling	B10	m²	6.70	1.34	11.97	15.95	4.19	32.10
Carried forward				55.36	714.24	645.16	203.91	1,563.31

	Ref	Unit	Qty	Hours	Hours £	Mat'ls £	O & P £	Total £
Brought forward				55.36	714.24	645.16	203.91	1,563.31
Damp-proof membrane	B11	m²	20.63	0.83	10.73	47.86	8.79	67.38
Reinforcement ref A193 in foundation	B12	m²	6.70	0.80	10.45	13.53	3.60	27.58
Steel fabric reinforcement ref A193 in slab	B13	m²	19.98	3.00	38.96	39.96	11.84	90.76
Solid blockwork 140mm thick in cavity wall	B14	m²	8.71	11.32	192.49	145.28	50.67	388.44
Common bricks 112.5mm thick in cavity wall	B15	m²	6.70	11.39	193.63	110.22	45.58	349.42
Facing bricks in 112.5mm thick in skin of cavity wall	B16	m²	2.01	3.62	61.51	28.55	13.51	103.56
Form cavity 50mm wide in cavity wall	B17	m²	8.71	0.26	4.44	13.24	2.65	20.33
DPC 112mm wide	B18	m	13.40	0.67	11.39	13.67	3.76	28.82
DPC 140mm wide	B19	m	13.40	0.80	13.67	18.49	4.82	36.98
Bond in block wall	B20	m	1.30	0.57	9.72	3.78	2.03	15.53
Bond in half brick wall	B21	m	0.30	0.11	1.79	0.80	0.39	2.98
50mm thick insulation board	B22	m²	19.98	5.99	101.90	143.86	36.86	282.62
Carried to summary				94.72	1,364.91	1,224.41	388.40	2,977.72

PART C
EXTERNAL WALLS

	Ref	Unit	Qty	Hours	Hours £	Mat'ls £	O & P £	Total £
Solid blockwork 140mm thick in cavity wall	C1	m²	51.87	67.43	1,146.33	865.19	301.73	2,313.25
Facing brickwork 112.5mm thick in cavity wall	C2	m²	51.87	93.37	1,587.22	1,480.89	460.22	3,528.33
75mm thick insulation in cavity wall	C3	m²	51.87	11.41	193.99	276.99	70.65	541.63
Carried forward				172.21	2,927.54	2,623.07	832.59	6,383.20

	Ref	Unit	Qty	Hours	Hours £	Mat'ls £	O & P £	Total £
Brought forward				172.21	2,927.54	2,623.07	832.59	6,383.20
Steel lintel 2400mm long	C4	nr	2.00	0.50	8.50	296.58	45.76	350.84
Steel lintel 1500mm long	C5	nr	8.00	1.60	27.20	970.88	149.71	1,147.79
Steel lintel 1150mm long	C6	nr	1.00	0.15	2.55	98.72	15.19	116.46
Close cavity wall at jambs	C7	m	29.32	1.47	24.92	82.39	16.10	123.41
Close cavity wall at cills	C8	m	11.71	0.59	9.95	32.91	6.43	49.29
Close cavity wall at top	C9	m	13.40	0.67	11.39	37.65	7.36	56.40
DPC 112mm wide at jambs	C10	m	29.32	1.47	24.92	29.91	8.22	63.05
DPC 112mm wide at cills	C11	m	11.71	0.59	9.95	11.94	3.28	25.18
Carried to summary				297.77	3,142.01	4,221.66	1,188.55	8,312.22
PART D **FLAT ROOF**				N/A	N/A	N/A	N/A	N/A
PART E **PITCHED ROOF**								
100 x 75mm sawn softwood wall plate	E1	m	6.00	1.80	30.60	15.66	6.94	53.20
200 x 50mm sawn softwood pole plate plugged	E2	nr	6.30	1.89	32.13	20.92	7.96	61.00
100 x 50mm sawn softwood rafters	E3	m	49.50	9.90	168.30	122.76	43.66	334.72
100 x 50mm sawn softwood purlin	E4	m	6.30	1.26	21.42	18.27	5.95	45.64
150 x 50mm softwood joists	E5	m	40.50	8.10	137.70	126.36	39.61	303.67
150 x 50mm sawn softwood sprockets	E6	nr	20.00	2.40	40.80	31.20	10.80	82.80
100mm layer of insulation quilt laid over and between joists	E7	m²	24.00	11.52	195.84	278.40	71.14	545.38
Carried forward				36.87	626.79	613.57	186.05	1,426.41

	Ref	Unit	Qty	Hours	Hours £	Mat'ls £	O & P £	Total £
Brought forward				36.87	626.79	613.57	186.05	1,426.41
6mm softwood soffit 150mm wide	E8	m	15.30	6.12	104.04	32.28	20.45	156.77
19mm wrought softwood fascia/ barge board 200mm high	E9	m	6.30	3.15	53.55	17.07	10.59	81.22
Marley Plain roof tiles on felt and battens	E10	m²	36.65	69.64	1,183.80	1,328.20	376.80	2,888.79
Double eaves course	E11	m	6.30	2.21	37.49	26.52	9.60	73.61
Verge with tile undercloak	E12	m	11.00	2.75	46.75	72.71	17.92	137.38
Lead flashing code 5, 200mm girth	E13	m	6.30	3.78	64.26	54.31	17.78	136.35
Rake out joint for flashing	E14	m	6.30	2.21	37.49	1.51	5.85	44.85
112mm diameter PVC-U gutter	E15	m	6.30	1.64	27.85	20.41	7.24	55.50
Stop end	E16	nr	1.00	0.14	2.38	2.12	0.68	5.18
Stop end outlet	E17	nr	1.00	0.25	4.25	4.19	1.27	9.71
68mm diameter PVC-U down pipe	E18	m	2.50	0.63	10.63	14.25	3.73	28.61
Shoe	E19	nr	1.00	0.30	5.10	3.15	1.24	9.49
Paint fascia and soffit	E20	m²	4.59	0.92	0.05	4.87	0.74	5.65
Carried to summary				128.38	2,166.92	2,168.63	650.33	4,985.88

**PART F
WINDOWS AND
EXTERNAL DOORS**

	Ref	Unit	Qty	Hours	Hours £	Mat'ls £	O & P £	Total £
PVC-U door size 840 x 1980mm complete (B)	F1	nr	2.00	5.00	85.00	278.67	54.55	418.22
PVC-U sliding patio door size 1700 x 2075mm (C)	F2	nr	2.00	14.00	238.00	616.94	128.24	983.18
Carried forward				19.00	323.00	895.61	182.79	1,401.40

	Ref	Unit	Qty	Hours	Hours £	Mat'ls £	O & P £	Total £
Brought forward				19.00	323.00	895.61	182.79	1,401.40
PVC-U window size 1200 x 1200mm complete (A)	F3	nr	8.00	16.00	272.00	1,457.12	259.37	1,988.49
25 x 225mm wrought softwood window board	F4	m	9.60	2.88	48.96	38.59	13.13	100.68
Paint window board	F5	m	9.60	0.14	2.38	8.83	1.68	12.89
Carried to summary				38.02	646.34	2,400.15	456.97	3,503.47

**PART G
INTERNAL
PARTITIONS AND
DOORS**

	Ref	Unit	Qty	Hours	Hours £	Mat'ls £	O & P £	Total £
50 x 75mm sawn softwood sole plate	G1	m	3.70	0.81	13.84	4.96	2.82	21.62
50 x 75mm softwood head	G2	m	3.70	0.81	13.84	4.96	2.82	21.62
50 x 75mm softwood studs	G3	m	17.15	4.80	81.63	22.98	15.69	120.31
50 x 75mm sawn softwood noggings	G4	m	3.70	1.04	17.61	4.96	3.39	25.96
Plasterboard	G5	m²	14.80	5.33	90.58	39.52	19.51	149.61
35mm thick veneered internal door size 762 x 1981mm	G6	nr	1.00	1.25	21.25	69.91	13.67	104.83
35 x 150mm wrought softwood lining	G7	m	4.87	1.07	18.21	22.50	6.11	46.82
13 x 38mm wrought softwood stop	G8	m	4.87	0.97	16.56	331.16	52.16	399.88
19 x 50mm wrought softwood architrave	G9	m	9.74	1.46	24.84	13.05	5.68	43.57
19 x 100mm wrought softwood skirting	G10	m	5.88	1.00	16.99	14.76	4.76	36.51
Carried forward				18.55	315.35	528.75	126.62	970.72

	Ref	Unit	Qty	Hours	Hours £	Mat'ls £	O & P £	Total £
Brought forward				18.55	315.35	528.75	126.62	970.72
100mm rising steel butts	G11	pr	1.00	0.30	5.10	1.92	1.05	8.07
SAA mortice latch with lever furniture	G12	nr	1.00	0.80	13.60	16.22	4.47	34.29
Two coats emulsion on plaster– board walls	G13	m²	14.80	3.85	65.42	1,420.80	222.93	1,709.15
Paint general surfaces	G14	m²	3.29	0.66	11.19	5.26	2.47	18.92
Carried to summary				24.16	410.65	1,972.95	357.54	2,741.15

PART H
WALL FINISHES

	Ref	Unit	Qty	Hours	Hours £	Mat'ls £	O & P £	Total £
19 x 100mm wrought softwood skirting	H1	m	28.90	4.91	83.52	72.54	23.41	179.47
12mm plasterboard fixed to walls with dabs	H2	m²	46.22	18.03	306.44	131.73	65.72	503.89
12mm plasterboard fixed to walls less than 300mm wide	H3	m	41.03	7.39	125.55	37.75	24.49	187.79
Two coats emulsion paint to walls	H4	m²	67.91	17.66	300.16	65.19	54.80	420.16
Paint skirting	H5	m	28.90	5.78	98.26	26.59	18.73	143.58
Carried to summary				53.76	913.93	333.80	187.16	1,434.89

PART J
FLOOR FINISHES

	Ref	Unit	Qty	Hours	Hours £	Mat'ls £	O & P £	Total £
Cement and sand floor screed 40mm thick	J1	m²	19.98	5.00	84.92	95.10	27.00	207.02
Vinyl floor tiles, size 300 x 300mm	J2	m²	19.98	3.40	57.74	155.04	31.92	244.71
Carried forward				8.39	142.66	250.15	58.92	451.73

	Ref	Unit	Qty	Hours	Hours £	Mat'ls £	O & P £	Total £
Brought forward				8.39	142.66	250.15	58.92	451.73
25mm thick tongued and grooved boarding	J3	m²	19.98	14.79	251.35	249.35	75.10	575.80
150 x 50mm softwood joists	J4	m	39.00	8.58	145.86	121.68	40.13	307.67
Cut and pin end of joists to existing brick wall	J5	nr	10.00	1.80	30.60	3.00	5.04	38.64
Build in ends of joists to blockwork	J6	nr	10.00	1.00	17.00	2.40	2.91	22.31
Carried to summary				34.56	587.47	626.58	182.11	1,396.15

PART K
CEILING FINISHES

	Ref	Unit	Qty	Hours	Hours £	Mat'ls £	O & P £	Total £
Plasterboard with taped butt joints fixed to joists	K1	m²	39.96	14.39	244.56	113.89	53.77	412.21
5mm skim coat to plasterboard ceilings	K2	m²	39.96	19.98	339.66	69.13	61.32	470.11
Two coats emulsion paint to ceilings	K3	m²	39.96	10.39	176.62	38.36	32.25	247.23
Carried to summary				44.76	760.84	221.38	147.33	1,129.55

PART L
ELECTRICAL WORK

	Ref	Unit	Qty	Hours	Hours £	Mat'ls £	O & P £	Total £
13 amp double switched socket outlet with neon	L1	nr	10.00	8.00	152.00	88.10	36.02	276.12
Lighting point	L2	nr	6.00	4.20	79.80	44.04	18.58	142.42
Lighting switch	L3	nr	5.00	3.50	66.50	22.10	13.29	101.89
Lighting wiring	L4	m	22.00	4.40	83.60	40.92	18.68	143.20
Power cable	L5	m	44.00	13.20	250.80	119.24	55.51	425.55
Carried to summary				33.30	632.70	314.40	142.07	1,089.17

	Ref	Unit	Qty	Hours	Hours £	Mat'ls £	O & P £	Total £
PART M **HEATING WORK**								
15mm copper pipe	M1	m	23.00	10.12	182.16	49.45	34.74	266.35
Elbow	M2	nr	8.00	4.48	80.64	19.52	15.02	115.18
Tee	M3	nr	2.00	1.12	20.16	4.88	3.76	28.80
Radiator, double convector size 1400 x 520mm	M4	nr	6.00	7.80	140.40	834.66	146.26	1,121.32
Break into existing pipe and insert tee	M5	nr	1.00	0.75	13.50	4.42	2.69	20.61
Carried to summary				24.27	436.86	912.93	202.47	1,552.26
PART N **ALTERATION WORK**								
Take out existing window size 1500 x 1000mm and lintel over, adapt opening to receive 1770 x 2000mm patio door and insert new lintel over (both measured separately) and make good	N1	nr	1.00	20.00	340.00	32.10	55.82	427.92
Take out existing window size 1500 x 1000mm, enlarge opening to receive new PVC–U door (measured separately)	N2	nr	1.00	24.00	408.00	58.16	69.92	536.08
Carried to summary				44.00	748.00	90.26	125.74	964.00

SUMMARY

	Hours	Hours £	Mat'ls £	O & P £	Total £
PART A **PRELIMINARIES**	0.00	0.00	0.00	0.00	4,026.00
PART B SUBSTRUCTURE TO **DPC LEVEL**	94.72	1,364.91	1,224.41	388.40	2,977.72
PART C **EXTERNAL WALLS**	297.77	3,142.01	4,221.66	1,188.55	8,312.22
PART D **FLAT ROOF**	0.00	0.00	0.00	0.00	0.00
PART E **PITCHED ROOF**	128.38	2,166.92	2,168.63	650.33	4,985.88
PART F WINDOWS AND **EXTERNAL DOORS**	38.02	646.34	2,400.15	456.97	3,503.47
PART G INTERNAL **PARTITIONS AND DOORS**	24.16	410.65	1,972.95	357.54	2,741.15
PART H **WALL FINISHES**	53.76	913.93	333.80	187.16	1,434.89
PART J **FLOOR FINISHES**	34.56	587.47	626.58	182.11	1,396.15
PART K **CEILING FINISHES**	44.76	760.84	221.38	147.33	1,129.55
PART L **ELECTRICAL WORK**	33.30	632.70	314.40	142.07	1,089.17
PART M **HEATING WORK**	24.27	436.86	912.93	202.47	1,552.26
PART N **ALTERATION WORK**	44.00	748.00	90.26	125.74	964.00
Final total	817.70	11,810.63	14,487.15	4,028.67	34,112.46

SUMMARY OF EXTENSION COSTS

One storey flat roof

	2 x 3m £	2 x 3m %	2 x 4m £	2 x 4m %	2 x 5m £	2 x 5m %	3 x 3m £	3 x 3m %	3 x 4m £	3 x 4m %
PART A: PRELIMINARIES	1132	13	1418	14	1704	15	1554	14	1788	15
PART B: SUBSTRUCTURE TO DPC LEVEL	1079	13	1319	13	1532	13	1483	14	1755	14
PART C: EXTERNAL WALLS	2037	24	2349	23	2661	23	2661	25	2973	24
PART D: FLAT ROOF	1227	14	1524	15	1906	17	1598	15	2061	17
PART E: PITCHED ROOF	0	0	0	0	0	0	0	0	0	0
PART F: WINDOWS AND EXTERNAL DOORS	1511	18	1511	15	1511	13	1510	14	1510	12
PART G: INTERNAL PARTITIONS AND DOORS	0	0	0	0	0	0	0	0	0	0
PART H: WALL FINISHES	274	3	328	3	382	3	382	4	436	4
PART J: FLOOR FINISHES	92	1	131	1	169	1	146	1	208	2
PART K: CEILING FINISHES	115	1	163	2	211	2	184	2	260	2
PART L: ELECTRICAL WORK	292	3	318	3	395	3	294	3	365	3
PART M: HEATING WORK	326	4	527	5	538	5	527	5	538	4
PART N: ALTERATION WORK	428	5	428	4	428	4	428	4	428	3
Final total £	8513	100	10016	100	11437	100	10767	100	12322	100
Floor area m²	4.08		6.29		7.48		6.48		9.18	
£ per m² £	2,087		1,592		1,529		1,662		1,342	

SUMMARY OF EXTENSION COSTS

One storey flat roof

	3 x 5 m		3 x 6m		4 x 4m		4 x 5m		4 x 6m	
	£	%	£	%	£	%	£	%	£	%
PART A: PRELIMINARIES	2074	15	2420	15	2194	14	2480	14	2740	1
PART B: SUBSTRUCTURE TO DPC LEVEL	2062	15	2357	15	2273	15	2585	15	2954	1
PART C: EXTERNAL WALLS	3285	24	3425	22	3425	23	3582	21	3914	2
PART D: FLAT ROOF	2462	18	2862	18	2598	17	3101	18	3605	1
PART E: PITCHED ROOF	0	0	0	0	0	0	0	0	0	
PART F: WINDOWS AND EXTERNAL DOORS	1510	11	2133	13	2133	14	2395	14	2395	1
PART G: INTERNAL PARTITIONS AND DOORS	0	0	0	0	0	0	0	0	0	
PART H: WALL FINISHES	496	4	538	3	534	4	589	3	638	
PART J: FLOOR FINISHES	269	2	330	2	284	2	368	2	449	
PART K: CEILING FINISHES	336	2	412	3	355	2	460	3	562	
PART L: ELECTRICAL WORK	390	3	458	3	418	3	495	3	555	
PART M: HEATING WORK	550	4	561	4	573	4	772	4	783	
PART N: ALTERATION WORK	428	3	428	3	428	3	428	2	428	
Final total £	13862	100	15924	100	15215	100	17255	100	19023	10
Floor area m²	11.88		14.58		12.58		16.28		19.98	
£ per m² £	1,167		1,092		1,209		1,060		952	

SUMMARY OF EXTENSION COSTS

One storey pitched roof

		2 x 3m		2 x 4m		2 x 5m		3 x 3m		3 x 4m	
		£	%	£	%	£	%	£	%	£	%
PART A: PRELIMINARIES		1132	12	1418	13	1704	14	1554	13	1788	13
PART B: SUBSTRUCTURE TO DPC LEVEL		1079	11	1310	12	1542	12	1471	12	1776	13
PART C: EXTERNAL WALLS		2406	25	2718	24	3030	24	3399	28	3711	27
PART D: FLAT ROOF		0	0	0	0	0	0	0	0	0	0
PART E: PITCHED ROOF		1844	19	2265	20	2686	21	2422	20	2959	21
PART F: WINDOWS AND EXTERNAL DOORS		1511	16	1511	14	1511	12	1511	12	1511	11
PART G: INTERNAL PARTITIONS AND DOORS		0	0	0	0	0	0	0	0	0	0
PART H: WALL FINISHES		274	3	328	3	382	3	382	3	436	3
PART J: FLOOR FINISHES		92	1	131	1	169	1	146	1	207	1
PART K: CEILING FINISHES		115	1	163	1	211	2	183	1	259	2
PART L: ELECTRICAL WORK		292	3	318	3	395	3	294	2	365	3
PART M: HEATING WORK		325	3	524	5	536	4	524	4	536	4
PART N: ALTERATION WORK		428	5	428	4	428	3	428	3	428	3
Final total	£	9499	100	11114	100	12594	100	12314	100	13976	100
Floor area	m²	4.08		6.29		7.48		6.48		9.18	
per m²	£	2,328		1,767		1,684		1,900		1,522	

SUMMARY OF EXTENSION COSTS

One storey pitched roof

	3 x 5 m £	3 x 5 m %	3 x 6m £	3 x 6m %	4 x 4m £	4 x 4m %	4 x 5m £	4 x 5m %	4 x 6m £	4 x 6m %
PART A: PRELIMINARIES	2074	13	2420	14	2220	13	2480	12	2740	1
PART B: SUBSTRUCTURE TO DPC LEVEL	2075	13	2375	13	2291	13	2605	13	2978	1
PART C: EXTERNAL WALLS	4023	26	4401	25	4901	28	5202	26	5389	2
PART D: FLAT ROOF	0	0	0	0	0	0	0	0	0	
PART E: PITCHED ROOF	3468	22	3773	21	3525	20	4136	21	4863	2
PART F: WINDOWS AND EXTERNAL DOORS	1511	9.7	2143	12	2143	12	2405	12	2405	1
PART G: INTERNAL PARTITIONS AND DOORS	0	0	0	0	0	0	0	0	0	
PART H: WALL FINISHES	494	3	538	3	538	3	639	3	637	
PART J: FLOOR FINISHES	268	2	330	2	284	2	368	2	449	
PART K: CEILING FINISHES	336	2	412	2	356	2	460	2	562	
PART L: ELECTRICAL WORK	391	3	464	3	418	2	495	2	555	
PART M: HEATING WORK	547	4	559	3	571	3	769	4	780	
PART N: ALTERATION WORK	428	3	428	2	428	2	428	2	428	
Final total £	15615	100	17843	100	17675	100	19987	100	21786	10
Floor area m²	11.88		14.58		12.58		16.28		19.98	
£ per m² £	1,314		1,224		1,405		1,228		1,090	

SUMMARY OF EXTENSION COSTS

Two storey flat roof

	2 x 3m		2 x 4m		2 x 5m		3 x 3m		3 x 4m	
	£	%	£	%	£	%	£	%	£	%
PART A: PRELIMINARIES	2054	14	2400	15	2746	15	2570	15	2916	14
PART B: SUBSTRUCTURE TO DPC LEVEL	1079	7.5	1310	8	1541	8.4	1471	8.4	1758	8.7
PART C: EXTERNAL WALLS	4017	28	4644	28	5269	29	5269	30	5894	29
PART D: FLAT ROOF	1253	8.7	1550	9.5	1849	10	1687	9.6	2395	12
PART E: PITCHED ROOF	0	0	0	0	0	0	0	0	0	0
PART F: WINDOWS AND EXTERNAL DOORS	2405	17	2405	15	2405	13	2405	14	2405	12
PART G: INTERNAL PARTITIONS AND DOORS	0	0	0	0	0	0	0	0	0	0
PART H: WALL FINISHES	559	4	659	4	771	4	771	4	884	4
PART J: FLOOR FINISHES	315	2	445	3	553	3	478	3	675	3
PART K: CEILING FINISHES	231	2	327	2	423	2	366	2	519	3
PART L: ELECTRICAL WORK	537	4	636	4	743	4	588	3	730	4
PART M: HEATING WORK	1016	7	1039	6	1063	6	1040	6	1062	5
PART N: ALTERATION WORK	964	7	964	6	964	5	964	5	964	5
Final total £	14430	100	16379	100	18327	100	17609	100	20202	100
Floor area m²	8.16		12.58		14.96		12.96		18.36	
per m² £	1,768		1,302		1,225		1,359		1,100	

SUMMARY OF EXTENSION COSTS

Two storey flat roof

	3 x 5 m		3 x 6m		4 x 4m		4 x 5m		4 x 6m	
	£	%	£	%	£	%	£	%	£	%
PART A: PRELIMINARIES	3236	15	3676	15	3476	14	3796	14	4026	1
PART B: SUBSTRUCTURE TO DPC LEVEL	2074	9.4	2375	9.5	2291	9.5	2606	9.5	2978	
PART C: EXTERNAL WALLS	6520	30	7332	29	7086	29	7690	28	8316	2
PART D: FLAT ROOF	2488	11	2889	12	2624	11	3128	11	3631	1
PART E: PITCHED ROOF	0	0	0	0	0	0	0	0	0	
PART F: WINDOWS AND EXTERNAL DOORS	2405	11	2979	12	2979	12	3503	13	3947	1
PART G: INTERNAL PARTITIONS AND DOORS	0	0	0	0	0	0	0	0	720	
PART H: WALL FINISHES	984	4	873	3	1089	5	1184	4	1273	
PART J: FLOOR FINISHES	843	4	1040	4	905	4	1132	4	1379	
PART K: CEILING FINISHES	672	3	824	3	711	3	949	3	1130	
PART L: ELECTRICAL WORK	781	4	909	4	814	3	969	4	1089	
PART M: HEATING WORK	1086	5	1120	4	1132	5	1529	6	1552	
PART N: ALTERATION WORK	964	4	964	4	964	4	964	4	964	
Final total £	22053	100	24981	100	24071	100	27450	100	31005	9
Floor area m²	23.76		29.16		25.16		32.56		39.96	
£ per m² £	928		857		957		843		776	

SUMMARY OF EXTENSION COSTS

Two storey pitched roof

	2 x 3m		2 x 4m		2 x 5m		3 x 3m		3 x 4m	
	£	%	£	%	£	%	£	%	£	%
PART A: PRELIMINARIES	2054	14	2400	14	2746	14	2570	14	2916	14
PART B: SUBSTRUCTURE TO DPC LEVEL	1079	7.2	1310	7.7	1542	8.1	1471	8.1	1767	8.5
PART C: EXTERNAL WALLS	4017	27	4644	27	5269	28	5269	29	5894	28
PART D: FLAT ROOF	0	0	0	0	0	0	0	0	0	0
PART E: PITCHED ROOF	1853	12	2273	13	2636	14	2302	13	2895	14
PART F: WINDOWS AND EXTERNAL DOORS	2405	16	2405	14	2405	13	2405	13	2405	16
PART G: INTERNAL PARTITIONS AND DOORS	0	0	0	0	0	0	0	0	0	0
PART H: WALL FINISHES	559	4	661	4	771	4	771	4	884	4
PART J: FLOOR FINISHES	315	2	445	3	553	3	478	3	675	3
PART K: CEILING FINISHES	231	2	327	2	423	2	366	2	519	3
PART L: ELECTRICAL WORK	537	4	636	4	743	4	636	3	730	4
PART M: HEATING WORK	1016	7	1039	6	1063	6	1040	6	1063	5
PART N: ALTERATION WORK	964	6	964	6	964	5	964	5	964	5
Final total £	15030	100	17105	100	19115	100	18272	100	20712	104
Floor area m²	8.16		12.58		14.96		12.96		18.36	
per m² £	1,842		1,360		1,278		1,410		1,128	

SUMMARY OF EXTENSION COSTS

Two storey pitched roof

	3 x 5m		3 x 6m		4 x 4m		4 x 5m		4 x 6m	
	£	%	£	%	£	%	£	%	£	%
PART A: PRELIMINARIES	3236	14	3676	14	3476	14	3796	13	4026	1
PART B: SUBSTRUCTURE TO DPC LEVEL	2075	9.1	2375	9.1	2311	9.1	2606	9.1	2978	8.
PART C: EXTERNAL WALLS	6284	28	7294	28	7349	29	7687	27	8312	2
PART D: FLAT ROOF	0	0	0	0	0	0	0	0	0	
PART E: PITCHED ROOF	3389	15	3896	15	3606	14	4240	15	4986	1
PART F: WINDOWS AND EXTERNAL DOORS	2405	11	2979	11	2978	12	3503	12	3503	1
PART G: INTERNAL PARTITIONS AND DOORS	0	0	0	0	0	0	0	0	2741	
PART H: WALL FINISHES	984	4	1088	4	1088	4	1184	4	1435	
PART J: FLOOR FINISHES	843	4	1040	4	905	4	1132	4	1396	
PART K: CEILING FINISHES	672	3	825	3	711	3	948	3	1130	
PART L: ELECTRICAL WORK	781	3	909	3	815	3	969	3	1089	
PART M: HEATING WORK	1086	5	1120	4	1132	4	1529	5	1552	
PART N: ALTERATION WORK	964	4	964	4	964	4	964	3	964	
Final total £	22719	100	26166	100	25335	100	28558	100	34112	10
Floor area m²	23.76		29.16		25.16		32.56		39.96	
£ per m² £	956		897		1,007		877		854	

	Ref	Qty	Hours	Hours £	Plant £	O & P £	Total £

ALTERNATIVE ITEMS

The following items are alternatives to those listed in the previous extension take-offs and can be substituted to suit the needs of any particular project.

	Ref	Qty	Hours	Hours £	Plant £	O & P £	Total £
Excavate topsoil 150mm thick by machine and deposit on site		m²	0.00	0.00	0.52	0.08	0.60
Excavate to reduce levels by machine and deposit on site		m³	0.00	0.00	2.10	0.32	2.42
Load excavated material by machine into skips and remove to tip		m³	0.00	0.00	18.20	2.73	20.93

	Ref	Qty	Hours	Hours £	Mat'ls £	O & P £	Total £
Common bricks (£175 per 1,000) in leaf of cavity wall		m²	1.70	28.90	16.05	6.74	51.69
Common bricks (£200 per 1,000) in leaf of cavity wall		m²	1.70	28.90	17.36	6.94	53.20
Common bricks (£225 per 1,000) in leaf of cavity wall		m²	1.70	28.90	17.82	7.01	53.73
Facing bricks (£300 per 1,000) in leaf of cavity wall		m²	1.85	31.45	26.04	8.62	66.11
Facing bricks (£350 per 1,000) in leaf of cavity wall		m²	1.85	31.45	29.12	9.09	69.66
Facing bricks (£400 per 1,000) in leaf of cavity wall		m²	1.85	31.45	32.41	9.58	73.44

	Qty	Hours	Hours £	Mat'ls £	O & P £	Total £
Marley Modern roofing tiles smooth finish, size 420 x 330mm 75mm lap, type 1F reinforced roofing felt and 38 x 19mm sawn softwood battens	m²	0.59	10.03	11.89	3.29	25.21
Marley Mendip roofing tiles smooth finish, size 420 x 330mm 75mm lap, type 1F reinforced roofing felt and 25 x 19mm sawn softwood battens	m²	0.59	10.03	11.27	3.20	24.50
100mm diameter cast iron gutter	m	0.36	6.12	11.92	2.71	20.75
Stop end	nr	0.18	3.06	2.87	0.89	6.82
Stop end outlet	nr	0.18	3.06	9.04	1.82	13.92
65mm diameter cast iron down pipe	m	0.25	4.25	21.80	3.91	29.96
Shoe	nr	0.25	4.25	17.92	3.33	25.50
Quarry floor tiles, size 150 x 150 x 12.5mm	m²	1.00	17.00	31.82	7.32	56.14
Ceramic floor tiles, size 150 x 150 x 12.5mm	m²	0.90	15.30	24.07	5.91	45.28
Vinyl floor floor sheeting 2.5mm thick with welded joints	m²	0.40	6.80	14.12	3.14	24.06
Acrylic reinforced bath1700mm long complete with chromium plated grip handles, pair of taps, waste fitting, overflow, chain, plug, trap and bath panels	nr	3.50	63.00	251.67	47.20	361.87
Vitreous china wash basin size 540 x 430mm, complete with pair of taps, waste fitting, overflow, chain, plug, trap and pedestal	nr	2.00	36.00	132.41	25.26	193.67

	Qty	Hours	Hours £	Mat'ls £	O & P £	Total £
Vitreous china bidet complete	nr	1.50	27.00	287.08	47.11	361.19
Vitreous china WC suite comprising pan, plastic seat and cover, 9 litre cistern and brackets, ball valve and plastic connecting pipe, complete	nr	1.50	25.50	251.16	41.50	318.16
Shower cubicle with plastic tray, aluminium and acrylic sides with opening door and surface-mounted thermostatically-controlled shower fitting	nr	1.50	25.50	742.11	115.14	882.75

Prepare, size and hang wallpaper to walls with adhesive

	Qty	Hours	Hours £	Mat'ls £	O & P £	Total £
lining paper, £1.00 per roll	m²	0.25	4.25	0.24	0.67	5.16
lining paper, £1.10 per roll	m²	0.25	4.25	0.26	0.68	5.19
lining paper, £1.20 per roll	m²	0.25	4.25	0.28	0.68	5.21
lining paper, £1.30 per roll	m²	0.25	4.25	0.30	0.68	5.23
washable paper, £3.50 per roll	m²	0.30	5.10	0.82	0.89	6.81
washable paper, £4.50 per roll	m²	0.30	5.10	1.06	0.92	7.08
washable paper, £5.50 per roll	m²	0.30	5.10	1.30	0.96	7.36
embossed paper, £4.00 per roll	m²	0.32	5.44	1.06	0.98	7.48
embossed paper, £5.00 per roll	m²	0.32	5.44	1.18	0.99	7.61
embossed paper, £6.00 per roll	m²	0.32	5.44	1.30	1.01	7.75
embossed paper, £7.00 per roll	m²	0.32	5.44	1.42	1.03	7.89

Prepare, size and hang wallpaper to ceilings with adhesive

	Qty	Hours	Hours £	Mat'ls £	O & P £	Total £
lining paper, £1.00 per roll	m²	0.30	5.10	0.24	0.80	6.14
lining paper, £1.10 per roll	m²	0.30	5.10	0.26	0.80	6.16
lining paper, £1.20 per roll	m²	0.30	5.10	0.28	0.81	6.19
lining paper, £1.30 per roll	m²	0.30	5.10	0.30	0.81	6.21
washable paper, £3.50 per roll	m²	0.35	5.95	0.82	1.02	7.79
washable paper, £4.50 per roll	m²	0.35	5.95	1.06	1.05	8.06
washable paper, £5.50 per roll	m²	0.35	5.95	1.30	1.09	8.34
embossed paper, £4.00 per roll	m²	0.37	6.29	1.06	1.10	8.45
embossed paper, £5.00 per roll	m²	0.37	6.29	1.18	1.12	8.59
embossed paper, £6.00 per roll	m²	0.37	6.29	1.30	1.14	8.73
embossed paper, £7.00 per roll	m²	0.37	6.29	1.42	1.16	8.87

Part Two

LOFT CONVERSIONS

Standard items

Drawings

Loft conversions

      4.5 x 4.5m

      4.5 x 5.5m

      4.5 x 6.5m

Summary of loft conversion costs

	Ref	Unit	Hours	Hours £	Mat'ls £	O & P £	Total £

PART A
PRELIMINARIES

	Ref	Unit	Hours	Hours £	Mat'ls £	O & P £	Total £
Small tools	A1	wk					40.00
Scaffolding (m²/weeks)	A2	m²/wk					6.00
Skip	A3	wk					110.00
Clean up	A4	hour					13.00

PART B
PREPARATION

	Ref	Unit	Hours	Hours £	Mat'ls £	O & P £	Total £
Clear out roof space	B1	item	4.00	52.00	0.00	7.80	59.80
Remove carpet from bedroom	B2	item	2.00	26.00	0.00	3.90	29.90
Erect temporary screen (10 m²) consisting of 75 x 50mm covered both sides with polythene sheeting	B3	item	6.00	102.00	14.72	17.51	134.23
Disconnect pipes from cold water storage tank, move tank to new position and reconnect pipes	B4	item	1.90	34.20	5.21	5.91	45.32

PART C
DORMER WINDOW

	Ref	Unit	Hours	Hours £	Mat'ls £	O & P £	Total £
Remove clay tiles or slates, felt and battens and remove	C1	m²	0.75	12.75	0.00	1.91	14.66
Cut into rafters and purlin to form new opening size 2000 x 3000mm, trim with 150 x 50mm sawn softwood bearers	C2	nr	14.00	238.00	42.92	42.14	323.06
Cut into rafters and purlin to form new opening size 2600 x 3000mm, trim with 150 x 50mm sawn softwood bearers	C3	nr	16.00	272.00	47.66	47.95	367.61

	Ref	Unit	Hours	Hours £	Mat'ls £	O & P £	Total £
200 x 75mm sawn softwood purlin	C4	m	0.20	3.40	2.99	0.96	7.35
Cut and pin end of purlin to existing brickwork	C5	nr	1.00	17.00	2.91	2.99	22.90
50 x 75mm sawn softwood sole plate	C6	m	0.22	3.74	1.34	0.76	5.84
50 x 75mm sawn softwood head	C7	m	0.22	3.74	1.34	0.76	5.84
50 x 75mm sawn softwood studs	C8	m	0.28	4.76	1.34	0.92	7.02
50 x 75mm sawn joists	C9	m	0.20	3.40	1.34	0.71	5.45
Mild steel bolt, M10 x 100mm	C10	nr	0.15	2.55	1.72	0.64	4.91
18mm thick WPB grade plywood roof decking fixed to roof joists	C11	m²	0.90	15.30	10.47	3.87	29.64
50mm wide sawn softwood tapered firring pieces average depth 50mm	C12	m	0.18	3.06	1.91	0.75	5.72
Crown Wool insulation or similar 100mm thick fixed between joists with chicken wire and 150mm thick layer laid over joists	C13	m²	0.48	8.16	11.60	2.96	22.72
19mm thick wrought softwood fascia board 200mm high	C14	m	0.50	8.50	2.71	1.68	12.89
6mm thick asbestos-free insulation board soffit 150mm wide	C15	m	0.40	6.80	2.11	1.34	10.25
Three-layer polyester-base mineral-surfaced roofing felt	C16	m²	0.55	9.35	13.04	3.36	25.75
Turn down to edge of roof	C17	m	0.10	1.70	1.40	0.47	3.57
Lead flashing code 5, 200mm girth dressing under existing tiles and over new dormer roof	C18	m	0.60	10.20	8.62	6.68	25.50

	Ref	Unit	Hours	Hours £	Mat'ls £	O & P £	Total £
Lead stepped flashing code 5, 200mm girth dressing under vertical tiles and over existing roof	C19	m	0.90	15.30	8.62	6.68	30.60
Marley Plain roofing tiles size 278 x 165mm, 65mm lap, type 1F reinforced fel, 38 x 19mm battens hung, vertically	C20	m²	1.90	32.30	36.24	10.28	78.82
Raking cutting on tiling	C21	m	0.15	2.55	0.00	0.38	2.93
Make good existing tiling up to new dormer	C22	m	0.18	3.06	0.00	0.46	3.52
PVC-U window size 1800 x 1200mm complete	C23	nr	2.00	34.00	721.18	113.28	868.46
PVC-U window size 2400 x 1200mm complete	C24	nr	2.30	39.10	801.36	126.07	966.53
25 x 225mm wrought softwood window board	C25	m	0.30	5.10	4.02	1.37	10.49
25 x 150mm wrought softwood lining	C26	m	0.24	4.08	3.64	1.16	8.88
PVC-U gutter, 112mm half round with gutter union joints, fixed to wrought softwood fascia board with support brackets at 1m maximum centres	C27	m	0.26	4.42	3.24	1.15	8.81
Extra for stop end	C28	nr	0.14	2.38	2.12	0.68	5.18
Extra for stop end outlet	C29	nr	0.14	2.38	4.19	0.99	7.56
PVC-U down pipe 68mm diameter, loose spigot and socket joints, plugged to faced brickwork with pipe clips at 2m centres	C30	m	0.25	4.25	5.70	1.49	11.44
Extra for shoe	C31	nr	0.30	5.10	3.15	1.24	9.49

	Ref	Unit	Hours	Hours £	Mat'ls £	O & P £	Total £
Plasterboard 9.5mm thick fixed to softwood studding, joints filled with filler and taped to receive decoration	C32	m²	0.36	6.12	2.67	1.32	10.11
Plasterboard 9.5mm thick fixed to softwood joists, joints filled with filler and taped to receive decoration	C33	m²	0.36	6.12	2.67	1.32	10.11
One coat skim plaster to plaster-board ceiling including scrimming joints	C34	m²	0.50	8.50	1.74	1.54	11.78
Two coats emulsion paint to plasterboard walls and ceilings	C35	m²	0.36	6.12	0.96	1.06	8.14
Apply one coat primer, one oil-based undercoat and one coat gloss paint on surfaces not exceeding 300mm girth	C36	m	0.20	3.40	1.60	0.75	5.75

PART D
ROOF WINDOW

	Ref	Unit	Hours	Hours £	Mat'ls £	O & P £	Total £
Remove clay tiles or slates, felt and battens and remove	D1	m²	0.75	12.75	0.00	1.91	14.66
Cut into rafters and purlin to form new opening size 880 x 1080mm, trim with 150 x 50mm sawn softwood bearers	D2	nr	3.00	51.00	39.14	13.52	103.66
Make good existing tiling up to new dormer	D3	m	0.18	3.06	0.00	0.46	3.52
Velux window size 780 x 980mm complete with flashings	D4	nr	3.25	55.25	510.22	84.82	650.29
25 x 150mm wrought softwood lining	D5	m	0.24	4.08	3.64	1.16	8.88

	Ref	Unit	Hours	Hours £	Mat'ls £	O & P £	Total £
Apply one coat primer, one oil-based undercoat and one coat gloss paint on surfaces not exceeding 300mm girth	D6	m	0.20	3.40	1.60	0.75	5.75

PART E
STAIRS

	Ref	Unit	Hours	Hours £	Mat'ls £	O & P £	Total £
Break into existing ceiling joists and plasterboard ceiling, trim with 150 x 50mm trimmer including temporary supports, to form new opening size 2750 x 1200mm	E1	nr	3.00	51.00	18.29	10.39	79.68
Make good plasterboard ceiling up to new opening	E2	m	0.20	3.40	0.74	0.62	4.76
Wrought softwood straight-flight staircase with 13 close treads, 2700mm going, 2600mm rise complete with 38 x 200mm strings, 66 x 66 x 1350mm newel post, 63 x 44mm handrail, balusters 848mm high and spacers	E3	nr	32.00	544.00	647.11	178.67	1,369.78
25 x 150mm wrought softwood lining	E4	m	0.24	4.08	3.64	1.16	8.88
Apply one coat primer, one oil-based undercoat and one coat gloss paint on surfaces exceeding 300mm girth	E5	m²	0.70	11.90	1.60	2.03	15.53

PART F
FLOORING

	Ref	Unit	Hours	Hours £	Mat'ls £	O & P £	Total £
25mm thick tongued and grooved wrought softwood flooring	F1	m²	0.74	12.58	12.48	3.76	28.82
19 x 100mm wrought softwood skirting	F2	m	0.17	2.89	2.51	0.81	6.21

	Ref	Unit	Hours	Hours £	Mat'ls £	O & P £	Total £
Apply one coat primer, one oil-based undercoat and one coat gloss paint on skirting not exceeding 300mm girth	F3	m	0.20	3.40	0.92	0.65	4.97

PART G
INTERNAL
PARTITIONS AND
DOORS

	Ref	Unit	Hours	Hours £	Mat'ls £	O & P £	Total £
50 x 75mm sawn softwood sole plate	G1	m	0.22	3.74	1.34	0.76	5.84
50 x 75mm sawn softwood head	G2	m	0.22	3.74	1.34	0.76	5.84
50 x 75mm sawn softwood studs	G3	m	0.28	4.76	1.34	0.92	7.02
50 x 75mm sawn softwood noggings	G4	m	0.28	4.76	1.34	0.92	7.02
Plasterboard 9.5mm thick fixed to softwood studding, filled joints and taped to receive decoration	G5	m²	0.36	6.12	2.67	1.32	10.11
Flush door 35mm thick, size 762 x 1981mm, internal quality, half hour fire check, veneered finish both sides	G6	nr	1.25	21.25	69.91	13.67	104.83
38 x 150mm wrought softwood lining	G7	m	0.22	3.74	4.62	1.25	9.61
13 x 38mm wrought softwood door stop	G8	m	0.20	3.40	0.68	0.61	4.69
19 x 50mm wrought softwood chamfered architrave	G9	m	0.15	2.55	1.34	0.58	4.47
19 x 100mm wrought softwood chamfered skirting	G10	m	0.17	2.89	2.51	0.81	6.21
100mm rising steel butts	G11	pair	0.30	5.10	1.92	1.05	8.07

	Ref	Unit	Hours	Hours £	Mat'ls £	O & P £	Total £
Silver anodised aluminium mortice latch with lever furniture	G12	nr	0.80	13.60	16.22	4.47	34.29
Two coats emulsion paint to plasterboard walls	G13	m²	0.26	4.42	0.96	0.81	6.19
Apply one coat primer, one oil-based undercoat and one coat gloss paint to general surfaces exceeding 300mm girth	G14	m²	0.70	11.90	1.60	2.03	15.53

PART H
CEILINGS AND SOFFITS

	Ref	Unit	Hours	Hours £	Mat'ls £	O & P £	Total £
Plasterboard 9.5mm thick fixed to ceiling joists, joints filled with filler and taped to receive decoration	H1	m²	0.36	6.12	2.67	1.32	10.11
Plasterboard 9.5mm thick fixed to sloping ceiling softwood joists, joints filled with filler and taped to receive decoration	H2	m²	0.36	6.12	2.67	1.32	10.11
One coat skim plaster to plasterboard ceiling and scrim joints	H3	m²	0.50	8.50	1.74	1.54	11.78
Two coats emulsion paint to plasterboard walls and ceilings	H4	m²	0.26	4.42	0.96	0.81	6.19

PART J
WALL FINISHES

	Ref	Unit	Hours	Hours £	Mat'ls £	O & P £	Total £
Plaster, first coat 11mm bonding, second coat 2mm finish to walls	J1	m²	0.39	6.63	2.31	1.34	10.28
Two coats emulsion paint to plasterboard walls and ceilings	J2	m²	0.26	4.42	0.96	0.81	6.19

	Ref	Unit	Hours	Hours £	Mat'ls £	O & P £	Total £
PART K **ELECTRICAL WORK**							
13 amp double switched socket outlet with neon	K1	nr	0.40	7.60	8.81	2.46	18.87
Lighting point	K2	nr	0.35	6.65	7.34	2.10	16.09
Lighting switch 1 way	K3	nr	0.35	6.65	4.42	1.66	12.73
Lighting switch 2 way	K4	nr	0.40	7.60	7.27	2.23	17.10
Lighting wiring	K5	m	0.10	1.90	1.86	0.56	4.32
Power cable	K6	m	0.15	2.85	2.71	0.83	6.39
Three-floor linked smoke alarm system	K7	nr	1.00	19.00	47.20	9.93	76.13
PART L **HEATING WORK**							
15mm copper pipe	L1	m	0.22	3.96	2.15	0.92	7.03
Elbow	L2	nr	0.28	5.04	2.44	1.12	8.60
Tee	L3	nr	0.34	6.12	2.90	1.35	10.37
Radiator, double convector size 1400 x 520mm	L4	nr	1.30	23.40	139.11	24.38	186.89
Break into existing pipe, insert tee	L5	nr	0.75	11.25	4.42	2.35	18.02

Loft conversion size 4.5 × 4.5m (not to scale)

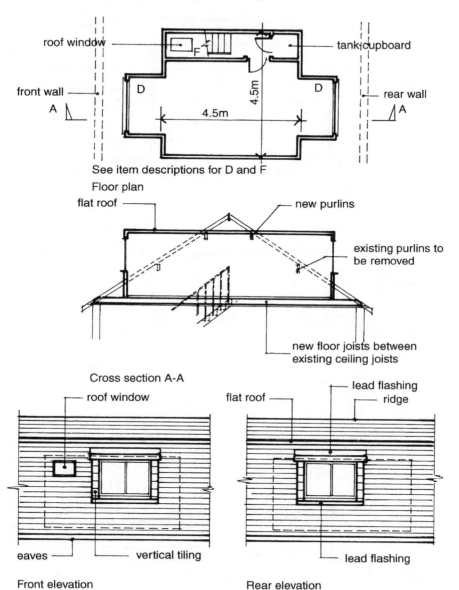

roof window

tank cupboard

front wall

rear wall

A

A

4.5m

4.5m

D

D

F

See item descriptions for D and F

Floor plan

flat roof

new purlins

existing purlins to be removed

new floor joists between existing ceiling joists

Cross section A-A

roof window

flat roof

lead flashing

ridge

eaves

vertical tiling

lead flashing

Front elevation

Rear elevation

Loft conversion size 4.5×5.5m (not to scale)

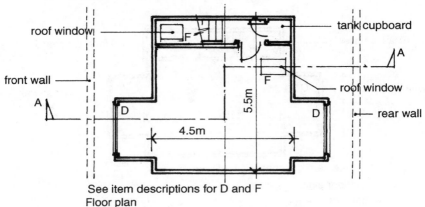

roof window

front wall

A

tank cupboard

A

roof window

rear wall

5.5m

4.5m

D   D

See item descriptions for D and F
Floor plan

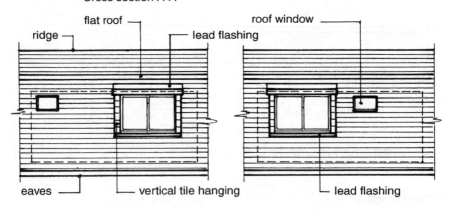

flat roof

existing purlins to
be removed

new purlins

roof window

new floor joists between
existing ceiling joists

Cross section A-A

flat roof

ridge

lead flashing

roof window

eaves   vertical tile hanging

lead flashing

Front elevation   Rear elevation

Loft conversion size 4.5 × 6.5m (not to scale)

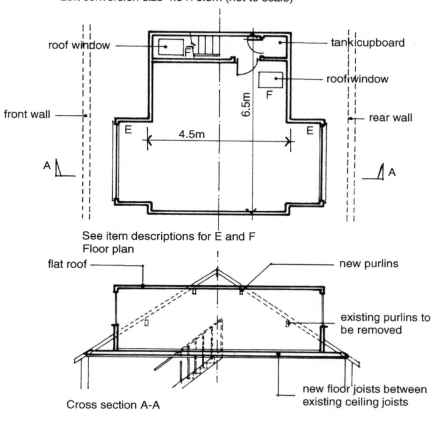

See item descriptions for E and F
Floor plan

Cross section A-A

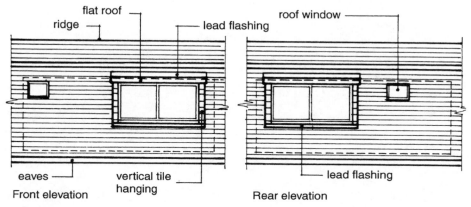

Front elevation

Rear elevation

	Ref	Unit	Qty	Hours	Hours £	Mat'ls £	O & P £	Total £

For full item descriptioms
see pages 367 to 374

PART A
PRELIMINARIES

	Ref	Unit	Qty	Hours	Hours £	Mat'ls £	O & P £	Total £
Small tools	A1	wks	6					300.00
Scaffolding (m²/weeks)	A2		120					720.00
Skip	A3	wks	6					660.00
Clean up	A4	hrs	8					104.00
Carried to summary								1,784.00

PART B
PREPARATION

	Ref	Unit	Qty	Hours	Hours £	Mat'ls £	O & P £	Total £
Clear out roof space	B1		item	4.0	52.00	0.00	7.80	59.80
Remove carpet from bedroom	B2		item	2.0	26.00	0.00	3.90	29.90
Erect temporary screen (10 m²) in bedroom	B3		item	6.0	102.00	147.22	37.38	286.60
Disconnect pipes from cold water storage tank	B4		item	1.9	34.20	5.21	5.91	45.32
Carried to summary				13.9	214.20	152.43	54.99	421.62

PART C
DORMER WINDOW

	Ref	Unit	Qty	Hours	Hours £	Mat'ls £	O & P £	Total £
Remove clay tiles or slates, felt and battens	C1	m²	17.5	12.8	217.43	0.00	32.61	250.04
Cut into rafters and purlin to form new opening size 2000 x 3000mm	C2	nr	1.0	14.0	238.00	42.92	42.14	323.06
Carried forward				26.8	455.43	42.92	74.75	573.10

	Ref	Unit	Qty	Hours	Hours £	Mat'ls £	O & P £	Total £
Brought forward				26.8	455.43	42.92	74.75	573.10
Cut into rafters and purlin to form new opening size 2600 x 3000mm	C3	m	0.0	0.0	0.00	0.00	0.00	0.00
200 x 75mm sawn softwood purlin	C4	m	9.6	2.4	40.80	28.70	10.43	79.93
Cut and pin end of purlin to existing brickwork	C5	nr	4.0	4.0	68.00	11.64	11.95	91.59
50 x 75mm sawn softwood sole plate	C6	m	9.4	2.1	35.36	12.60	7.19	55.15
50 x 75mm sawn softwood head	C7	m	9.4	2.1	35.36	12.60	7.19	55.15
50 x 75mm sawn softwood studs	C8	m	36.0	2.8	47.94	48.24	14.43	110.61
50 x 75mm sawn joists	C9	m	21.0	4.2	71.40	28.14	14.93	114.47
Mild steel bolt, M10 x 100mm	C10	nr	14.0	2.1	35.70	24.08	8.97	68.75
18mm thick WPB grade plywood roof decking	C11	m²	13.0	11.7	198.90	136.11	50.25	385.26
50mm wide sawn softwood tapered firring pieces	C12	m	21.0	3.8	64.26	40.11	15.66	120.03
Crown Wool insulation or similar100mm thick	C13	m²	23.0	20.7	351.90	266.80	92.81	711.51
19mm thick wrought softwood fascia board	C14	m	15.2	7.6	129.20	36.63	24.87	190.71
Carried forward				90.3	1,534.25	688.57	333.42	2,556.24

	Ref	Unit	Qty	Hours	Hours £	Mat'ls £	O & P £	Total £
Brought forward				90.3	1,534.25	688.57	333.42	2,556.24
6mm thick asbestos-free insulation board soffit	C15	m	15.2	6.1	103.36	32.07	20.31	155.75
Three-layer polyester-base roofing felt	C16	m²	13.0	7.2	121.55	169.52	43.66	334.73
Turn down to edge of roof	C17	m	15.2	1.5	25.84	21.28	7.07	54.19
Lead flashing code 5, 200mm girth dressing under existing tiles	C18	m	5.0	3.0	51.00	43.10	14.12	108.22
Lead flashing code 5, 200mm girth dressing under vertical tiles	C19	m	12.8	11.5	195.84	110.34	45.93	352.10
Marley Plain roofing tiles, hung vertically, size 278 x 165mm	C20	m²	10.0	19.0	323.00	362.40	102.81	788.21
Raking cutting on tiling	C21	m	12.8	1.9	32.64	0.00	4.90	37.54
Make good existing tiling up to new dormer	C22	m	15.3	1.9	32.64	0.00	4.90	37.54
PVC-U window size 1800 x 1200mm complete	C23	nr	2.0	4.0	68.00	1,442.36	226.55	1,736.91
PVC-U window size 2400 x 1200mm complete	C24	nr	0.0	0.0	0.00	0.00	0.00	0.00
25 x 225mm wrought softwood window board	C25	m	5.0	1.5	25.50	20.10	6.84	52.44
25 x 150mm wrought softwood lining	C26	m	20.0	4.8	81.60	72.80	23.16	177.56
Carried forward				152.7	2,595.22	2,962.54	833.66	6,391.42

	Ref	Unit	Qty	Hours	Hours £	Mat'ls £	O & P £	Total £
Brought forward				152.7	2,595.22	2,962.54	833.66	6,391.42
PVC-U gutter, 112mm half round	C27	m	5.0	1.3	22.10	16.20	5.75	44.05
Extra for stop end	C28	nr	4.0	0.6	9.52	8.48	2.70	20.70
Extra for stop end outlet	C29	nr	2.0	0.3	4.76	8.38	1.97	15.11
PVC-U down pipe 68mm diameter	C30	m	5.0	1.3	21.25	28.50	7.46	57.21
Extra for shoe	C31	nr	2.0	0.6	10.20	6.30	2.48	18.98
Plasterboard 9.5mm thick fixed to studding	C32	m²	5.4	5.5	94.18	14.42	16.29	124.89
Plasterboard 9.5mm thick fixed to softwood joists	C33	m²	3.0	4.7	79.56	8.01	13.14	100.71
One coat skim plaster to plasterboard ceiling	C34	m²	8.0	14.0	238.00	13.92	37.79	289.71
Two coats emulsion paint to plasterboard walls	C35	m²	8.0	7.3	123.76	7.68	19.72	151.16
One coat primer, one undercoat and one coat gloss paint on surfaces not exceeding 300mm	C36	m²	20.0	4.0	340.00	32.00	55.80	427.80
Carried to summary				192.2	3,538.55	3,106.42	996.75	7,641.72

PART D
ROOF WINDOW

	Ref	Unit	Qty	Hours	Hours £	Mat'ls £	O & P £	Total £
Remove clay tiles or slates, felt and battens	D1	m²	1.2	0.9	14.79	0.00	2.22	17.01
Carried forward				0.9	14.79	0.00	2.22	17.01

	Ref	Unit	Qty	Hours	Hours £	Mat'ls £	O & P £	Total £
Brought forward				0.9	14.79	0.00	2.22	17.01
Cut into rafters and purlin to form new opening size 880 x 1080mm	D2	nr	2.0	6.0	102.00	78.28	27.04	207.32
Make good existing tiling up to new dormer	D3	m	2.1	0.4	6.63	0.00	0.99	7.62
Velux window size 780 x 980mm	D4	nr	2.0	6.5	110.50	1,020.44	169.64	1300.58
25 x 150mm wrought softwood lining	D5	m	3.5	0.8	14.28	12.74	4.05	31.07
One coat primer, one undercoat and one coat gloss paint on surfaces not exceeding 300mm	D6	m	3.5	0.7	11.90	5.63	2.63	20.16
Carried to summary				15.3	260.10	1,117.09	206.58	1,583.77

PART E
STAIRS

	Ref	Unit	Qty	Hours	Hours £	Mat'ls £	O & P £	Total £
Break into existing ceiling joists and ceiling to form new opening size 2750 x 1200mm	E1	nr	1.0	3.0	51.00	18.29	10.39	79.68
Make good plasterboard ceiling up to opening	E2	nr	7.9	0.2	3.40	5.85	1.39	10.63
Softwood staircase 2700mm going, 2600mm	E3	nr	1.0	32.0	544.00	647.11	178.67	1369.78
25 x 150mm wrought softwood lining	E4	m	7.9	1.9	32.30	28.76	9.16	70.21
Carried forward				37.1	630.7	700.0	199.61	1530.31

	Ref	Unit	Qty	Hours	Hours £	Mat'ls £	O & P £	Total £
Brought forward				37.1	630.70	700.00	199.61	1,530.31
One coat primer, one undercoat and one coat gloss paint on surfaces not exceeding 300mm	E5	m	23.0	16.1	273.70	36.80	46.58	357.08
Carried to summary				53.2	904.40	736.80	246.18	1,887.38

PART F
FLOORING

	Ref	Unit	Qty	Hours	Hours £	Mat'ls £	O & P £	Total £
25mm thick tongued and grooved softwood flooring	F1	m²	21.0	15.5	263.33	262.08	78.81	604.22
19 x 100mm wrought softwood skirting	F2	m	22.0	3.7	63.58	55.22	17.82	136.62
One coat primer, one undercoat and one coat gloss paint on surfaces not exceeding 300mm	F3	m	22.0	4.4	74.80	20.24	14.26	109.30
Carried to summary				23.6	401.71	337.54	110.89	850.14

PART G
INTERNAL
PARTITIONS AND
DOORS

	Ref	Unit	Qty	Hours	Hours £	Mat'ls £	O & P £	Total £
50 x 75mm sawn softwood soleplate	G1	m	9.2	2.0	34.34	12.33	7.00	53.67
50 x 75mm sawn softwood head	G2	m	10.1	2.2	37.74	13.53	7.69	58.97
50 x 75mm sawn softwood studs	G3	m	29.8	8.3	141.78	39.93	27.26	208.97
Carried forward				12.6	213.86	65.79	41.95	321.60

	Ref	Unit	Qty	Hours	Hours £	Mat'ls £	O & P £	Total £
Brought forward				12.6	213.86	65.79	41.95	321.60
50 x 75mm sawn softwood noggings	G4	m	9.2	2.6	43.86	12.33	8.43	64.62
Plasterboard 9.5mm thick fixed to softwood studdimg	G5	m²	17.6	6.3	107.44	23.53	19.65	150.62
Flush door 35mm thick size 762 x 1981mm	G6	nr	2.0	2.5	42.50	2.68	6.78	51.96
38 x 150mm wrought softwood lining	G7	m	9.5	2.1	35.53	12.76	7.24	55.53
13 x 38mm wrought softwood door stop	G8	m	9.5	1.4	24.31	12.76	5.56	42.63
19 x 50mm wrought softwood architrave	G9	m	9.5	0.2	2.55	12.76	2.30	17.60
19 x 50mm wrought softwood skirting	G10	m	8.2	1.6	27.54	20.58	7.22	55.34
100mm rising steel butts	G11	pair	2.0	0.6	10.20	3.84	2.11	16.15
Silver anodised aluminium mortice latch	G12	nr	2.0	1.6	27.20	32.44	8.95	68.59
Two coats emulsion paint to plasterboard walls	G13	m²	17.6	4.6	77.69	16.86	14.18	108.73
One coat primer, one undercoat and one coat gloss paint on surfaces not exceeding 300mm	G14	m	6.0	0.7	11.90	9.66	3.23	24.80
Carried to summary				36.7	624.58	225.99	127.58	978.15

	Ref	Unit	Qty	Hours	Hours £	Mat'ls £	O & P £	Total £
PART H **CEILINGS AND SOFFITS**								
Plasterboard 9.5mm thick fixed to ceiling joists	H1	m²	7.7	2.8	46.75	20.43	10.08	77.25
Plasterboard 9.5mm thick fixed to sloping ceiling	H2	m²	30.6	12.2	208.08	81.70	43.47	333.25
One coat skim plaster to plasterboard ceiling	H3	m²	38.3	19.1	325.21	36.72	54.29	416.22
Two coats emulsion paint to plasterboard walls	H4	m²	38.3	10.0	169.15	66.56	35.36	271.06
Carried to summary				44.1	749.19	205.40	143.19	1,097.78
PART J **WALL FINISHES**								
Plaster, first coat 11mm bonding and 2mm finish coat	J1	m²	19.8	7.7	131.41	45.74	26.57	203.72
Two coats emulsion paint to plasterboard	J2	m²	19.8	5.2	87.55	19.01	15.98	122.54
Carried to summary				12.9	218.96	64.75	42.56	326.26
PART K **ELECTRICAL WORK**								
13 amp double switched socket outlet with neon	K1	nr	3.0	1.2	22.80	26.43	7.38	56.61
Lighting point	K2	nr	3.0	1.2	22.80	22.02	6.72	51.54
Carried forward				2.4	45.60	48.45	14.11	108.16

	Ref	Unit	Qty	Hours	Hours £	Mat'ls £	O & P £	Total £
Brought forward				2.4	45.60	48.45	14.11	108.16
Lighting switch 1 way	K3	nr	2.0	0.7	13.30	8.84	3.32	25.46
Lighting switch 2 way	K4	nr	2.0	0.8	15.20	14.54	4.46	34.20
Lighting wiring	K5	m	16.0	1.6	30.40	29.76	9.02	69.18
Power cable	K6	m	12.0	1.8	34.20	32.52	10.01	76.73
Three-floor linked smoke alarm system	K7	nr	1.0	1.0	19.00	47.20	9.93	76.13
Carried to summary				8.3	157.70	181.31	50.85	389.86

PART L
HEATING WORK

	Ref	Unit	Qty	Hours	Hours £	Mat'ls £	O & P £	Total £
15mm copper pipe	L1	m	9.0	2.0	35.64	24.39	9.00	69.03
Elbow	L2	nr	5.0	1.4	25.20	13.55	5.81	44.56
Tee	L3	nr	1.0	0.3	6.12	2.71	1.32	10.15
Radiator, double convector size 1400 x 520mm	L4	nr	2.0	2.6	46.80	5.42	7.83	60.05
Break into existing pipe, insert tee	L5	nr	1.0	0.8	13.50	2.71	2.43	18.64
Carried to summary				7.1	127.26	48.78	26.41	202.45

SUMMARY

	Hours	Hours £	Mat'ls £	O & P £	Total £
PART A **PRELIMINARIES**	0.0	0.00	0.00	0.00	1,784.00
PART B **PREPARATION**	13.9	214.20	152.43	54.99	421.62
PART C **DORMER WINDOW**	192.2	3,538.55	3,106.42	996.75	7,641.72
PART D **ROOF WINDOW**	15.3	260.10	1,117.09	206.58	1,583.77
PART E **STAIRS**	53.2	904.40	736.80	246.18	1,887.38
PART F **FLOORING**	23.6	401.71	337.54	110.89	850.14
PART G **INTERNAL PARTITIONS AND DOORS**	36.7	624.58	225.99	127.58	978.15
PART H **CEILINGS AND SOFFITS**	44.1	749.19	205.40	143.19	1,097.78
PART J **WALL FINISHES**	12.9	218.96	64.75	42.56	326.26
PART K **ELECTRICAL WORK**	8.3	157.70	181.31	50.85	389.86
PART L **HEATING WORK**	7.1	127.26	48.78	26.41	202.45
Final total	407.2	7,196.65	6,176.51	2,005.98	17,163.13

	Ref	Unit	Qty	Hours	Hours £	Mat'ls £	O & P £	Total £

*For full item descriptioms
see pages 367 to 374*

PART A
PRELIMINARIES

	Ref	Unit	Qty	Hours	Hours £	Mat'ls £	O & P £	Total £
Small tools	A1	wks	7					350.00
Scaffolding (m²/weeks)	A2		140					840.00
Skip	A3	wks	7					770.00
Clean up	A4	hrs	8					104.00
Carried to summary								2,064.00

PART B
PREPARATION

	Ref	Unit	Qty	Hours	Hours £	Mat'ls £	O & P £	Total £
Clear out roof space	B1		item	4.0	52.00	0.00	7.80	59.80
Remove carpet from bedroom	B2		item	2.0	26.00	0.00	3.90	29.90
Erect temporary screen (10 m²) in bedroom	B3		item	6.0	102.00	147.22	37.38	286.60
Disconnect pipes from cold water storage tank	B4		item	1.9	34.20	5.21	5.91	45.32
Carried to summary				13.9	214.20	152.43	54.99	421.62

PART C
DORMER WINDOW

	Ref	Unit	Qty	Hours	Hours £	Mat'ls £	O & P £	Total £
Remove clay tiles or slates, felt and battens	C1	m²	17.5	12.8	217.43	0.00	32.61	250.04
Cut into rafters and purlin to form new opening size 2000 x 3000mm	C2	nr	1.0	14.0	238.00	42.92	42.14	323.06
Carried forward				26.8	455.43	42.92	74.75	573.10

390 Loft, size 4.5 x 5.5m

	Ref	Unit	Qty	Hours	Hours £	Mat'ls £	O & P £	Total £
Brought forward				26.8	455.43	42.92	74.75	573.10
Cut into rafters and purlin to form new opening size 2600 x 3000mm	C3	m	0.0	0.0	0.00	0.00	0.00	0.00
200 x 75mm sawn softwood purlin	C4	m	9.6	2.4	40.80	28.70	10.43	79.93
Cut and pin end of purlin to existing brickwork	C5	nr	4.0	4.0	68.00	11.64	11.95	91.59
50 x 75mm sawn softwood sole plate	C6	m	9.4	2.1	35.36	12.60	7.19	55.15
50 x 75mm sawn softwood head	C7	m	9.4	2.1	35.36	12.60	7.19	55.15
50 x 75mm sawn softwood studs	C8	m	36.0	2.8	47.94	48.24	14.43	110.61
50 x 75mm sawn joists	C9	m	21.0	4.2	71.40	28.14	14.93	114.47
Mild steel bolt, M10 x 100mm	C10	nr	14.0	2.1	35.70	24.08	8.97	68.75
18mm thick WPB grade plywood roof decking	C11	m²	13.0	11.7	198.90	136.11	50.25	385.26
50mm wide sawn softwood tapered firring pieces	C12	m	21.0	3.8	64.26	40.11	15.66	120.03
Crown Wool insulation or similar100mm thick	C13	m²	23.0	20.7	351.90	266.80	92.81	711.51
19mm thick wrought softwood fascia board	C14	m	15.2	7.6	129.20	36.63	24.87	190.71
Carried forward				90.3	1,534.25	688.57	333.42	2,556.24

	Ref	Unit	Qty	Hours	Hours £	Mat'ls £	O & P £	Total £
Brought forward				90.3	1,534.25	688.57	333.42	2,556.24
6mm thick asbestos-free insulation board soffit	C15	m	15.2	6.1	103.36	32.07	20.31	155.75
Three-layer polyester-base roofing felt	C16	m²	13.0	7.2	121.55	169.52	43.66	334.73
Turn down to edge of roof	C17	m	15.2	1.5	25.84	21.28	7.07	54.19
Lead flashing code 5, 200mm girth dressing under existing tiles	C18	m	5.0	3.0	51.00	43.10	14.12	108.22
Lead flashing code 5, 200mm girth dressing under vertical tiles	C19	m	12.8	11.5	195.84	110.34	45.93	352.10
Marley Plain roofing tiles, hung vertically, size 278 x 165mm	C20	m²	10.0	19.0	323.00	362.40	102.81	788.21
Raking cutting on tiling	C21	m	12.8	1.9	32.64	0.00	4.90	37.54
Make good existing tiling up to new dormer	C22	m	15.3	1.9	32.64	0.00	4.90	37.54
PVC-U window size 1800 x 1200mm complete	C23	nr	2.0	4.0	68.00	1442.36	226.55	1,736.91
PVC-U window size 2400 x 1200mm complete	C24	nr	0.0	0.0	0.00	0.00	0.00	0.00
25 x 225mm wrought softwood window board	C25	m	5.0	1.5	25.50	20.10	6.84	52.44
25 x 150mm wrought softwood lining	C26	m	20.0	4.8	81.60	72.80	23.16	177.56
Carried forward				152.7	2595.22	2962.54	833.66	6,391.42

	Ref	Unit	Qty	Hours	Hours £	Mat'ls £	O & P £	Total £
Brought forward				152.7	2595.22	2,962.54	833.66	6,391.42
PVC-U gutter, 112mm half round	C27	m	5.0	1.3	22.10	16.20	5.75	44.05
Extra for stop end	C28	nr	4.0	0.6	9.52	8.48	2.70	20.70
Extra for stop end outlet	C29	nr	2.0	0.3	4.76	8.38	1.97	15.11
PVC-U down pipe 68mm diameter	C30	m	5.0	1.3	21.25	28.50	7.46	57.21
Extra for shoe	C31	nr	2.0	0.6	10.20	6.30	2.48	18.98
Plasterboard 9.5mm thick fixed to studding	C32	m²	5.4	5.5	94.18	14.42	16.29	124.89
Plasterboard 9.5mm thick fixed to softwood joists	C33	m²	3.0	4.7	79.56	8.01	13.14	100.71
One coat skim plaster to plasterboard ceiling	C34	m²	8.0	14.0	238.00	13.92	37.79	289.71
Two coats emulsion paint to plasterboard walls	C35	m²	8.0	7.3	123.76	7.68	19.72	151.16
One coat primer, one undercoat and one coat gloss paint on surfaces not exceeding 300mm	C36	m²	20.0	4.0	340.00	32.00	55.80	427.80
Carried to summary				192.2	3538.55	3,106.42	996.75	7,641.72

PART D
ROOF WINDOW

	Ref	Unit	Qty	Hours	Hours £	Mat'ls £	O & P £	Total £
Remove clay tiles or slates, felt and battens	D1	m²	1.2	0.9	14.79	0.00	2.22	17.01
Carried forward				0.9	14.79	0.00	2.22	17.01

	Ref	Unit	Qty	Hours	Hours £	Mat'ls £	O & P £	Total £
Brought forward				0.9	14.79	0.00	2.22	17.01
Cut into rafters and purlin to form new opening size 880 x 1080mm	D2	nr	2.0	6.0	102.00	78.28	27.04	207.32
Make good existing tiling up to new dormer	D3	m	2.1	0.4	6.63	0.00	0.99	7.62
Velux window size 780 x 980mm	D4	nr	2.0	6.5	110.50	1,020.44	169.64	1300.58
25 x 150mm wrought softwood lining	D5	m	3.5	0.8	14.28	12.74	4.05	31.07
One coat primer, one undercoat and one coat gloss paint on surfaces not exceeding 300mm	D6	m	3.5	0.7	11.90	5.63	2.63	20.16
Carried to summary				15.3	260.10	1,117.09	206.58	1,583.77

**PART E
STAIRS**

	Ref	Unit	Qty	Hours	Hours £	Mat'ls £	O & P £	Total £
Break into existing ceiling joists and ceiling to form new opening size 2750 x 1200mm	E1	nr	1.0	3.0	51.00	18.29	10.39	79.68
Make good plasterboard ceiling up to new opening	E2	nr	7.9	0.2	3.40	5.85	1.39	10.63
Softwood staircase 2700mm going, 2600mm	E3	nr	1.0	32.0	544.00	647.11	178.67	1369.78
25 x 150mm wrought softwood lining	E4	m	7.9	1.9	32.30	28.76	9.16	70.21
Carried forward				37.1	630.7	700.0	199.61	1530.31

394   Loft, size 4.5 x 5.5m

	Ref	Unit	Qty	Hours	Hours £	Mat'ls £	O & P £	Total £
Brought forward				37.1	630.70	700.00	199.61	1,530.31
One coat primer, one undercoat and one coat gloss paint on surfaces not exceeding 300mm	E5	m	23.0	16.1	273.70	36.80	46.58	357.08
Carried to summary				53.2	904.40	736.80	246.18	1,887.38

PART F
FLOORING

	Ref	Unit	Qty	Hours	Hours £	Mat'ls £	O & P £	Total £
25mm thick tongued and grooved softwood flooring	F1	m²	25.4	15.5	263.33	316.99	87.05	667.37
19 x 100mm wrought softwood skirting	F2	m	24.0	3.7	63.58	60.24	18.57	142.39
One coat primer, one undercoat and one coat gloss paint on surfaces not exceeding 300mm	F3	m	24.0	4.4	74.80	22.08	14.53	111.41
Carried to summary				23.6	401.71	399.31	120.15	921.18

PART G
INTERNAL
PARTITIONS AND
DOORS

	Ref	Unit	Qty	Hours	Hours £	Mat'ls £	O & P £	Total £
50 x 75mm sawn softwood soleplate	G1	m	11.2	2.0	34.34	15.01	7.40	56.75
50 x 75mm sawn softwood head	G2	m	12.1	2.2	37.74	16.21	8.09	62.05
50 x 75mm sawn softwood studs	G3	m	32.8	8.3	141.78	43.95	27.86	213.59
Carried forward				12.6	213.86	75.17	43.36	332.39

	Ref	Unit	Qty	Hours	Hours £	Mat'ls £	O & P £	Total £
Brought forward				12.6	213.86	75.17	43.36	332.39
50 x 75mm sawn softwood noggings	G4	m	11.2	2.6	43.86	15.01	8.83	67.70
Plasterboard 9.5mm thick fixed to softwood studdimg	G5	m²	18.8	6.3	107.44	25.14	19.89	152.47
Flush door 35mm thick size 762 x 1981mm	G6	nr	2.0	2.5	42.50	2.68	6.78	51.96
38 x 150mm wrought softwood lining	G7	m	9.5	2.1	35.53	12.76	7.24	55.53
13 x 38mm wrought softwood door stop	G8	m	9.5	1.4	24.31	12.76	5.56	42.63
19 x 50mm wrought softwood architrave	G9	m	9.5	0.2	2.55	12.76	2.30	17.60
19 x 50mm wrought softwood skirting	G10	m	8.2	1.6	27.54	20.58	7.22	55.34
100mm rising steel butts	G11	pair	2.0	0.6	10.20	3.84	2.11	16.15
Silver anodised aluminium mortice latch	G12	nr	2.0	1.6	27.20	32.44	8.95	68.59
Two coats emulsion paint to plasterboard walls	G13	m²	18.8	4.6	77.69	18.01	14.35	110.05
One coat primer, one undercoat and one coat gloss paint on surfaces not exceeding 300mm	G14	m	6.0	0.7	11.90	9.66	3.23	24.80
Carried to summary				36.7	624.58	240.81	129.81	995.19

	Ref	Unit	Qty	Hours	Hours £	Mat'ls £	O & P £	Total £
PART H								
CEILINGS AND SOFFITS								
Plasterboard 9.5mm thick fixed to ceiling joists	H1	m²	9.4	2.8	46.75	24.96	10.76	82.47
Plasterboard 9.5mm thick fixed to sloping ceiling	H2	m²	37.4	12.2	208.08	99.86	46.19	354.13
One coat skim plaster to plasterboard ceiling	H3	m²	47.8	19.1	325.21	45.84	55.66	426.71
Two coats emulsion paint to plasterboard walls	H4	m²	47.8	10.0	169.15	83.09	37.84	290.07
Carried to summary				44.1	749.19	253.75	150.44	1,153.38
PART J								
WALL FINISHES								
Plaster, first coat 11mm bonding and 2mm finish coat	J1	m²	19.8	7.7	131.41	45.74	26.57	203.72
Two coats emulsion paint to plasterboard	J2	m²	19.8	5.2	87.55	19.01	15.98	122.54
Carried to summary				12.9	218.96	64.75	42.56	326.26
PART K								
ELECTRICAL WORK								
13 amp double switched socket outlet with neon	K1	nr	3.0	1.2	22.80	26.43	7.38	56.61
Lighting point	K2	nr	3.0	1.2	22.80	22.02	6.72	51.54
Carried forward				2.4	45.60	48.45	14.11	108.16

	Ref	Unit	Qty	Hours	Hours £	Mat'ls £	O & P £	Total £
Brought forward				2.4	45.60	48.45	14.11	108.16
Lighting switch 1 way	K3	nr	2.0	0.7	13.30	8.84	3.32	25.46
Lighting switch 2 way	K4	nr	2.0	0.8	15.20	14.54	4.46	34.20
Lighting wiring	K5	m	16.0	1.6	30.40	29.76	9.02	69.18
Power cable	K6	m	12.0	1.8	34.20	32.52	10.01	76.73
Three-floor linked smoke alarm system	K7	nr	1.0	1.0	19.00	47.20	9.93	76.13
Carried to summary				8.3	157.70	181.31	50.85	389.86

PART L
HEATING WORK

	Ref	Unit	Qty	Hours	Hours £	Mat'ls £	O & P £	Total £
15mm copper pipe	L1	m	9.0	2.0	35.64	24.39	9.00	69.03
Elbow	L2	nr	5.0	1.4	25.20	13.55	5.81	44.56
Tee	L3	nr	1.0	0.3	6.12	2.71	1.32	10.15
Radiator, double convector size 1400 x 520mm	L4	nr	2.0	2.6	46.80	5.42	7.83	60.05
Break into existing pipe, insert tee	L5	nr	1.0	0.8	13.50	2.71	2.43	18.64
Carried to summary				7.10	127.26	48.78	26.39	202.43

SUMMARY

	Hours	Hours £	Mat'ls £	O & P £	Total £
PART A **PRELIMINARIES**	0.0	0.00	0.00	0.00	2,064.00
PART B **PREPARATION**	13.9	214.20	152.43	54.99	421.62
PART C **DORMER WINDOW**	192.2	3,538.55	3,106.42	996.75	7,641.72
PART D **ROOF WINDOW**	15.3	260.10	1,117.09	206.58	1,583.77
PART E **STAIRS**	53.2	904.40	736.80	246.18	1,887.38
PART F **FLOORING**	23.6	401.71	399.31	120.15	921.18
PART G **INTERNAL PARTITIONS AND** **DOORS**	36.7	624.58	240.81	129.81	995.19
PART H **CEILINGS AND SOFFITS**	44.1	749.19	253.75	150.44	1,153.38
PART J **WALL FINISHES**	12.9	218.96	64.75	42.56	326.26
PART K **ELECTRICAL WORK**	8.3	157.70	181.31	50.85	389.86
PART L **HEATING WORK**	7.1	127.26	48.78	26.39	202.43
Final total	407.2	7,196.65	6,301.45	2,024.70	17586.79

	Ref	Unit	Qty	Hours	Hours £	Mat'ls £	O & P £	Total £

*For full item descriptioms
see pages 367 to 374*

PART A
PRELIMINARIES

	Ref	Unit	Qty	Hours	Hours £	Mat'ls £	O & P £	Total £
Small tools	A1	wks	8					400.00
Scaffolding (m²/weeks)	A2		160					960.00
Skip	A3	wks	8					880.00
Clean up	A4	hrs	8					104.00
Carried to summary								2,344.00

PART B
PREPARATION

	Ref	Unit	Qty	Hours	Hours £	Mat'ls £	O & P £	Total £
Clear out roof space	B1		item	4.0	52.00	0.00	7.80	59.80
Remove carpet from bedroom	B2		item	2.0	26.00	0.00	3.90	29.90
Erect temporary screen (10 m²) in bedroom	B3		item	6.0	102.00	147.22	37.38	286.60
Disconnect pipes from cold water storage tank	B4		item	1.9	34.20	5.21	5.91	45.32
Carried to summary				13.9	214.20	152.43	54.99	421.62

PART C
DORMER WINDOW

	Ref	Unit	Qty	Hours	Hours £	Mat'ls £	O & P £	Total £
Remove clay tiles or slates, felt and battens	C1	m²	21.7	12.8	217.43	0.00	32.61	250.04
Cut into rafters and purlin to form new opening size 2000 x 3000mm	C2	nr	0.0	0.0	0.00	0.00	0.00	0.00
Carried forward				12.8	217.43	0.00	32.61	250.04

400 Loft, size 4.5 x 6.5m

	Ref	Unit	Qty	Hours	Hours £	Mat'ls £	O & P £	Total £
Brought forward				12.8	217.43	0.00	32.61	250.04
Cut into rafters and purlin to form new opening size 2600 x 3000mm	C3	m	1.0	2.4	40.80	2.99	6.57	50.36
200 x 75mm sawn softwood purlin	C4	m	13.6	2.4	40.80	40.66	12.22	93.68
Cut and pin end of purlin to existing brickwork	C5	nr	4.0	4.0	68.00	11.64	11.95	91.59
50 x 75mm sawn softwood sole plate	C6	m	9.4	2.1	35.36	12.60	7.19	55.15
50 x 75mm sawn softwood head	C7	m	9.4	2.1	35.36	12.60	7.19	55.15
50 x 75mm sawn softwood studs	C8	m	36.0	2.8	47.94	48.24	14.43	110.61
50 x 75mm sawn joists	C9	m	27.0	4.2	71.40	36.18	16.14	123.72
Mild steel bolt, M10 x 100mm	C10	nr	18.0	2.1	35.70	30.96	10.00	76.66
18mm thick WPB grade plywood roof decking	C11	m²	16.6	11.7	198.90	174.22	55.97	429.09
50mm wide sawn softwood tapered firring pieces	C12	m	27.0	3.8	64.26	51.57	17.37	133.20
Crown Wool insulation or similar100mm thick	C13	m²	26.6	20.7	351.90	309.02	99.14	760.06
19mm thick wrought softwood fascia board	C14	m	18.0	7.6	129.20	43.38	25.89	198.47
Carried forward				78.7	1,337.05	774.06	316.67	2,427.78

	Ref	Unit	Qty	Hours	Hours £	Mat'ls £	O & P £	Total £
Brought forward				78.7	1,337.05	774.06	316.67	2,427.78
6mm thick asbestos-free insulation board soffit	C15	m	18.0	6.1	103.36	37.98	21.20	162.54
Three-layer polyester-base roofing felt	C16	m²	16.6	7.2	121.55	216.99	50.78	389.32
Turn down to edge of roof	C17	m	18.0	1.5	25.84	25.20	7.66	58.70
Lead flashing code 5, 200mm girth dressing under existing tiles	C18	m	6.4	3.0	51.00	55.17	15.93	122.09
Lead flashing code 5, 200mm girth dressing under vertical tiles	C19	m	12.8	11.5	195.84	110.34	45.93	352.10
Marley Plain roofing tiles, hung vertically, size 278 x 165mm	C20	m²	10.0	19.0	323.00	362.40	102.81	788.21
Raking cutting on tiling	C21	m	12.8	1.9	32.64	0.00	4.90	37.54
Make good existing tiling up to new dormer	C22	m	16.0	1.9	32.64	0.00	4.90	37.54
PVC-U window size 1800 x 1200mm complete	C23	nr	0.0	0.0	0.00	0.00	0.00	0.00
PVC-U window size 2400 x 1200mm complete	C24	nr	2.0	1.9	32.64	1,602.72	245.30	1,880.66
25 x 225mm wrought softwood window board	C25	m	6.4	1.5	25.50	25.73	7.68	58.91
25 x 150mm wrought softwood lining	C26	m	22.4	4.8	81.60	81.54	24.47	187.61
Carried forward				139.0	2,362.66	3,292.11	848.22	6,502.99

	Ref	Unit	Qty	Hours	Hours £	Mat'ls £	O & P £	Total £
Brought forward				139.0	2,362.66	3,292.11	848.22	6,502.99
PVC-U gutter, 112mm half round	C27	m	6.4	1.3	22.10	20.74	6.43	49.26
Extra for stop end	C28	nr	4.0	0.6	9.52	8.48	2.70	20.70
Extra for stop end outlet	C29	nr	2.0	0.3	4.76	8.38	1.97	15.11
PVC-U down pipe 68mm diameter	C30	m	5.0	1.3	21.25	28.50	7.46	57.21
Extra for shoe	C31	nr	2.0	0.6	10.20	6.30	2.48	18.98
Plasterboard 9.5mm thick fixed to studding	C32	m²	15.4	5.5	94.18	41.12	20.29	155.59
Plasterboard 9.5mm thick fixed to softwood joists	C33	m²	16.4	4.7	79.56	43.79	18.50	141.85
One coat skim plaster to plasterboard ceiling	C34	m²	31.6	14.0	238.00	55.05	43.96	337.01
Two coats emulsion paint to plasterboard walls	C35	m²	31.6	7.3	123.76	30.37	23.12	177.25
One coat primer, one undercoat and one coat gloss paint on surfaces not exceeding 300mm	C36	m²	22.4	4.0	380.80	35.84	62.50	479.14
Carried to summary				178.5	3,346.79	3,570.68	1,037.62	7,955.10

PART D
ROOF WINDOW

	Ref	Unit	Qty	Hours	Hours £	Mat'ls £	O & P £	Total £
Remove clay tiles or slates, felt and battens	D1	m²	1.2	0.9	14.79	0.00	2.22	17.01
Carried forward				0.9	14.79	0.00	2.22	17.01

	Ref	Unit	Qty	Hours	Hours £	Mat'ls £	O & P £	Total £
Brought forward				0.9	14.79	0.00	2.22	17.01
Cut into rafters and purlin to form new opening size 880 x 1080mm	D2	nr	2.0	6.0	102.00	78.28	27.04	207.32
Make good existing tiling up to new dormer	D3	m	2.2	0.4	6.63	0.00	0.99	7.62
Velux window size 780 x 980mm	D4	nr	2.0	6.5	110.50	1,020.44	169.64	1300.58
25 x 150mm wrought softwood lining	D5	m	3.5	0.8	14.28	12.81	4.06	31.16
One coat primer, one undercoat and one coat gloss paint on surfaces not exceeding 300mm	D6	m	3.5	0.7	11.90	5.63	2.63	20.16
Carried to summary				15.3	260.10	1,117.16	206.59	1,583.85

PART E
STAIRS

	Ref	Unit	Qty	Hours	Hours £	Mat'ls £	O & P £	Total £
Break into existing ceiling joists and ceiling to form new opening size 2750 x 1200mm	E1	nr	1.0	3.0	51.00	18.29	10.39	79.68
Make good plasterboard ceiling up to new opening	E2	nr	7.9	0.2	3.40	5.85	1.39	10.63
Softwood staircase 2700mm going, 2600mm	E3	nr	1.0	32.0	544.00	647.11	178.67	1369.78
25 x 150mm wrought softwood lining	E4	m	7.9	1.9	32.30	28.76	9.16	70.21
Carried forward				37.1	630.7	700.0	199.61	1,530.31

404   Loft, size 4.5 x 6.5m

	Ref	Unit	Qty	Hours	Hours £	Mat'ls £	O & P £	Total £
Brought forward				37.1	630.70	700.00	199.61	1,530.31
One coat primer, one undercoat and one coat gloss paint on surfaces not exceeding 300mm	E5	m	23.0	16.1	273.70	36.80	46.58	357.08
Carried to summary				53.2	904.40	736.80	246.18	1,887.38

PART F
FLOORING

	Ref	Unit	Qty	Hours	Hours £	Mat'ls £	O & P £	Total £
25mm thick tongued and grooved softwood flooring	F1	m²	31.9	15.5	263.33	398.24	99.24	760.80
19 x 100mm wrought softwood skirting	F2	m	26.0	3.7	63.58	65.26	19.33	148.17
One coat primer, one undercoat and one coat gloss paint on surfaces not exceeding 300mm	F3	m	26.0	4.4	74.80	23.92	14.81	113.53
Carried to summary				23.6	401.71	487.42	133.37	1,022.50

PART G
INTERNAL
PARTITIONS AND
DOORS

	Ref	Unit	Qty	Hours	Hours £	Mat'ls £	O & P £	Total £
50 x 75mm sawn softwood soleplate	G1	m	13.2	2.0	34.34	17.69	7.80	59.83
50 x 75mm sawn softwood head	G2	m	14.1	2.2	37.74	18.89	8.50	65.13
50 x 75mm sawn softwood studs	G3	m	33.5	8.3	141.78	44.89	28.00	214.67
Carried forward				12.6	213.86	81.47	44.30	339.63

	Ref	Unit	Qty	Hours	Hours £	Mat'ls £	O & P £	Total £
Brought forward				12.6	213.86	81.47	44.30	339.63
50 x 75mm sawn softwood noggings	G4	m	13.2	2.6	43.86	17.69	9.23	70.78
Plasterboard 9.5mm thick fixed to softwood studdimg	G5	m²	20.0	6.3	107.44	26.75	20.13	154.31
Flush door 35mm thick size 762 x 1981mm	G6	nr	2.0	2.5	42.50	2.68	6.78	51.96
38 x 150mm wrought softwood lining	G7	m	9.5	2.1	35.53	12.76	7.24	55.53
13 x 38mm wrought softwood door stop	G8	m	9.5	1.4	24.31	12.76	5.56	42.63
19 x 50mm wrought softwood architrave	G9	m	9.5	0.2	2.55	12.76	2.30	17.60
19 x 50mm wrought softwood skirting	G10	m	8.2	1.6	27.54	20.58	7.22	55.34
100mm rising steel butts	G11	pair	2.0	0.6	10.20	3.84	2.11	16.15
Silver anodised aluminium mortice latch	G12	nr	2.0	1.6	27.20	32.44	8.95	68.59
Two coats emulsion paint to plasterboard walls	G13	m²	20.0	4.6	77.69	19.16	14.53	111.38
One coat primer, one undercoat and one coat gloss paint on surfaces not exceeding 300mm	G14	m	6.0	0.7	11.90	9.66	3.23	24.80
Carried to summary				36.7	624.58	252.54	131.57	1,008.69

	Ref	Unit	Qty	Hours	Hours £	Mat'ls £	O & P £	Total £
PART H **CEILINGS AND SOFFITS**								
Plasterboard 9.5mm thick fixed to ceiling joists	H1	m²	11.1	2.8	46.75	29.50	11.44	87.69
Plasterboard 9.5mm thick fixed to sloping ceiling	H2	m²	44.2	12.2	208.08	118.01	48.91	375.01
One coat skim plaster to plasterboard ceiling	H3	m²	55.3	19.1	325.21	53.04	56.74	434.99
Two coats emulsion paint to plasterboard walls	H4	m²	55.3	10.0	169.15	96.14	39.79	305.08
Carried to summary				44.1	749.19	296.69	156.88	1,202.76
PART J **WALL FINISHES**								
Plaster, first coat 11mm bonding and 2mm finish coat	J1	m²	19.8	7.7	131.41	45.74	26.57	203.72
Two coats emulsion paint to plasterboard	J2	m²	19.8	5.2	87.55	19.01	15.98	122.54
Carried to summary				12.9	218.96	64.75	42.56	326.26
PART K **ELECTRICAL WORK**								
13 amp double switched socket outlet with neon	K1	nr	3.0	1.2	22.80	26.43	7.38	56.61
Lighting point	K2	nr	4.0	1.2	22.80	29.36	7.82	59.98
Carried forward				2.4	45.60	55.79	15.21	116.60

	Ref	Unit	Qty	Hours	Hours £	Mat'ls £	O & P £	Total £
Brought forward				2.4	45.60	55.79	15.21	116.60
Lighting switch 1 way	K3	nr	2.0	0.7	13.30	8.84	3.32	25.46
Lighting switch 2 way	K4	nr	2.0	0.8	15.20	14.54	4.46	34.20
Lighting wiring	K5	m	18.0	1.6	30.40	33.48	9.58	73.46
Power cable	K6	m	14.0	1.8	34.20	37.94	10.82	82.96
Three-floor linked smoke alarm system	K7	nr	1.0	1.0	19.00	47.20	9.93	76.13
Carried to summary				8.3	157.70	197.79	53.32	408.81

PART L
HEATING WORK

	Ref	Unit	Qty	Hours	Hours £	Mat'ls £	O & P £	Total £
15mm copper pipe	L1	m	9.0	2.0	35.64	24.39	9.00	69.03
Elbow	L2	nr	5.0	1.4	25.20	13.55	5.81	44.56
Tee	L3	nr	1.0	0.3	6.12	2.71	1.32	10.15
Radiator, double convector size 1400 x 520mm	L4	nr	2.0	2.6	46.80	5.42	7.83	60.05
Break into existing pipe, insert tee	L5	nr	1.0	0.8	13.50	2.71	2.43	18.64
Carried to summary				7.1	127.26	48.78	26.41	202.45

SUMMARY

	Hours	Hours £	Mat'ls £	O & P £	Total £
PART A **PRELIMINARIES**	0.0	0.00	0.00	0.00	2,344.00
PART B **PREPARATION**	13.9	214.20	152.43	54.99	421.62
PART C **DORMER WINDOW**	178.5	3,346.79	3,570.68	1,037.62	7,955.10
PART D **ROOF WINDOW**	15.3	260.10	1,117.16	206.59	1,583.85
PART E **STAIRS**	53.2	904.40	736.80	246.18	1,887.38
PART F **FLOORING**	23.6	401.71	487.42	133.37	1,022.50
PART G **INTERNAL PARTITIONS AND DOORS**	36.7	624.58	252.54	131.57	1,008.69
PART H **CEILINGS AND SOFFITS**	44.1	749.19	296.69	156.88	1,202.76
PART J **WALL FINISHES**	12.9	218.96	64.75	42.56	326.26
PART K **ELECTRICAL WORK**	8.3	157.70	197.79	53.32	408.81
PART L **HEATING WORK**	7.1	127.26	48.78	26.41	202.45
Final total	393.6	7,004.89	6,925.04	2,089.49	18363.42

SUMMARY OF LOFT CONVERSION COSTS

Loft conversion size

	4.5 x 4.5m		4.5 x 5.5m		4.5 x 6.5m	
	£	%	£	%	£	%
PART A **PRELIMINARIES**	1,784	10	2,064	18	2,344	13
PART B **PREPARATION**	422	2	422	2	422	2
PART C **DORMER WINDOW**	7,642	45	7,642	37	7,955	42
PART D **ROOF WINDOW**	1,584	9	1,584	9	1,584	9
PART E **STAIRS**	1,887	11	1,887	11	1,887	10
PART F **FLOORING**	850	5	921	5	1,022	6
PART G **INTERNAL PARTITIONS** **AND DOORS**	978	7	995	6	1,009	6
PART H **CEILINGS AND SOFFITS**	1,098	6	1,153	7	1,203	7
PART J **WALL FINISHES**	326	2	326	2	326	2
PART K **ELECTRICAL WORK**	390	2	390	2	409	2
PART L **HEATING WORK**	202	1	202	1	202	1
Final total	17,163	100	17,586	100	18,363	100
Floor area (inc. dormers) sq. m	24.21		28.75		34.85	
Cost per square metre £	709		612		527	

Part Three

INSULATION WORK

Standard items

	Unit	Hours	Hours £	Mat'ls £	O & P £	Total £

INSULATION

Quilt insulation

Lightweight glasswool insulation
quilt laid between joists

	Unit	Hours	Hours £	Mat'ls £	O & P £	Total £
60mm thick	m²	0.12	2.04	2.81	0.73	5.58
80mm thick	m²	0.12	2.04	3.50	0.83	6.37
100mm thick	m²	0.12	2.04	4.67	1.01	7.72
150mm thick	m²	0.14	2.38	6.22	1.29	9.89
200mm thick	m²	0.16	2.72	8.02	1.61	12.35

Cavity wall insulation

Semi-rigid glass mineral wool
insulation in cavity walls

	Unit	Hours	Hours £	Mat'ls £	O & P £	Total £
30mm thick	m²	0.12	2.04	2.34	0.66	5.04
40mm thick	m²	0.12	2.04	3.12	0.77	5.93
50mm thick	m²	0.12	2.04	3.61	0.85	6.50
60mm thick	m²	0.14	2.38	3.98	0.95	7.31
75mm thick	m²	0.14	2.38	4.81	1.08	8.27
80mm thick	m²	0.14	2.38	3.37	0.86	6.61
85mm thick	m²	0.16	2.72	5.49	1.23	9.44
100mm thick	m²	0.18	3.06	6.53	1.44	11.03

Cavity closer system 47.5mm
thick in 3m lengths including
stainless steel clips

	Unit	Hours	Hours £	Mat'ls £	O & P £	Total £
50mm wide	m	0.10	1.70	2.96	0.70	5.36
65mm thick	m	0.10	1.70	3.37	0.76	5.83
75mm thick	m	0.10	1.70	3.92	0.84	6.46
85mm thick	m	0.10	1.70	4.18	0.88	6.76
90mm thick	m	0.12	2.04	4.39	0.96	7.39
100mm thick	m	0.12	2.04	4.88	1.04	7.96

	Qty	Hours	Hours £	Mat'ls £	O & P £	Total £

Acoustic insulation

Semi-rigid glass mineral wool
insulation in stud partitions and
roofs

80mm thick	m²	0.22	3.74	4.90	1.30	9.94
90mm thick	m²	0.24	4.08	5.41	1.42	10.91
100mm thick	m²	0.26	4.42	5.80	1.53	11.75
140mm thick	m²	0.28	4.76	9.07	2.07	15.90
150mm thick	m²	0.18	3.06	9.38	1.87	14.31

Mineral wool acoustic blanket
30mm thick fixed to

stud partitions	m²	0.14	2.38	3.17	0.83	6.38
ceilings	m²	0.18	3.06	3.17	0.93	7.16

Rigid rock wool acoustic slab
25mm thick fixed to timber

floors	m²	0.18	3.06	6.70	1.46	11.22

Mineral wool sound-deadening
quilt 25mm thick fixed to

floors	m²	0.12	2.04	3.29	0.80	6.13

Polyethylene sound-deadening
foam blanket quilt under screed

2mm thick	m²	0.08	1.36	2.30	0.55	4.21
5mm thick	m²	0.10	1.70	2.47	0.63	4.80

Thermal insulation

Non-combustible glass mineral
wool quilt fitted between timber
rafters in roof space

60mm thick	m²	0.12	2.04	4.21	0.94	7.19
80mm thick	m²	0.14	2.38	5.42	1.17	8.97
90mm thick	m²	0.14	2.38	5.90	1.24	9.52
100mm thick	m²	0.16	2.72	6.37	1.36	10.45
140mm thick	m²	0.16	2.72	8.95	1.75	13.42
160mm thick	m²	0.18	3.06	10.04	1.97	15.07

	Unit	Hours	Hours £	Mat'ls £	O & P £	Total £

Lagging

Jute felt lagging in strips 100mm
wide secured with galvanised
steel wire to pipe diameter

15mm	m	0.10	1.70	0.38	0.31	2.39
22mm	m	0.11	1.87	0.45	0.35	2.67
28mm	m	0.12	2.04	0.61	0.40	3.05
35mm	m	0.13	2.21	0.64	0.43	3.28
42mm	m	0.14	2.38	0.86	0.49	3.73

Expanded polystryrene lagging
fixed with aluminium bands to pipe
diameter

15mm	m	0.10	1.70	1.12	0.42	3.24
22mm	m	0.11	1.87	1.54	0.51	3.92
28mm	m	0.12	2.04	1.86	0.59	4.49
35mm	m	0.13	2.21	2.80	0.75	5.76
42mm	m	0.14	2.38	3.17	0.83	6.38

Expanded polystryrene lagging
jacket set to bottom and sides of
galvanised steel cisterns, size

10 gallon	nr	0.50	8.50	4.22	1.91	14.63
25 gallon	nr	0.60	10.20	7.80	2.70	20.70

PVC-U insulating jacket 80mm
thick, filled with expanded
polystyrene securing with fixing
bands to hot water cylinder, size

400 x 1050mm	nr	0.80	13.60	16.10	4.46	34.16
450 x 900mm	nr	0.80	13.60	17.32	4.64	35.56
400 x 1200mm	nr	0.80	13.60	18.09	4.75	36.44

Part Four

DAMAGE REPAIRS

Emergency measures

Fire damage

Flood damage

Gale damage

Theft damage

	Unit	Hours	Hours £	Plant £	O & P £	Total £

EMERGENCY MEASURES

Flooding

	Unit	Hours	Hours £	Plant £	O & P £	Total £
Install hired pump and hoses in position	Item	1.00	13.00	0.00	1.95	14.95
Hire diaphragm pump and hoses						
50mm	day	1.00	0.00	55.00	8.25	63.25
75mm	day	1.00	0.00	65.00	9.75	74.75

	Unit	Hours	Hours £	Mat'ls £	O & P £	Total £

Hoardings and screens

	Unit	Hours	Hours £	Mat'ls £	O & P £	Total £
Temporary screens and hoardings 2m high consisting of 22mm thick exterior quality plywood fixed to 100 x 50mm posts and rails	m	2.60	44.20	35.80	12.00	92.00
Hire tarpaulin sheeting size 5 x 4m fixed in position	week	1.00	12.00	18.00	4.50	34.50
Plywood sheeting 18mm thick blocking up window or door opening	m²	1.00	17.00	15.17	4.83	37.00

Shoring

	Unit	Hours	Hours £	Mat'ls £	O & P £	Total £
Timber dead shores consisting of 200 x 200mm shores, 250 x 50mm plates and 200 x 50mm braces at centres of						
2m	m²	3.80	64.60	73.58	20.73	158.91
3m	m²	3.00	51.00	59.64	16.60	127.24
4m	m²	2.20	37.40	41.05	11.77	90.22
Timber raking and flying shores consisting of 200 x 200mm shores, 250 x 50mm plates and 200 x 50mm braces to gable end of two-storey house	Item	60.00	1020.00	527.61	232.14	1779.75

	Unit	Hours	Hours £	Mat'ls £	O & P £	Total £

Access towers

Hire narrow width access tower size
0.85 x 1.8m, height

	Unit	Hours	Hours £	Mat'ls £	O & P £	Total £
7.20m	week	1.00	0.00	200.00	30.00	230.00
6.20m	week	1.00	0.00	180.00	27.00	207.00
5.20m	week	1.00	0.00	160.00	24.00	184.00
4.20m	week	1.00	0.00	140.00	21.00	161.00

Hire full width access tower size
1.45 x 2.5m, height

	Unit	Hours	Hours £	Mat'ls £	O & P £	Total £
7.20m	week	1.00	0.00	280.00	42.00	322.00
6.20m	week	1.00	0.00	260.00	39.00	299.00
5.20m	week	1.00	0.00	220.00	33.00	253.00
4.20m	week	1.00	0.00	200.00	30.00	230.00

	Unit	Hours	Hours £	Mat'ls £	O & P £	Total £

FIRE DAMAGE

Window replacement

Take out existing window, prepare
jambs and cill to receive new
  PVC-U

	Unit	Hours	Hours £	Mat'ls £	O & P £	Total £
size 600 x 900mm	nr	2.00	34.00	238.29	40.84	313.13
size 600 x 1200mm	nr	2.00	34.00	271.15	45.77	350.92
size 1200 x 1200mm	nr	2.50	42.50	332.01	56.18	430.69
size 1800 x 1200mm	nr	3.00	51.00	370.32	63.20	484.52
softwood painted						
size 630 x 900mm	nr	2.00	34.00	155.91	28.49	218.40
size 915 x 900mm	nr	2.00	34.00	165.15	29.87	229.02
size 915 x 1200mm	nr	2.50	42.50	282.01	48.68	373.19
size 1200 x 1200mm	nr	3.00	51.00	301.15	52.82	404.97
hardwood stained						
size 915 x 1050mm	nr	2.00	34.00	224.64	38.80	297.44
size 915 x 1500mm	nr	2.00	34.00	264.27	44.74	343.01
size 1200 x 1500mm	nr	2.50	42.50	302.28	51.72	396.50
size 1770 x 1200mm	nr	3.00	51.00	430.00	72.15	553.15

Take out existing bay window,
prepare jambs and cill to receive
new
  PVC-U

	Unit	Hours	Hours £	Mat'ls £	O & P £	Total £
size 1800 x 900mm	nr	3.00	51.00	308.92	53.99	413.91
size 1800 x 1200mm	nr	3.00	51.00	331.02	57.30	439.32
size 2400 x 900mm	nr	3.50	59.50	461.68	78.18	599.36
size 2400 x 1200mm	nr	3.50	59.50	489.47	82.35	631.32
softwood painted						
size 1800 x 900mm	nr	3.00	51.00	221.03	40.80	312.83
size 1800 x 1200mm	nr	3.00	51.00	250.88	45.28	347.16
size 2400 x 900mm	nr	3.50	59.50	308.87	55.26	423.63
size 2400 x 1200mm	nr	3.50	59.50	341.78	60.19	461.47

	Unit	Hours	Hours £	Mat'ls £	O & P £	Total £

Take out window (cont'd)

hardwood stained

	Unit	Hours	Hours £	Mat'ls £	O & P £	Total £
size 1800 x 900mm	nr	3.00	51.00	306.16	53.57	410.73
size 1800 x 1200mm	nr	3.00	51.00	339.79	58.62	449.41
size 2400 x 900mm	nr	3.50	59.50	361.27	63.12	483.89
size 2400 x 1200mm	nr	3.50	59.50	401.59	69.16	530.25

Window repairs

Take off and replace defective
ironmongery to softwood windows

	Unit	Hours	Hours £	Mat'ls £	O & P £	Total £
casement fastener	nr	0.20	3.40	5.13	1.28	9.81
casement stay	nr	0.20	3.40	6.79	1.53	11.72
hinges	pair	0.25	4.25	3.92	1.23	9.40
cockspur fastener	nr	0.20	3.40	11.07	2.17	16.64
sash fastener	nr	0.25	4.25	5.75	1.50	11.50

Take off and replace defective
ironmongery to hardwood windows

	Unit	Hours	Hours £	Mat'ls £	O & P £	Total £
casement fastener	nr	0.30	5.10	5.13	1.53	11.76
casement stay	nr	0.30	5.10	6.79	1.78	13.67
hinges	pair	0.35	5.95	3.92	1.48	11.35
cockspur fastener	nr	0.30	5.10	11.07	2.43	18.60
sash fastener	nr	0.35	5.95	5.75	1.76	13.46

Take off and replace defective
ironmongery to PVC-U windows

	Unit	Hours	Hours £	Mat'ls £	O & P £	Total £
casement fastener	nr	0.20	3.40	5.13	1.28	9.81
casement stay	nr	0.20	3.40	6.79	1.53	11.72
hinges	pair	0.25	4.25	3.92	1.23	9.40
cockspur fastener	nr	0.20	3.40	11.07	2.17	16.64
sash fastener	nr	0.25	4.25	5.75	1.50	11.50

Take out existing window cill and
replace

	Unit	Hours	Hours £	Mat'ls £	O & P £	Total £
softwood	m	0.55	9.35	10.21	2.93	22.49
hardwood	m	0.65	11.05	27.50	5.78	44.33

	Unit	Hours	Hours £	Mat'ls £	O & P £	Total £
Take out existing window board and replace						
softwood	m	0.45	7.65	8.21	2.38	18.24
hardwood	m	0.60	10.20	21.70	4.79	36.69

Door replacement

	Unit	Hours	Hours £	Mat'ls £	O & P £	Total £
Take off existing external door and frame and replace with new flush door						
softwood	nr	2.20	37.40	98.65	20.41	156.46
hardwood	nr	2.50	42.50	350.76	58.99	452.25
PVC-U	nr	2.20	37.40	341.97	56.91	436.28
panelled door						
softwood	nr	2.20	37.40	124.92	24.35	186.67
hardwood	nr	2.50	42.50	380.39	63.43	486.32
PVC-U	nr	2.20	37.40	358.02	59.31	454.73
half glazed door						
softwood	nr	2.20	37.40	124.92	24.35	186.67
hardwood	nr	2.50	42.50	380.39	63.43	486.32
PVC-U	nr	2.20	37.40	358.02	59.31	454.73
fully glazed door						
softwood	nr	2.80	47.60	101.36	22.34	171.30
hardwood	nr	3.20	54.40	360.15	62.18	476.73
PVC-U	nr	2.80	47.60	342.26	58.48	448.34
fully glazed patio doors						
softwood	pair	3.80	64.60	240.39	45.75	350.74
hardwood	pair	4.30	73.10	510.64	87.56	671.30
PVC-U	pair	3.80	64.60	480.28	81.73	626.61
galvanised steel up-and-over garage doors						
2135 x 1980mm	nr	6.00	102.00	360.51	69.38	531.89
3965 x 2135mm	nr	7.50	127.50	900.30	154.17	1181.97

	Unit	Hours	Hours £	Mat'ls £	O & P £	Total £
Door repairs						
Take off and replace defective ironmongery to softwood doors						
bolts						
barrel	nr	0.20	3.40	6.04	1.42	10.86
flush	nr	0.20	3.40	8.67	1.81	13.88
tower	nr	0.30	5.10	7.31	1.86	14.27
butts						
light	pair	0.20	3.40	3.95	1.10	8.45
medium	pair	0.25	4.25	4.10	1.25	9.60
heavy	pair	0.30	5.10	4.97	1.51	11.58
locks						
cupboard	nr	0.30	5.10	7.93	1.95	14.98
mortice dead lock	nr	0.85	14.45	13.80	4.24	32.49
rim lock	nr	0.45	7.65	6.10	2.06	15.81
cylinder	nr	1.10	18.70	25.10	6.57	50.37
Take off and replace defective ironmongery to hardwood doors						
bolts						
barrel	nr	0.30	5.10	6.04	1.67	12.81
flush	nr	0.30	5.10	8.67	2.07	15.84
indicating	nr	0.30	5.10	8.15	1.99	15.24
tower	nr	0.40	6.80	7.31	2.12	16.23
butts						
light	pair	0.40	6.80	3.95	1.61	12.36
medium	pair	0.35	5.95	4.10	1.51	11.56
heavy	pair	0.40	6.80	4.97	1.77	13.54
locks						
cupboard	nr	0.30	5.10	7.93	1.95	14.98
mortice dead lock	nr	0.85	14.45	13.80	4.24	32.49
rim lock	nr	0.45	7.65	6.10	2.06	15.81
cylinder	nr	1.10	18.70	25.10	6.57	50.37

	Unit	Hours	Hours £	Mat'ls £	O & P £	Total £

Partitions, walls and ceilings

Pull down existing damaged
partitions and walls and rebuild

	Unit	Hours	Hours £	Mat'ls £	O & P £	Total £
stud partition plasterboard both sides	m²	1.50	25.50	18.08	6.54	50.12
brickwork 112mm thick plastered both sides	m²	4.00	68.00	23.32	13.70	105.02
blockwork 75mm thick plastered both sides	m²	3.10	52.70	15.86	10.28	78.84
blockwork 100mm thick plastered both sides	m²	3.30	56.10	19.98	11.41	87.49
Hack off scorched plaster to walls and renew	m²	1.50	25.50	2.17	4.15	31.82
Pull down plasterboard and skim ceilings and renew	m²	1.56	26.52	4.03	4.58	35.13
Hack off damaged wall tiles and renew	m²	1.45	24.65	18.08	6.41	49.14

Take off damaged skirting and
replace

	Unit	Hours	Hours £	Mat'ls £	O & P £	Total £
19 x 75mm softwood	m	0.27	4.59	2.48	1.06	8.13
19 x 100mm softwood	m	0.30	5.10	2.63	1.16	8.89
19 x 100mm hardwood	m	0.32	5.44	4.71	1.52	11.67
25 x 150mm hardwood	m	0.38	6.46	7.39	2.08	15.93

Take up damaged flooring and
renew

	Unit	Hours	Hours £	Mat'ls £	O & P £	Total £
19mm tongued and grooved softwood flooring	m²	1.14	19.38	10.89	4.54	34.81
25mm tongued and grooved softwood flooring	m²	1.15	19.55	12.41	4.79	36.75
12mm tongued and grooved chipboard flooring	m²	0.80	13.60	5.05	2.80	21.45
18mm tongued and grooved chipboard flooring	m²	1.00	17.00	6.08	3.46	26.54
2mm vinyl sheeting	m²	0.50	8.50	12.61	3.17	24.28

	Unit	Hours	Hours £	Mat'ls £	O & P £	Total £
Take up damaged flooring (cont'd)						
2.5mm vinyl sheeting	m²	0.55	9.35	13.08	3.36	25.79
3mm vinyl sheeting	m²	0.60	10.20	13.91	3.62	27.73
2mm vinyl tiling	m²	0.40	6.80	12.88	2.95	22.63
3mm laminated wood flooring	m²	0.90	15.30	11.11	3.96	30.37
12.5mm quarry tiling	m²	1.15	19.55	28.17	7.16	54.88
Cut out and replace floor joists and roof members						
38 x 100mm sawn softwood	m	0.22	3.74	1.51	0.79	6.04
50 x 75mm sawn softwood	m	0.24	4.08	1.53	0.84	6.45
50 x 100mm sawn softwood	m	0.28	4.76	1.62	0.96	7.34
50 x 125mm sawn softwood	m	0.30	5.10	1.98	1.06	8.14
50 x 150mm sawn softwood	m	0.32	5.44	2.11	1.13	8.68
75 x 125mm sawn softwood	m	0.34	5.78	3.18	1.34	10.30
75 x 200mm sawn softwood	m	0.36	6.12	3.47	1.44	11.03
Take down damaged softwood staircase 2600mm rise and renew						
straight flight 900mm wide	nr	12.20	207.40	607.42	122.22	937.04
straight flight 900mm wide with balustrade	nr	13.30	226.10	640.24	129.95	996.29
two flights 900mm wide with landing and balustrade	nr	14.50	246.50	800.43	157.04	1203.97
Take down damaged hardwood staircase 2600mm rise and renew						
straight flight 900mm wide	nr	12.40	210.80	1007.79	182.79	1401.38
straight flight 900mm wide with balustrade	nr	13.90	236.30	1211.03	217.10	1664.43
two flights 900mm wide with landing and balustrade	nr	15.00	255.00	1401.26	248.44	1904.70

	Unit	Hours	Hours £	Mat'ls £	O & P £	Total £

Furniture and fittings

Take down damaged fittings and
replace (material prices not
included due to the wide variation
in the quality and costs of the
fittings)

	Unit	Hours	Hours £	Mat'ls £	O & P £	Total £
wall units 2000mm high						
300 x 600mm	nr	1.00	17.00	0.00	2.55	19.55
300 x 1000mm	nr	1.10	18.70	0.00	2.81	21.51
300 x 1200mm	nr	1.20	20.40	0.00	3.06	23.46
base units 750mm high						
600 x 900mm	nr	1.00	17.00	0.00	2.55	19.55
600 x 1000mm	nr	1.00	17.00	0.00	2.55	19.55
600 x 1200mm	nr	1.10	18.70	0.00	2.81	21.51
900 x 900mm	nr	1.10	18.70	0.00	2.81	21.51
900 x 1000mm	nr	1.20	20.40	0.00	3.06	23.46
900 x 1200mm	nr	1.20	20.40	0.00	3.06	23.46
sink units 750mm high						
600 x 900mm	nr	1.40	23.80	0.00	3.57	27.37
600 x 1000mm	nr	1.40	23.80	0.00	3.57	27.37
600 x 1200mm	nr	1.40	23.80	0.00	3.57	27.37
worktops						
600 x 900mm	nr	0.80	13.60	0.00	2.04	15.64
600 x 1000mm	nr	0.90	15.30	0.00	2.30	17.60
600 x 1200mm	nr	1.00	17.00	0.00	2.55	19.55

Remove smoke damaged furniture
and place in skip

	Unit	Hours	Hours £	Mat'ls £	O & P £	Total £
easy chair	nr	0.15	2.55	0.00	0.38	2.93
settee	nr	0.18	3.06	0.00	0.46	3.52
single bed and mattress	nr	0.20	3.40	0.00	0.51	3.91
double bed and mattress	nr	0.22	3.74	0.00	0.56	4.30
set of six dining chairs	nr	0.12	2.04	0.00	0.31	2.35
dining table	nr	0.18	3.06	0.00	0.46	3.52
sideboard	nr	0.20	3.40	0.00	0.51	3.91

	Unit	Hours	Hours £	Mat'ls £	O & P £	Total £

Remove smoke damaged furniture (cont'd)

	Unit	Hours	Hours £	Mat'ls £	O & P £	Total £
wardrobe	nr	0.15	2.55	0.00	0.38	2.93
chest of drawers	nr	0.20	3.40	0.00	0.51	3.91
carpet	nr	0.25	4.25	0.00	0.64	4.89
TV and hi-fi equipment	nr	0.25	4.25	0.00	0.64	4.89

Plumbing and heating work

Take out damaged sanitary fittings and associated pipework and replace

	Unit	Hours	Hours £	Mat'ls £	O & P £	Total £
lavatory basin	nr	3.30	59.40	126.37	27.87	213.64
low level WC	nr	3.40	61.20	298.27	53.92	413.39
shower cubicle	nr	4.20	75.60	907.43	147.45	1130.48
sink with single drainer	nr	2.30	41.40	188.04	34.42	263.86
sink with double drainer	nr	2.90	52.20	241.93	44.12	338.25
bath, cast iron	nr	4.40	79.20	401.53	72.11	552.84
bath, acrylic	nr	4.20	75.60	190.62	39.93	306.15
bidet	nr	3.50	63.00	371.82	65.22	500.04

Take out damaged cisterns and cylinders and associated pipework and replace
polyethylene cold water cisterns

	Unit	Hours	Hours £	Mat'ls £	O & P £	Total £
68 litres	nr	2.00	36.00	88.55	18.68	143.23
86 litres	nr	2.00	36.00	100.91	20.54	157.45
191 litres	nr	2.50	45.00	119.34	24.65	188.99
327 litres	nr	3.80	68.40	138.75	31.07	238.22

copper cylinders, indirect pattern

	Unit	Hours	Hours £	Mat'ls £	O & P £	Total £
114 litres	nr	3.50	63.00	149.79	31.92	244.71
117 litres	nr	3.50	63.00	158.74	33.26	255.00
140 litres	nr	4.00	72.00	168.52	36.08	276.60
162 litres	nr	4.10	73.80	181.23	38.25	293.28

	Unit	Hours	Hours £	Mat'ls £	O & P £	Total £
copper cylinders, direct pattern						
116 litres	nr	3.80	68.40	137.19	30.84	236.43
120 litres	nr	3.80	68.40	141.63	31.50	241.53
144 litres	nr	4.20	75.60	149.34	33.74	258.68
166 litres	nr	4.30	77.40	161.74	35.87	275.01
Disconnect damaged central heating boiler and associated pipe work and replace						
floor-mounted gas boiler						
40,000 Btu	nr	9.00	162.00	629.47	118.72	910.19
50,000 Btu	nr	9.00	162.00	675.08	125.56	962.64
60,000 Btu	nr	9.20	165.60	700.23	129.87	995.70
70,000 Btu	nr	9.20	165.60	762.11	139.16	1066.87
wall-mounted gas boiler						
40,000 Btu	nr	9.00	162.00	581.65	111.55	855.20
50,000 Btu	nr	9.00	162.00	620.98	117.45	900.43
60,000 Btu	nr	9.20	165.60	658.32	123.59	947.51
70,000 Btu	nr	9.20	165.60	712.14	131.66	1009.40
floor-mounted oil-fired boiler						
52,000 Btu	nr	9.20	165.60	1310.50	221.42	1697.52
70,000 Btu	nr	9.40	169.20	1398.06	235.09	1802.35
Disconnect damaged radiator and associated pipework and and valves and replace						
single panel, 450mm high						
length, 1000mm	nr	1.65	29.70	74.67	15.66	120.03
length, 1600mm	nr	1.85	33.30	97.24	19.58	150.12
length, 2000mm	nr	2.10	37.80	198.68	35.47	271.95
single panel, 600mm high						
length, 1000mm	nr	1.85	33.30	98.76	19.81	151.87
length, 1600mm	nr	2.05	36.90	116.90	23.07	176.87
length, 2000mm	nr	2.30	41.40	228.86	40.54	310.80
double panel, 450mm high						
length, 1000mm	nr	1.85	33.30	94.76	19.21	147.27
length, 1400mm	nr	1.35	24.30	115.51	20.97	160.78
length, 2000mm	nr	1.45	26.10	218.86	36.74	281.70

	Unit	Hours	Hours £	Mat'ls £	O & P £	Total £

Disconnect damaged radiator (cont'd)

double panel, 600mm high

	Unit	Hours	Hours £	Mat'ls £	O & P £	Total £
length, 1000mm	nr	2.05	36.90	116.75	23.05	176.70
length, 1400mm	nr	2.25	40.50	135.15	26.35	202.00
length, 2000mm	nr	2.45	44.10	246.59	43.60	334.29

Electrics

Disconnect power supply, remove
damaged electric fittings and
replace

	Unit	Hours	Hours £	Mat'ls £	O & P £	Total £
single power point	nr	0.60	11.40	7.05	2.77	21.22
double power point	nr	0.60	11.40	8.81	3.03	23.24
light switch	nr	0.60	11.40	6.25	2.65	20.30
light point	nr	0.60	11.40	7.35	2.81	21.56
wall light	nr	0.90	17.10	11.41	4.28	32.79
cooker control unit	nr	1.50	28.50	27.21	8.36	64.07

Painting and decorating

Burn off damaged woodwork and
leave ready to receive new paintwork

	Unit	Hours	Hours £	Mat'ls £	O & P £	Total £
general surfaces	m²	1.10	18.70	0.00	2.81	21.51
general surfaces up to 300mm wide	m	0.30	5.10	0.00	0.77	5.87

Burn off damaged metalwork and
leave ready to receive new paintwork

	Unit	Hours	Hours £	Mat'ls £	O & P £	Total £
general surfaces	m²	1.10	18.70	0.00	2.81	21.51
general surfaces up to 300mm wide	m	0.30	5.10	0.00	0.77	5.87

Wash down existing plastered
surfaces, stop cracks and rub
down and leave ready to receive
new paintwork

	Unit	Hours	Hours £	Mat'ls £	O & P £	Total £
walls	m²	0.14	2.38	0.00	0.36	2.74
ceilings	m²	0.05	0.85	0.00	0.13	0.98

	Unit	Hours	Hours £	Mat'ls £	O & P £	Total £
Apply one mist coat and two coats emulsion paint to plastered surfaces						
walls	m²	0.30	5.10	0.81	0.89	6.80
ceilings	m²	0.32	5.44	0.81	0.94	7.19
Apply two undercoats and one coat gloss paint to plastered surfaces						
walls	m²	0.30	5.10	1.71	1.02	7.83
ceilings	m²	0.36	6.12	1.71	1.17	9.00
Prepare, size, apply adhesive, supply and hang paper to plastered walls						
lining paper						
£1.50 per roll	m²	0.30	5.10	0.36	0.82	6.28
£2.00 per roll	m²	0.30	5.10	0.48	0.84	6.42
£2.50 per roll	m²	0.30	5.10	0.59	0.85	6.54
washable paper						
£2.50 per roll	m²	0.30	5.10	0.59	0.85	6.54
£4.00 per roll	m²	0.30	5.10	0.94	0.91	6.95
£5.00 per roll	m²	0.30	5.10	1.18	0.94	7.22
vinyl paper						
£4.00 per roll	m²	0.30	5.10	0.94	0.91	6.95
£5.00 per roll	m²	0.30	5.10	1.18	0.94	7.22
£6.00 per roll	m²	0.30	5.10	1.41	0.98	7.49
washable paper						
£5.00 per roll	m²	0.30	5.10	1.18	0.94	7.22
£6.00 per roll	m²	0.30	5.10	1.41	0.98	7.49
£7.00 per roll	m²	0.30	5.10	1.64	1.01	7.75
hessian paper						
£7.00 per m²	m²	0.50	8.50	7.70	2.43	18.63
£8.00 per m²	m²	0.50	8.50	8.80	2.60	19.90
£9.00 per m²	m²	0.50	8.50	9.90	2.76	21.16

	Unit	Hours	Hours £	Plant £	O & P £	Total £

FLOOD DAMAGE

Pumping

	Unit	Hours	Hours £	Plant £	O & P £	Total £
Instal pump and hoses in position	Item	1.00	13.00	0.00	1.95	14.95
Hire diaphragm pump and hoses						
50mm	day	1.00	0.00	28.00	4.20	32.20
75mm	day	1.00	0.00	34.50	5.18	39.68
Hire portable dryer, 350W	week	1.00	0.00	30.00	4.50	34.50

	Unit	Hours	Hours £	Mat'ls £	O & P £	Total £

Repairs

	Unit	Hours	Hours £	Mat'ls £	O & P £	Total £
Clean up and remove debris from basement and ground floor	m²	0.20	3.40	0.00	0.51	3.91
Hack off defective wall plaster and replace	m²	1.50	25.50	2.17	4.15	31.82
Pull down plasterboard and skim ceilings and renew	m²	1.56	26.52	4.03	4.58	35.13
Hack off damaged wall tiles and renew	m²	1.45	24.65	22.80	7.12	54.57
Take off damaged skirting and replace						
19 x 75mm softwood	m	0.27	4.59	2.48	1.06	8.13
19 x 100mm softwood	m	0.30	5.10	2.63	1.16	8.89
19 x 100mm hardwood	m	0.32	5.44	4.71	1.52	11.67
25 x 150mm hardwood	m	0.38	6.46	7.39	2.08	15.93
Take up damaged flooring and renew						
19mm tongued and grooved softwood flooring	m²	1.14	19.38	10.89	4.54	34.81
25mm tongued and grooved softwood flooring	m²	1.15	19.55	12.41	4.79	36.75

	Unit	Hours	Hours £	Mat'ls £	O & P £	Total £

Take up damaged flooring (cont'd)

	Unit	Hours	Hours £	Mat'ls £	O & P £	Total £
12mm tongued and grooved chipboard flooring	m²	0.80	13.60	5.05	2.80	21.45
18mm tongued and grooved chipboard flooring	m²	1.00	17.00	6.08	3.46	26.54

Electrics

Disconnect power supply, remove damaged electric fittings and replace

	Unit	Hours	Hours £	Mat'ls £	O & P £	Total £
single power point	nr	0.60	11.40	7.05	2.77	21.22
double power point	nr	0.60	11.40	8.81	3.03	23.24
light switch	nr	0.60	11.40	6.25	2.65	20.30
cooker control unit	nr	1.50	28.50	27.21	8.36	64.07

Painting and decorating

Burn off damaged woodwork and leave ready to receive new paint- work

	Unit	Hours	Hours £	Mat'ls £	O & P £	Total £
general surfaces	m²	1.10	18.70	0.00	2.81	21.51
general surfaces up to 300mm wide	m	0.30	5.10	0.00	0.77	5.87

Burn off damaged metalwork and leave ready to receive new paint- work

	Unit	Hours	Hours £	Mat'ls £	O & P £	Total £
general surfaces	m²	1.10	18.70	0.00	2.81	21.51
general surfaces up to 300mm wide	m	0.30	5.10	0.00	0.77	5.87

Wash down existing plastered surfaces, stop cracks and rub down and leave ready to receive new paintwork

	Unit	Hours	Hours £	Mat'ls £	O & P £	Total £
walls	m²	0.05	0.85	0.00	0.13	0.98
ceilings	m²	0.07	1.19	0.00	0.18	1.37

	Unit	Hours	Hours £	Mat'ls £	O & P £	Total £

GALE DAMAGE

Hoardings and screens

	Unit	Hours	Hours £	Mat'ls £	O & P £	Total £
Temporary screens and hoardings 2m high consisting of 22mm thick exterior quality plywood fixed to 100 x 50mm posts and rails	m	2.60	44.20	35.80	12.00	92.00
Hire tarpaulin sheeting size 5 x 4m fixed in position	week	0.00	0.00	18.00	2.70	20.70
Plywood sheeting blocking up window or door opening	m²	1.00	17.00	15.17	4.83	37.00

Shoring

	Unit	Hours	Hours £	Mat'ls £	O & P £	Total £
Timber dead shores consisting of 200 x 200mm shores, 200 x 50mm plates and 200 x 50mm braces at centres of						
2m	m²	3.80	64.60	73.58	20.73	158.91
3m	m²	3.00	51.00	59.64	16.60	127.24
4m	m²	2.20	37.40	41.05	11.77	90.22
Timber raking and flying shores consisting of 200 x 200mm shores, 250 x 50mm plates and 200 x 50mm braces to gable end of two-storey house	Item	60.00	1020.00	527.61	232.14	1779.75

Roof repairs

	Unit	Hours	Hours £	Mat'ls £	O & P £	Total £
Take up roof coverings from pitched roof						
tiles	m²	0.80	13.60	0.00	2.04	15.64
slates	m²	0.80	13.60	0.00	2.04	15.64
timber boarding	m²	1.00	17.00	0.00	2.55	19.55
metal sheeting	m²	0.20	3.40	0.00	0.51	3.91
flat sheeting	m²	0.30	5.10	0.00	0.77	5.87
corrugated sheeting	m²	0.30	5.10	0.00	0.77	5.87
underfelt	m²	0.10	1.70	0.00	0.26	1.96

	Unit	Hours	Hours £	Mat'ls £	O & P £	Total £
Take up roof coverings from flat roof						
bituminous felt	m²	0.25	4.25	0.00	0.64	4.89
metal sheeting	m²	0.30	5.10	0.00	0.77	5.87
woodwool slabs	m²	0.50	8.50	0.00	1.28	9.78
firrings	m²	0.20	3.40	0.00	0.51	3.91
Take up roof coverings from pitched roof, carefully lay aside for reuse						
tiles	m²	1.10	18.70	0.00	2.81	21.51
slates	m²	1.10	18.70	0.00	2.81	21.51
metal sheeting	m²	0.50	8.50	0.00	1.28	9.78
flat sheeting	m²	0.60	10.20	0.00	1.53	11.73
corrugated sheeting	m²	0.60	10.20	0.00	1.53	11.73
Take up roof coverings from flat roof, carefully lay aside for reuse						
metal sheeting	m²	0.60	10.20	0.00	1.53	11.73
woodwool slabs	m²	0.80	13.60	0.00	2.04	15.64
Inspect roof battens, refix loose and replace with new, size 38 x 25mm						
25% of area						
250mm centres	m²	0.14	2.38	0.50	0.43	3.31
450mm centres	m²	0.12	2.04	0.44	0.37	2.85
600mm centres	m²	0.10	1.70	0.37	0.31	2.38
50% of area						
250mm centres	m²	0.26	4.42	0.84	0.79	6.05
450mm centres	m²	0.16	2.72	0.72	0.52	3.96
600mm centres	m²	0.18	3.06	0.60	0.55	4.21
75% of area						
250mm centres	m²	0.36	6.12	1.27	1.11	8.50
450mm centres	m²	0.26	4.42	1.08	0.83	6.33
600mm centres	m²	0.22	3.74	0.88	0.69	5.31

	Unit	Hours	Hours £	Mat'ls £	O & P £	Total £
100% of area						
250mm centres	m²	0.44	7.48	1.60	1.36	10.44
450mm centres	m²	0.32	5.44	1.40	1.03	7.87
600mm centres	m²	0.38	6.46	1.20	1.15	8.81
Remove single slipped slate and replace	nr	1.00	17.00	1.20	2.73	20.93
Remove single broken slate, renew with new Welsh blue slate						
405 x 255mm	nr	1.20	20.40	1.65	3.31	25.36
510 x 255mm	nr	1.20	20.40	3.17	3.54	27.11
610 x 305mm	nr	1.20	20.40	6.51	4.04	30.95
Remove slates in area approximately 1m² and replace with Welsh blue slates previously laid aside	nr	1.60	27.20	0.00	4.08	31.28
Remove slates in area approximately 1m² and replace with Welsh blue slates previously laid aside						
405 x 255mm	nr	1.80	30.60	0.00	4.59	35.19
510 x 255mm	nr	1.70	28.90	0.00	4.34	33.24
610 x 305mm	nr	1.60	27.20	0.00	4.08	31.28
Remove single slipped tile and refix	nr	0.30	5.10	0.00	0.77	5.87
Remove single broken tile and renew						
Marley plain tile	nr	0.30	5.10	0.92	0.90	6.92
Marley Ludlow Plus tile	nr	0.30	5.10	1.12	0.93	7.15
Marley Modern tile	nr	0.30	5.10	1.78	1.03	7.91
Redland Renown tile	nr	0.30	5.10	1.78	1.03	7.91
Redland Norfolk tile	nr	0.30	5.10	1.42	0.98	7.50
Remove tiles in area approximately 1m² and replace with tiles previously laid aside						
Marley plain tile	nr	1.80	30.60	0.00	4.59	35.19
Marley Ludlow Plus tile	nr	1.20	20.40	0.00	3.06	23.46
Marley Modern tile	nr	1.10	18.70	0.00	2.81	21.51

	Unit	Hours	Hours £	Mat'ls £	O & P £	Total £
Remove tiles in and replace (cont'd)						
Redland Renown tile	nr	1.10	18.70	0.00	2.81	21.51
Redland Norfolk tile	nr	1.15	19.55	0.00	2.93	22.48
Take off defective ridge capping and refix including pointing in mortar	m	1.10	18.70	1.68	3.06	23.44
Chimney stack						
Erect and take down chimney scaffold	Item	4.00	68.00	0.00	10.20	78.20
Hire chimney scaffold and platform	week	1.00	17.00	115.00	19.80	151.80
Take down existing chimney stack to below roof level and remove debris						
single stack	m	2.50	42.50	0.00	6.38	48.88
double stack	m	3.50	59.50	0.00	8.93	68.43
Chimney stack in facing brick £450.00 per thousand in gauged mortar						
single stack	m	3.80	64.60	42.57	16.08	123.25
double stack	m	5.20	88.40	88.53	26.54	203.47
Terracotta chimney pot bedded and flaunched in gauged mortar						
185mm diameter x 300mm high	nr	1.25	21.25	25.98	7.08	54.31
185mm diameter x 600mm high	nr	1.80	30.60	43.55	11.12	85.27
External work						
Remove blown-down trees, trunk girth 1m above ground level						
600 to 1500mm	nr	22.00	374.00	0.00	56.10	430.10
1500 to 3000mm	nr	38.00	646.00	0.00	96.90	742.90

	Unit	Hours	Hours £	Mat'ls £	O & P £	Total £

THEFT DAMAGE

Window and door repairs

	Unit	Hours	Hours £	Mat'ls £	O & P £	Total £
Hack out glass and remove	m²	0.45	7.65	0.00	1.15	8.80
Clean rebates, remove sprigs or clips and prepare for reglazing	m	0.20	3.40	0.00	0.51	3.91

Reglaze existing softwood windows in clear float glass with putty and sprigs

under 0.15m², thickness

	Unit	Hours	Hours £	Mat'ls £	O & P £	Total £
3mm	m²	0.90	15.30	44.23	8.93	68.46
4mm	m²	0.90	15.30	46.02	9.20	70.52
5mm	m²	0.90	15.30	66.07	12.21	93.58
6mm	m²	1.00	17.00	67.10	12.62	96.72

over 0.15m², thickness

	Unit	Hours	Hours £	Mat'ls £	O & P £	Total £
3mm	m²	0.60	10.20	44.23	8.16	62.59
4mm	m²	0.60	10.20	46.02	8.43	64.65
5mm	m²	0.60	10.20	66.07	11.44	87.71
6mm	m²	0.65	11.05	67.10	11.72	89.87

Reglaze existing softwood windows in clear float glass with pinned beads

under 0.15m², thickness

	Unit	Hours	Hours £	Mat'ls £	O & P £	Total £
3mm	m²	1.10	18.70	44.23	9.44	72.37
4mm	m²	1.10	18.70	46.02	9.71	74.43
5mm	m²	1.10	18.70	66.07	12.72	97.49
6mm	m²	1.20	20.40	67.10	13.13	100.63

over 0.15m², thickness

	Unit	Hours	Hours £	Mat'ls £	O & P £	Total £
3mm	m²	0.80	13.60	44.23	8.67	66.50
4mm	m²	0.80	13.60	46.02	8.94	68.56
5mm	m²	0.80	13.60	66.07	11.95	91.62
6mm	m²	0.90	15.30	67.10	12.36	94.76

	Unit	Hours	Hours £	Mat'ls £	O & P £	Total £
Reglaze existing metal windows in clear float glass with clips and putty						
under 0.15m², thickness						
3mm	m²	0.95	16.15	44.23	9.06	69.44
4mm	m²	0.95	16.15	46.02	9.33	71.50
5mm	m²	0.95	16.15	66.07	12.33	94.55
6mm	m²	1.05	17.85	67.10	12.74	97.69
over 0.15m², thickness						
3mm	m²	0.65	11.05	44.23	8.29	63.57
4mm	m²	0.65	11.05	46.02	8.56	65.63
5mm	m²	0.65	11.05	66.07	11.57	88.69
6mm	m²	0.70	11.90	67.10	11.85	90.85
Reglaze existing metal windows in clear float glass with screwed metal beads						
under 0.15m², thickness						
3mm	m²	1.30	22.10	44.23	9.95	76.28
4mm	m²	1.30	22.10	46.02	10.22	78.34
5mm	m²	1.30	22.10	66.07	13.23	101.40
6mm	m²	1.40	23.80	67.10	13.64	104.54
over 0.15m², thickness						
3mm	m²	1.00	17.00	44.23	9.18	70.41
4mm	m²	1.00	17.00	46.02	9.45	72.47
5mm	m²	1.00	17.00	66.07	12.46	95.53
6mm	m²	1.05	17.85	67.10	12.74	97.69
Reglaze existing softwood windows in white patterned glass with putty and sprigs						
under 0.15m², thickness						
4mm	m²	0.90	15.30	25.92	6.18	47.40
6mm	m²	1.00	17.00	43.57	9.09	69.66

	Unit	Hours	Hours £	Mat'ls £	O & P £	Total £
over 0.15m², thickness						
4mm	m²	0.60	10.20	24.12	5.15	39.47
6mm	m²	0.65	11.05	42.07	7.97	61.09
Reglaze existing softwood windows in white patterned glass with pinned beads						
under 0.15m², thickness						
4mm	m²	1.10	18.70	25.92	6.69	51.31
6mm	m²	1.20	20.40	43.57	9.60	73.57
over 0.15m², thickness						
4mm	m²	0.80	13.60	24.12	5.66	43.38
6mm	m²	0.85	14.45	42.07	8.48	65.00
Reglaze existing softwood windows in white patterned glass with screwed beads						
under 0.15m², thickness						
4mm	m²	1.30	22.10	25.92	7.20	55.22
6mm	m²	1.40	23.80	43.57	10.11	77.48
over 0.15m², thickness						
4mm	m²	1.00	17.00	24.12	6.17	47.29
6mm	m²	1.50	25.50	42.07	10.14	77.71
Reglaze existing metal windows in white patterned glass with putty						
under 0.15m², thickness						
4mm	m²	0.90	15.30	25.92	6.18	47.40
6mm	m²	1.00	17.00	43.57	9.09	69.66
over 0.15m², thickness						
4mm	m²	0.60	10.20	24.12	5.15	39.47
6mm	m²	0.65	11.05	42.07	7.97	61.09

	Unit	Hours	Hours £	Mat'ls £	O & P £	Total £
Reglaze existing metal windows in white patterned glass with metal clips and putty						
under 0.15m², thickness						
4mm	m²	0.95	16.15	25.92	6.31	48.38
6mm	m²	1.05	17.85	43.57	9.21	70.63
over 0.15m², thickness						
4mm	m²	0.65	11.05	24.12	5.28	40.45
6mm	m²	0.70	11.90	42.07	8.10	62.07
Reglaze existing metal windows in white patterned glass with screwed beads						
under 0.15m², thickness						
4mm	m²	1.30	22.10	25.92	7.20	55.22
6mm	m²	1.40	23.80	43.57	10.11	77.48
over 0.15m², thickness						
4mm	m²	1.00	17.00	24.12	6.17	47.29
6mm	m²	1.05	17.85	42.07	8.99	68.91
Take off and replace damaged ironmongery to softwood windows						
casement fastener	nr	0.20	3.40	5.13	1.28	9.81
casement stay	nr	0.20	3.40	6.79	1.53	11.72
hinges	pair	0.25	4.25	3.92	1.23	9.40
lock	nr	0.60	10.20	14.12	3.65	27.97
Take off and replace damaged ironmongery to hardwood windows						
casement fastener	nr	0.30	5.10	5.13	1.53	11.76
casement stay	nr	0.30	5.10	6.79	1.78	13.67
hinges	pair	0.35	5.95	3.92	1.48	11.35
lock	nr	0.75	12.75	14.12	4.03	30.90

	Unit	Hours	Hours £	Mat'ls £	O & P £	Total £
Take off and replace defective ironmongery to PVC-U windows						
casement fastener	nr	0.20	3.40	5.13	1.28	9.81
casement stay	nr	0.20	3.40	6.79	1.53	11.72
hinges	pair	0.25	4.25	3.92	1.23	9.40
lock	nr	0.60	10.20	14.12	3.65	27.97
Take out existing window cill and replace						
softwood	m	0.55	9.35	10.21	2.93	22.49
hardwood	m	0.65	11.05	27.50	5.78	44.33
Take off existing external door and frame and replace with new						
flush door						
softwood	nr	2.20	37.40	98.65	20.41	156.46
hardwood	nr	2.50	42.50	350.76	58.99	452.25
PVC-U	nr	2.20	37.40	341.97	56.91	436.28
panelled door						
softwood	nr	2.20	37.40	124.92	24.35	186.67
hardwood	nr	2.50	42.50	380.39	63.43	486.32
PVC-U	nr	2.20	37.40	358.02	59.31	454.73
half glazed door						
softwood	nr	2.20	37.40	124.92	24.35	186.67
hardwood	nr	2.50	42.50	380.39	63.43	486.32
PVC-U	nr	2.20	37.40	358.02	59.31	454.73
fully glazed door						
softwood	nr	2.80	47.60	101.36	22.34	171.30
hardwood	nr	3.20	54.40	360.15	62.18	476.73
PVC-U	nr	2.80	47.60	342.26	58.48	448.34
fully glazed patio doors						
softwood	pair	3.80	64.60	240.39	45.75	350.74
hardwood	pair	4.30	73.10	510.64	87.56	671.30
PVC-U	pair	2.80	47.60	480.28	79.18	607.06

	Unit	Hours	Hours £	Mat'ls £	O & P £	Total £
Take off and replace damaged ironmongery to softwood doors						
bolts						
barrel	nr	0.20	3.40	6.04	1.42	10.86
flush	nr	0.20	3.40	8.67	1.81	13.88
tower	nr	0.30	5.10	7.31	1.86	14.27
butts						
light	pair	0.20	3.40	3.95	1.10	8.45
medium	pair	0.25	4.25	4.10	1.25	9.60
heavy	pair	0.30	5.10	4.97	1.51	11.58
locks						
cupboard	nr	0.30	5.10	7.93	1.95	14.98
mortice dead lock	nr	0.85	14.45	13.80	4.24	32.49
rim lock	nr	0.45	7.65	6.10	2.06	15.81
cylinder	nr	1.10	18.70	25.10	6.57	50.37
Take off and replace damaged ironmongery to hardwood doors						
bolts						
barrel	nr	0.30	5.10	6.04	1.67	12.81
flush	nr	0.30	5.10	8.67	2.07	15.84
tower	nr	0.40	6.80	7.31	2.12	16.23
butts						
light	pair	0.30	5.10	3.95	1.36	10.41
medium	pair	0.35	5.95	4.10	1.51	11.56
heavy	pair	0.40	6.80	4.97	1.77	13.54
locks						
cupboard	nr	0.30	5.10	7.93	1.95	14.98
mortice dead lock	nr	0.85	14.45	13.80	4.24	32.49
rim lock	nr	0.45	7.65	6.10	2.06	15.81
cylinder	nr	1.10	18.70	25.10	6.57	50.37

Part Five

KITCHENS

Standard items

	Unit	Hours	Hours £	Mat'ls £	O & P £	Total £

KITCHENS

Floor tiling

Red quarry tiles, bedded, jointed
and pointed in cement mortar (1:3),
butt jointed, straight both ways, to
floors

150 x 150 x 12.5mm thick

	Unit	Hours	Hours £	Mat'ls £	O & P £	Total £
over 300mm wide	m²	0.90	15.30	32.96	7.24	55.50
less than 300mm wide	m²	0.36	6.12	36.81	6.44	49.37

200 x 200 x 19mm thick

over 300mm wide	m²	0.80	13.60	58.48	10.81	82.89
less than 300mm wide	m²	0.32	5.44	65.68	10.67	81.79

Vitrified ceramic tiles, bedded,
jointed and pointed in cement mortar
mortar (1:3), butt jointed, straight
both ways. to floors

100 x 100 x 9mm thick

over 300mm wide	m²	1.00	17.00	23.47	6.07	46.54
less than 300mm wide	m²	0.44	7.48	26.50	5.10	39.08

150 x 150 x 9mm thick

over 300mm wide	m²	0.90	15.30	25.96	6.19	47.45
less than 300mm wide	m²	0.36	6.12	28.90	5.25	40.27

200 x 200 x 9mm thick

over 300mm wide	m²	0.80	13.60	29.07	6.40	49.07
less than 300mm wide	m²	0.32	5.44	32.81	5.74	43.99

Vinyl floor tiling, fixed with adhesive,
butt jointed, straight both ways, to
floors

300 x 300 x 2mm thick

over 300mm wide	m²	0.28	4.76	14.06	2.82	21.64
less than 300mm wide	m²	0.10	1.70	16.27	2.70	20.67

	Unit	Hours	Hours £	Mat'ls £	O & P £	Total £
300 x 300 x 2.5mm thick						
over 300mm wide	m²	0.28	4.76	15.18	2.99	22.93
less than 300mm wide	m²	0.10	1.70	16.97	2.80	21.47
300 x 300 x 2.5mm thick						
over 300mm wide	m²	0.28	4.76	16.12	3.13	24.01
less than 300mm wide	m²	0.10	1.70	17.06	2.81	21.57

Wall tiling

White glazed ceramic wall tiling
fixed with adhesive, pointing with
white grout

	Unit	Hours	Hours £	Mat'ls £	O & P £	Total £
100 x 100 x 4mm thick						
over 300mm wide	m²	0.95	16.15	19.80	5.39	41.34
less than 300mm wide	m²	0.38	6.46	22.12	4.29	32.87
150 x 150 x 6.5mm thick						
over 300mm wide	m²	0.80	13.60	16.62	4.53	34.75
less than 300mm wide	m²	0.35	5.95	19.72	3.85	29.52
200 x 200 x 6.5mm thick						
over 300mm wide	m²	0.75	12.75	71.74	12.67	97.16
less than 300mm wide	m²	0.32	5.44	35.08	6.08	46.60
200 x 250 x 6.5mm thick						
over 300mm wide	m²	0.70	11.90	32.94	6.73	51.57
less than 300mm wide	m²	0.30	5.10	36.19	6.19	47.48
250 x 265 x 6.5mm thick						
over 300mm wide	m²	0.70	11.90	33.15	6.76	51.81
less than 300mm wide	m²	0.30	5.10	37.54	6.40	49.04
250 x 230 x 6.5mm thick						
over 300mm wide	m²	0.65	11.05	33.15	6.63	50.83
less than 300mm wide	m²	0.28	4.76	37.54	6.35	48.65

	Unit	Hours	Hours £	Mat'ls £	O & P £	Total £

Kitchen fittings

Floor units with doors, size

300 x 870 x 565mm

prime cost £40	nr	1.00	17.00	41.00	8.70	66.70
prime cost £50	nr	1.05	17.85	51.00	10.33	79.18
prime cost £60	nr	1.10	18.70	61.00	11.96	91.66
prime cost £70	nr	1.15	19.55	71.00	13.58	104.13

600 x 870 x 565mm

prime cost £50	nr	1.10	18.70	61.00	2.81	21.51
prime cost £60	nr	1.15	19.55	51.00	10.58	81.13
prime cost £70	nr	1.20	20.40	71.00	13.71	105.11
prime cost £80	nr	1.25	21.25	81.00	15.34	117.59

1000 x 870 x 565mm

prime cost £80	nr	1.25	21.25	82.00	3.19	24.44
prime cost £90	nr	1.30	22.10	92.00	17.12	131.22
prime cost £100	nr	1.35	22.95	102.00	18.74	143.69
prime cost £110	nr	1.40	23.80	112.00	20.37	156.17

Floor units with drawers, size

300 x 870 x 575mm

prime cost £70	nr	1.20	20.40	72.00	13.86	106.26
prime cost £80	nr	1.25	21.25	82.00	15.49	118.74
prime cost £90	nr	1.25	21.25	92.00	16.99	130.24
prime cost £100	nr	1.35	22.95	102.00	18.74	143.69

600 x 870 x 575mm

prime cost £90	nr	1.30	22.10	92.00	3.32	25.42
prime cost £100	nr	1.35	22.95	102.00	18.74	143.69
prime cost £110	nr	1.40	23.80	112.00	20.37	156.17
prime cost £120	nr	1.45	24.65	122.00	22.00	168.65

1000 x 870 x 575mm

prime cost £140	nr	1.50	25.50	142.00	3.83	29.33
prime cost £150	nr	1.55	26.35	152.00	26.75	205.10
prime cost £160	nr	1.60	27.20	162.00	28.38	217.58
prime cost £170	nr	1.65	28.05	172.00	30.01	230.06

	Unit	Hours	Hours £	Mat'ls £	O & P £	Total £
Wall units, size						
300 x 704 x 330mm						
prime cost £40	nr	1.10	18.70	41.00	8.96	68.66
prime cost £50	nr	1.15	19.55	51.00	10.58	81.13
prime cost £60	nr	1.20	20.40	61.00	12.21	93.61
prime cost £70	nr	1.25	21.25	71.00	13.84	106.09
600 x 704 x 330mm						
prime cost £50	nr	1.20	20.40	61.00	3.06	23.46
prime cost £60	nr	1.25	21.25	51.00	10.84	83.09
prime cost £70	nr	1.30	22.10	71.00	13.97	107.07
prime cost £80	nr	1.35	22.95	81.00	15.59	119.54
1000 x 704 x 330mm						
prime cost £80	nr	1.35	22.95	82.00	3.44	26.39
prime cost £90	nr	1.40	23.80	92.00	17.37	133.17
prime cost £100	nr	1.45	24.65	102.00	19.00	145.65
prime cost £110	nr	1.50	25.50	112.00	20.63	158.13
Larder and broom cupboards, size						
300 x 2118 x 575mm						
prime cost £120	nr	1.35	22.95	122.00	21.74	166.69
prime cost £130	nr	1.40	23.80	132.00	23.37	179.17
prime cost £140	nr	1.45	24.65	142.00	25.00	191.65
prime cost £150	nr	1.50	25.50	150.00	26.33	201.83
600 x 2118 x 575mm						
prime cost £140	nr	1.55	26.35	142.00	25.25	193.60
prime cost £150	nr	1.60	27.20	152.00	26.88	206.08
prime cost £160	nr	1.65	28.05	162.00	28.51	218.56
prime cost £170	nr	1.70	28.90	172.00	30.14	231.04
Single oven housing, size 600 x 2118 x 575mm						
prime cost £210	nr	1.55	26.35	212.00	35.75	274.10
prime cost £220	nr	1.60	27.20	222.00	37.38	286.58
prime cost £230	nr	1.65	28.05	232.00	39.01	299.06
prime cost £240	nr	1.70	28.90	242.00	40.64	311.54

	Unit	Hours	Hours £	Mat'ls £	O & P £	Total £
Double oven housing, size 600 x 2118 x 575mm						
prime cost £190	nr	1.55	26.35	192.00	32.75	251.10
prime cost £200	nr	1.60	27.20	202.00	34.38	263.58
prime cost £210	nr	1.65	28.05	212.00	36.01	276.06
prime cost £220	nr	1.70	28.90	222.00	37.64	288.54

Painting

One mist coat and two coats white matt emulsion on surfaces over 300mm girth

	Unit	Hours	Hours £	Mat'ls £	O & P £	Total £
plastered walls	m²	0.30	5.10	0.81	0.89	6.80
plastered ceilings	m²	0.32	5.44	0.81	0.94	7.19

One mist coat and two coats coloured matt emulsion on surfaces over 300mm girth

	Unit	Hours	Hours £	Mat'ls £	O & P £	Total £
plastered walls	m²	0.30	5.10	0.85	0.89	6.84
plastered ceilings	m²	0.32	5.44	0.85	0.94	7.23

One coat wood primer, one white oil-based undercoat and one white oil-based finishing coat on wood

	Unit	Hours	Hours £	Mat'ls £	O & P £	Total £
general surfaces	m²	0.20	3.40	1.60	0.75	5.75
windows and glazed doors	m²	0.25	4.25	1.12	0.81	6.18
frames and linings, 150 to 300mm girth	m	0.20	3.40	0.92	0.65	4.97

Part Six

ALTERATIONS AND SPOT ITEMS

	Unit	Hours	Hours £	Mat'ls £	O & P £	Total £

ALTERATIONS

Shoring

Timber dead shores consisting of
200 x 200mm shores, 250 x
50mm plates and 200 x 50mm
braces at centres of

	Unit	Hours	Hours £	Mat'ls £	O & P £	Total £
2m	m²	3.80	64.60	73.58	20.73	158.91
3m	m²	3.00	51.00	59.64	16.60	127.24
4m	m²	2.20	37.40	41.05	11.77	90.22

Timber raking and flying shores
consisting of 200 x 200mm shores,
250 x 50mm plates and 200 x
50mm braces to gable end
of two-storey house

	Unit	Hours	Hours £	Mat'ls £	O & P £	Total £
	Item	60.00	1020.00	527.61	232.14	1779.75

Temporary screens

Erect, maintain and remove
temporary screens consisting of
50 x 50mm softwood framing
covered one side with

	Unit	Hours	Hours £	Mat'ls £	O & P £	Total £
chipboard (three uses)	m²	1.15	19.55	1.98	3.23	24.76
insulation board (three uses)	m²	1.15	19.55	2.55	3.32	25.42
polythene sheeting three uses)	m²	0.80	13.60	1.51	2.27	17.38

Underpinning

Excavate preliminary trench by
hand, maximum depth not
exceeding

	Unit	Hours	Hours £	Mat'ls £	O & P £	Total £
1.00m	m³	3.70	62.90	0.00	9.44	72.34
1.50m	m³	4.15	70.55	0.00	10.58	81.13
2.00m	m³	4.40	74.80	0.00	11.22	86.02
2.50m	m³	6.10	103.70	0.00	15.56	119.26
3.00m	m³	7.25	123.25	0.00	18.49	141.74

	Unit	Hours	Hours £	Mat'ls £	O & P £	Total £
Excavate trench by hand below existing foundations, maximum not exceeding						
1.00m	m³	4.25	72.25	0.00	10.84	83.09
1.50m	m³	4.75	80.75	0.00	12.11	92.86
2.00m	m³	5.45	92.65	0.00	13.90	106.55
2.50m	m³	7.30	124.10	0.00	18.62	142.72
3.00m	m³	8.35	141.95	0.00	21.29	163.24
Excavate and backfill working space, maximum depth not exceeding						
1.00m	m³	5.00	85.00	0.00	12.75	97.75
1.50m	m³	5.25	89.25	0.00	13.39	102.64
2.00m	m³	5.90	100.30	0.00	15.05	115.35
2.50m	m³	8.10	137.70	0.00	20.66	158.36
3.00m	m³	9.25	157.25	0.00	23.59	180.84
Open-boarded earthwork support to sides of preliminary trenches, distance between faces not exceeding 2.0m						
maximum depth not exceeding 1.0m	m²	0.30	5.10	1.98	1.06	8.14
maximum depth not exceeding 2.0m	m²	0.40	6.80	1.98	1.32	10.10
Close-boarded earthwork support to sides of preliminary trenches, distance between faces not exceeding 2.0m						
maximum depth not exceeding 1.0m	m²	0.85	14.45	3.39	2.68	20.52
maximum depth not exceeding 2.0m	m²	1.10	18.70	3.39	3.31	25.40

	Unit	Hours	Hours £	Mat'ls £	O & P £	Total £
Open-boarded earthwork support to sides of excavation trenches, distance between faces not exceeding 2.0m						
maximum depth not exceeding 1.0m	m²	0.35	5.95	2.36	1.25	9.56
maximum depth not exceeding 2.0m	m²	0.45	7.65	2.36	1.50	11.51
Close-boarded earthwork support to sides of excavation trenches, distance between faces not exceeding 2.0m						
maximum depth not exceeding 1.0m	m²	0.95	16.15	4.65	3.12	23.92
maximum depth not exceeding 2.0m	m²	1.20	20.40	4.65	3.76	28.81
Cut away projecting concrete foundations, size						
150 × 150mm	m	0.45	7.65	0.00	1.15	8.80
150 × 225mm	m	0.55	9.35	0.00	1.40	10.75
150 × 300mm	m	0.65	11.05	0.00	1.66	12.71
Cut away projecting brickwork in footings, one brick thick						
one course	m	1.10	18.70	0.00	2.81	21.51
two courses	m	1.90	32.30	0.00	4.85	37.15
three courses	m	2.60	44.20	0.00	6.63	50.83

	Unit	Hours	Hours £	Mat'ls £	O & P £	Total £
Prepare underside of existing foundations to receive the new work, width						
300mm	m	0.75	12.75	0.00	1.91	14.66
500mm	m	1.15	19.55	0.00	2.93	22.48
750mm	m	1.40	23.80	0.00	3.57	27.37
1000mm	m	1.60	27.20	0.00	4.08	31.28
1200mm	m	1.80	30.60	0.00	4.59	35.19
Load surplus excavated material into barrows, wheel and deposit in temporary spoil heaps, average distance						
15m	m³	1.25	21.25	0.00	3.19	24.44
25m	m³	1.45	24.65	0.00	3.70	28.35
50m	m³	1.70	28.90	0.00	4.34	33.24
Load surplus excavated material into barrows, wheel and deposit in temporary spoil heaps, average distance						
25m	m³	1.45	24.65	0.00	3.70	28.35
50m	m³	1.70	28.90	0.00	4.34	33.24
Load surplus excavated material into barrows, wheel and deposit in skips or lorries, average distance						
25m	m³	1.40	23.80	0.00	3.57	27.37
50m	m³	1.65	28.05	0.00	4.21	32.26
Level and compact bottom of excavation	m³	0.15	2.55	0.00	0.38	2.93

	Unit	Hours	Hours £	Mat'ls £	O & P £	Total £
Site-mixed concrete 1:3:6 40mm aggregate in foundations to under-pinning, thickness						
150 to 300mm	m³	4.35	73.95	85.62	23.94	183.51
300 to 450mm	m³	4.05	68.85	85.62	23.17	177.64
over 450mm	m³	3.65	62.05	85.62	22.15	169.82
Site-mixed concrete 1:2:4 20mm aggregate in foundations to under-pinning, thickness						
150 to 300mm	m³	4.35	73.95	91.44	24.81	190.20
300 to 450mm	m³	4.05	68.85	91.44	24.04	184.33
over 450mm	m³	3.65	62.05	91.44	23.02	176.51
Plain vertical formwork to sides of underpinned foundations, height						
over 1m	m²	2.30	39.10	9.87	7.35	56.32
not exceeding 250mm	m	0.75	12.75	3.46	2.43	18.64
250 to 500mm	m	1.25	21.25	5.98	4.08	31.31
500mm to 1m	m	1.70	28.90	9.87	5.82	44.59
Plain vertical formwork to sides of underpinned foundations, left in, height						
over 1m	m²	2.15	36.55	36.56	10.97	84.08
not exceeding 250mm	m	0.65	11.05	9.55	3.09	23.69
250 to 500mm	m	1.15	19.55	18.04	5.64	43.23
500mm to 1m	m	1.60	27.20	36.56	9.56	73.32
High yield deformed steel reinforcement bars, straight or bent						
10mm diameter	m	0.04	0.68	0.84	0.23	1.75
12mm diameter	m	0.05	0.85	0.86	0.26	1.97
16mm diameter	m	0.06	1.02	0.88	0.29	2.19
20mm diameter	m	0.07	1.19	0.90	0.31	2.40
25mm diameter	m	0.08	1.12	0.92	0.31	2.35

	Unit	Hours	Hours £	Mat'ls £	O & P £	Total £
Common bricks basic price £200 per thousand in cement mortar in underpinning						
one brick thick	m²	5.20	88.40	16.45	15.73	120.58
one and a half brick thick	m²	7.40	125.80	16.45	21.34	163.59
two brick thick	m²	9.35	158.95	16.45	26.31	201.71
Class A engineering bricks in basic price £350 per thousand in cement mortar in underpinning						
one brick thick	m²	5.40	91.80	28.55	18.05	138.40
one and a half brick thick	m²	7.60	129.20	28.55	23.66	181.41
two brick thick	m²	9.50	161.50	28.55	28.51	218.56
Hessian-based bitumen damp proof course, bedded in cement mortar, horizontal						
over 225mm wide	m²	0.35	5.95	11.05	2.55	19.55
112mm wide	m	0.05	0.85	1.98	0.42	3.25
Two courses of slates bedded in cement mortar, horizontal						
over 225mm wide	m²	2.90	49.30	35.95	12.79	98.04
225mm wide	m	0.85	14.45	3.47	2.69	20.61
Wedge and pin new work to soffit of existing with slates in cement mortar, width						
one brick thick	m	1.90	32.30	7.48	5.97	45.75
one and a half brick thick	m	2.40	40.80	10.42	7.68	58.90
two brick thick	m	2.80	47.60	15.37	9.45	72.42

	Unit	Hours	Hours £	Mat'ls £	O & P £	Total £

Forming openings

Form opening for windows or
doors in existing walls in cement
mortar and make good to sides
of openings

	Unit	Hours	Hours £	Mat'ls £	O & P £	Total £
75mm blockwork	m²	2.00	34.00	2.88	5.53	42.41
100mm blockwork	m²	2.22	37.74	2.88	6.09	46.71
140mm blockwork	m²	2.65	45.05	3.63	7.30	55.98
215mm blockwork	m²	2.80	47.60	4.24	7.78	59.62
half brick wall	m²	2.90	49.30	2.88	7.83	60.01
one brick wall	m²	3.78	64.26	3.69	10.19	78.14
one and a half brick wall	m²	4.80	81.60	4.59	12.93	99.12
two brick wall	m²	7.20	122.40	6.03	19.26	147.69

Form opening for windows or
doors in existing walls in lime
mortar and make good to sides
of openings

	Unit	Hours	Hours £	Mat'ls £	O & P £	Total £
75mm blockwork	m²	1.80	30.60	2.88	5.02	38.50
100mm blockwork	m²	2.00	34.00	2.88	5.53	42.41
140mm blockwork	m²	2.40	40.80	3.63	6.66	51.09
215mm blockwork	m²	2.55	43.35	4.24	7.14	54.73
half brick wall	m²	2.60	44.20	2.88	7.06	54.14
one brick wall	m²	3.40	57.80	3.69	9.22	70.71
one and a half brick wall	m²	4.80	81.60	4.59	12.93	99.12
two brick wall	m²	6.50	110.50	6.03	17.48	134.01

Form opening for lintels above
openings in existing walls in cement
mortar and make good to sides
of openings

	Unit	Hours	Hours £	Mat'ls £	O & P £	Total £
75mm blockwork	m²	2.98	50.66	1.27	7.79	59.72
100mm blockwork	m²	3.36	57.12	1.27	8.76	67.15
140mm blockwork	m²	3.87	65.79	2.05	10.18	78.02
215mm blockwork	m²	4.23	71.91	2.82	11.21	85.94
half brick wall	m²	6.30	107.10	2.05	16.37	125.52

	Unit	Hours	Hours £	Mat'ls £	O & P £	Total £

Form opening for lintels (cont'd)

	Unit	Hours	Hours £	Mat'ls £	O & P £	Total £
one brick wall	m²	10.50	178.50	3.69	27.33	209.52
one and a half brick wall	m²	12.60	214.20	4.59	32.82	251.61
two brick wall	m²	16.98	288.66	6.03	44.20	338.89

Form opening for lintels above
openings in existing walls in lime
mortar and make good to sides
of openings

	Unit	Hours	Hours £	Mat'ls £	O & P £	Total £
75mm blockwork	m²	2.75	46.75	1.27	7.20	55.22
100mm blockwork	m²	3.00	51.00	1.27	7.84	60.11
140mm blockwork	m²	3.45	58.65	2.05	9.11	69.81
215mm blockwork	m²	3.86	65.62	2.82	10.27	78.71
half brick wall	m²	5.75	97.75	2.05	14.97	114.77
one brick wall	m²	9.45	160.65	3.69	24.65	188.99
one and a half brick wall	m²	11.40	193.80	4.59	29.76	228.15
two brick wall	m²	15.30	260.10	6.03	39.92	306.05

Form opening in reinforced
concrete floor slab and make good
to edges, slab thickness

	Unit	Hours	Hours £	Mat'ls £	O & P £	Total £
100mm	m²	5.65	96.05	3.21	14.89	114.15
150mm	m²	7.86	133.62	3.53	20.57	157.72
200mm	m²	9.50	161.50	3.85	24.80	190.15
250mm	m²	11.57	196.69	4.17	30.13	230.99
300mm	m²	14.55	247.35	7.72	38.26	293.33

Form opening in reinforced
concrete walls and make good
to edges, wall thickness

	Unit	Hours	Hours £	Mat'ls £	O & P £	Total £
100mm	m²	6.50	110.50	3.21	17.06	130.77
150mm	m²	7.20	122.40	3.53	18.89	144.82
200mm	m²	10.33	175.61	3.85	26.92	206.38
250mm	m²	12.10	205.70	4.17	31.48	241.35
300mm	m²	15.32	260.44	7.72	40.22	308.38

	Unit	Hours	Hours £	Mat'ls £	O & P £	Total £

Filling openings

Take out existing door and frame
complete, make good to reveals
and piece up skirting to match
existing

single internal door	nr	6.00	102.00	12.71	17.21	131.92
single external door	nr	6.20	105.40	12.71	17.72	135.83
double internal door	nr	6.20	105.40	13.43	17.82	136.65
double external door	nr	6.40	108.80	13.43	18.33	140.56

Take out existing single door and
frame complete, fill in opening, fix
new skirting to match existing and
plaster both sides

100mm blockwork	nr	10.45	177.65	46.11	33.56	257.32
215mm blockwork	nr	12.47	211.99	67.41	41.91	321.31
half brick wall	nr	12.40	210.80	42.12	37.94	290.86
one brick wall	nr	14.60	248.20	75.98	48.63	372.81
one and a half brick wall	nr	16.37	278.29	104.69	57.45	440.43

Take out existing double doors and
frame complete, fill in opening, fix
new skirting to match existing and
plaster both sides

100mm blockwork	nr	14.59	248.03	54.37	45.36	347.76
215mm blockwork	nr	16.36	278.12	122	60.02	460.14
half brick wall	nr	15.54	264.18	70.05	50.13	384.36
one brick wall	nr	20.34	345.78	105.09	67.63	518.50
one and a half brick wall	nr	22.20	377.40	128.68	75.91	581.99

	Unit	Hours	Hours £	Mat'ls £	O & P £	Total £

SPOT ITEMS

Brickwork, blockwork and masonry

Take out existing fireplace, fill
opening with 100mm thick blockwork plastered one side, fix new
skirting to match existing, make
good flooring where hearth
removed

medium size fireplace	nr	12.00	204.00	17.78	33.27	255.05
large size fireplace	nr	14.00	238.00	28.58	39.99	306.57

Cut out existing projecting chimney
breast, size 1350 × 350mm, including
making good flooring and ceiling
to match existing and plastering
wall where breast removed

one storey	nr	36.00	612.00	45.47	98.62	756.09

Take down chimney stack to
100mm below roof level, seal
flue with slates in cement mortar,
make good roof timbers

slated roof	nr	46.00	782.00	115.44	134.62	1032.06
tiled roof	nr	46.00	782.00	74.62	128.49	985.11

Cut out single brick in half brick
wall and replace in gauged mortar

commons	nr	0.25	4.25	0.25	0.68	5.18
facings	nr	0.30	5.10	0.51	0.84	6.45

	Unit	Hours	Hours £	Mat'ls £	O & P £	Total £
Cut out decayed brickwork in half brick wall in areas 0.5 to 1m³ and replace in gauged mortar						
commons	nr	5.20	88.40	14.63	15.45	118.48
facings	nr	6.00	102.00	2.05	15.61	119.66
Cut out decayed brickwork in one brick wall in areas 0.5 to 1m³ and replace in gauged mortar						
commons	nr	7.60	129.20	29.27	23.77	182.24
facings	nr	8.80	149.60	51.15	30.11	230.86
Cut out vertical, horizontal or stepped cracks in half brick wall, replace average 350mm wide with new bricks in gauged mortar						
commons	m	2.75	46.75	5.45	7.83	60.03
facings	m	3.10	52.70	13.88	9.99	76.57
Cut out vertical, horizontal or stepped cracks in one brick wall, replace average 350mm wide with new bricks in gauged mortar						
commons	m	5.20	88.40	10.93	14.90	114.23
facings	m	5.80	98.60	27.73	18.95	145.28
Cut out defective brick-on-end soldier arch to half brick wall, and replace with new bricks in gauged mortar						
commons	m	3.00	51.00	8.29	8.89	68.18
facings	m	3.40	57.80	12.68	10.57	81.05

	Unit	Hours	Hours £	Mat'ls £	O & P £	Total £
Cut out defective brick-on-end soldier arch to one brick wall, and replace with new bricks in gauged mortar						
commons	m	3.80	64.60	16.59	12.18	93.37
facings	m	4.20	71.40	25.37	14.52	111.29
Cut out defective terracotta air brick and replace						
215 × 65mm	nr	0.35	5.95	2.56	1.28	9.79
215 × 140mm	nr	0.50	8.50	3.44	1.79	13.73
215 × 215mm	nr	0.65	11.05	9.54	3.09	23.68
Rake out joints in gauged mortar, refix loose flashing and point up on completion						
horizontal	m	0.45	7.65	0.51	1.22	9.38
stepped	m	0.65	11.05	0.75	1.77	13.57
Rake out joints and point up in gauged mortar	m²	0.65	11.05	0.75	1.77	13.57
Cut out one course of half brick wall, insert hessian-based damp course 112mm wide and replace with new bricks in gauged mortar						
commons	m	2.00	34.00	3.33	5.60	42.93
facings	m	2.00	34.00	3.78	5.67	43.45
Cut out one course of one brick wall, insert hessian-based damp course 112mm wide and replace with new bricks in gauged mortar						
commons	m	3.25	55.25	6.68	9.29	71.22
facings	m	3.25	55.25	7.59	9.43	72.27

	Unit	Hours	Hours £	Mat'ls £	O & P £	Total £
Cut out single block, replace with new in gauged mortar, thickness						
100mm	nr	0.30	5.10	1.44	0.98	7.52
140mm	nr	0.40	6.80	2.29	1.36	10.45
190mm	nr	0.50	8.50	2.89	1.71	13.10
255mm	nr	0.60	10.20	3.91	2.12	16.23
Rake out joints of random rubble walling and point up in gauged mortar						
flush pointing	m²	0.50	8.50	0.64	1.37	10.51
weather pointing	m²	0.55	9.35	0.64	1.50	11.49

Roofing

	Unit	Hours	Hours £	Mat'ls £	O & P £	Total £
Take up roof coverings from pitched roof						
tiles	m²	0.80	13.60	0.00	2.04	15.64
slates	m²	0.80	13.60	0.00	2.04	15.64
timber boarding	m²	1.00	17.00	0.00	2.55	19.55
metal sheeting	m²	0.20	3.40	0.00	0.51	3.91
flat sheeting	m²	0.30	5.10	0.00	0.77	5.87
corrugated sheeting	m²	0.30	5.10	0.00	0.77	5.87
underfelt	m²	0.10	1.70	0.00	0.26	1.96
Take up roof coverings from flat roof						
bituminous felt	m²	0.25	4.25	0.00	0.64	4.89
metal sheeting	m²	0.30	5.10	0.00	0.77	5.87
woodwool slabs	m²	0.50	8.50	0.00	1.28	9.78
firrings	m²	0.20	3.40	0.00	0.51	3.91

	Unit	Hours	Hours £	Mat'ls £	O & P £	Total £
Take up roof coverings from pitched roof, carefully lay aside for reuse						
tiles	m²	1.10	18.70	0.00	2.81	21.51
slates	m²	1.10	18.70	0.00	2.81	21.51
metal sheeting	m²	0.50	8.50	0.00	1.28	9.78
flat sheeting	m²	0.60	10.20	0.00	1.53	11.73
corrugated sheeting	m²	0.60	10.20	0.00	1.53	11.73
Take up roof coverings from flat roof, carefully lay aside for reuse						
metal sheeting	m²	0.60	10.20	0.00	1.53	11.73
woodwool slabs	m²	0.80	13.60	0.00	2.04	15.64
Inspect roof battens, refix loose and replace with new, size 38 × 25mm						
25% of area						
250mm centres	m²	0.14	2.38	0.38	0.41	3.17
450mm centres	m²	0.12	2.04	0.31	0.35	2.70
600mm centres	m²	0.10	1.70	0.25	0.29	2.24
50% of area						
250mm centres	m²	0.26	4.42	0.78	0.78	5.98
450mm centres	m²	0.16	2.72	0.64	0.50	3.86
600mm centres	m²	0.18	3.06	0.51	0.54	4.11
75% of area						
250mm centres	m²	0.36	6.12	1.16	1.09	8.37
450mm centres	m²	0.26	4.42	0.97	0.81	6.20
600mm centres	m²	0.22	3.74	0.78	0.68	5.20
100% of area						
250mm centres	m²	0.44	7.48	2.01	1.42	10.91
450mm centres	m²	0.32	5.44	1.30	1.01	7.75
600mm centres	m²	0.38	6.46	1.04	1.13	8.63

	Unit	Hours	Hours £	Mat'ls £	O & P £	Total £
Remove single slipped slate and refix	nr	1.00	17.00	1.04	2.71	20.75
Remove single broken slate, renew with new Welsh blue slate						
405 × 255mm	nr	1.20	20.40	1.57	3.30	25.27
510 × 255mm	nr	1.20	20.40	3.02	3.51	26.93
610 × 305mm	nr	1.20	20.40	6.39	4.02	30.81
Remove slates in area approximately 1m² and replace with Welsh blue slates previously laid aside	nr	1.60	27.20	0.00	4.08	31.28
Remove slates in area approximately 1m² and replace with Welsh blue slates previously laid aside						
405 × 255mm	nr	1.80	30.60	0.00	4.59	35.19
510 × 255mm	nr	1.70	28.90	0.00	4.34	33.24
610 × 305mm	nr	1.60	27.20	0.00	4.08	31.28
Remove double course at eaves and fix new Welsh blue slates						
405 × 255mm	m	0.70	11.90	56.85	10.31	79.06
510 × 255mm	m	0.70	11.90	58.41	10.55	80.86
610 × 305mm	m	0.70	11.90	64.20	11.42	87.52
Remove single verge undercloak course and renew						
405 × 255mm	m	0.90	15.30	21.84	5.57	42.71
510 × 255mm	m	0.90	15.30	23.17	5.77	44.24
610 × 305mm	m	0.90	15.30	25.39	6.10	46.79
Remove single slipped tile and refix	nr	0.30	5.10	0.00	0.77	5.87

	Unit	Hours	Hours £	Mat'ls £	O & P £	Total £

Remove single broken tile and renew

	Unit	Hours	Hours £	Mat'ls £	O & P £	Total £
Marley plain tile	nr	0.30	5.10	0.62	0.86	6.58
Marley Ludlow Plus tile	nr	0.30	5.10	0.93	0.90	6.93
Marley Modern tile	nr	0.30	5.10	1.44	0.98	7.52
Redland Renown tile	nr	0.30	5.10	1.44	0.98	7.52
Redland Norfolk tile	nr	0.30	5.10	1.44	0.98	7.52

Remove tiles in area approximately 1m² and replace with Welsh previously laid aside

	Unit	Hours	Hours £	Mat'ls £	O & P £	Total £
Marley Plain tile	nr	1.80	30.60	35.45	9.91	75.96
Marley Ludlow Plus tile	nr	1.20	20.40	12.34	4.91	37.65
Marley Modern tile	nr	1.10	18.70	19.50	5.73	43.93
Lafarge Renown tile	nr	1.10	18.70	18.50	5.58	42.78
Lafarge Norfolk tile	nr	1.15	19.55	14.50	5.11	39.16

Take off defective ridge capping and refix including pointing in mortar

	Unit	Hours	Hours £	Mat'ls £	O & P £	Total £
	m	1.10	18.70	1.80	3.08	23.58

Carpentry and joinery

Take down, cut out or demolish structural timbers and load into skips

structural timbers

	Unit	Hours	Hours £	Mat'ls £	O & P £	Total £
50 × 100mm	m	0.10	1.70	0.00	0.26	1.96
50 × 150mm	m	0.12	2.04	0.00	0.31	2.35
75 × 100mm	m	0.14	2.38	0.00	0.36	2.74
75 × 150mm	m	0.16	2.72	0.00	0.41	3.13
100 × 150mm	m	0.18	3.06	0.00	0.46	3.52
100 × 200mm	m	0.20	3.40	0.00	0.51	3.91
roof boarding	m²	0.28	4.76	0.00	0.71	5.47
floor boarding	m²	0.22	3.74	0.00	0.56	4.30

	Unit	Hours	Hours £	Mat'ls £	O & P £	Total £
stud partition plasterboard both sides	m²	0.75	12.75	0.00	1.91	14.66
skirtings and grounds						
100mm high	m	0.08	1.36	0.00	0.20	1.56
150mm high	m	0.09	1.53	0.00	0.23	1.76
200mm high	m	0.10	1.70	0.00	0.26	1.96
rails						
50mm high	m	0.05	0.85	0.00	0.13	0.98
75mm high	m	0.06	1.02	0.00	0.15	1.17
100mm high	m	0.07	1.19	0.00	0.18	1.37
fittings						
wall cupboards	nr	0.25	4.25	0.00	0.64	4.89
floor units	nr	0.20	3.40	0.00	0.51	3.91
sink units	nr	0.25	4.25	0.00	0.64	4.89
staircase, 900mm wide						
straight flight	nr	4.00	68.00	0.00	10.20	78.20
landing	nr	1.50	25.50	0.00	3.83	29.33
doors, frames and linings						
single, internal	nr	0.40	6.80	0.00	1.02	7.82
single, external	nr	0.60	10.20	0.00	1.53	11.73
double, internal	nr	0.60	10.20	0.00	1.53	11.73
double, external	nr	0.80	13.60	0.00	2.04	15.64
windows						
casement, 1200 × 900mm	nr	0.50	8.50	0.00	1.28	9.78
casement, 1800 × 900mm	nr	0.60	10.20	0.00	1.53	11.73
sash, 900 × 1500mm	nr	0.80	13.60	0.00	2.04	15.64
sash, 1800 × 900mm	nr	0.90	15.30	0.00	2.30	17.60
ironmongery						
bolt	nr	0.20	3.40	0.00	0.51	3.91
deadlock	nr	0.25	4.25	0.00	0.64	4.89
mortice lock	nr	0.35	5.95	0.00	0.89	6.84
mortice latch	nr	0.35	5.95	0.00	0.89	6.84
cylinder lock	nr	0.25	4.25	0.00	0.64	4.89
door closer	nr	0.35	5.95	0.00	0.89	6.84
casement stay	nr	0.15	2.55	0.00	0.38	2.93
casement fastener	nr	0.15	2.55	0.00	0.38	2.93
toilet roll holder	nr	0.15	2.55	0.00	0.38	2.93
shelf bracket	nr	0.15	2.55	0.00	0.38	2.93

	Unit	Hours	Hours £	Mat'ls £	O & P £	Total £

Cut out defective joists or rafters and replace with new

50 × 75mm	nr	0.35	5.95	1.51	1.12	8.58
50 × 100mm	nr	0.40	6.80	1.57	1.26	9.63
50 × 150mm	nr	0.60	10.20	2.35	1.88	14.43
75 × 100mm	nr	0.65	11.05	3.17	2.13	16.35
75 × 150mm	nr	0.75	12.75	3.87	2.49	19.11

Cut out defective skirting and renew

75mm high	m	0.30	5.10	2.48	1.14	8.72
100mm high	m	0.35	5.95	2.91	1.33	10.19
150mm high	m	0.40	6.80	3.77	1.59	12.16

Ease, adjust and oil

door	nr	0.60	10.20	0.00	1.53	11.73
casement window	nr	0.40	6.80	0.00	1.02	7.82
sash window including renewing cords	nr	1.50	25.50	5.48	4.65	35.63

Take up defective flooring and replace with 25mm thick plain edged boarding

areas less than 1m²	m²	1.50	25.50	13.78	5.89	45.17
areas more than 1m²	m²	1.20	20.40	13.78	5.13	39.31

Refix loose floorboards including punching in protruding nails

	m²	0.20	3.40	0.00	0.51	3.91

Take down existing door, lay aside, piece up frame or lining where butts removed, rehang door to opposite hand on existing butts and hardware

single, internal	nr	1.50	25.50	2.02	4.13	31.65
single, external	nr	1.70	28.90	2.02	4.64	35.56

	Unit	Hours	Hours £	Mat'ls £	O & P £	Total £

Finishings

Take down plasterboard sheeting
from

studded walls	m²	0.35	5.95	0.00	0.89	6.84
ceilings	m²	0.40	6.80	0.00	1.02	7.82

Hack off plaster from

walls	m²	0.25	4.25	0.00	0.64	4.89
ceiling	m²	0.30	5.10	0.00	0.77	5.87

Make good plaster to walls where
wall removed

100mm wide	m²	0.95	16.15	1.96	2.72	20.83
150mm wide	m²	1.05	17.85	2.10	2.99	22.94
200mm wide	m²	1.15	19.55	2.23	3.27	25.05

Make good plaster to ceilings
where wall removed

100mm wide	m²	1.10	18.70	1.96	3.10	23.76
150mm wide	m²	1.15	19.55	2.10	3.25	24.90
200mm wide	m²	1.20	20.40	2.23	3.39	26.02

Cut out cracks in plasterwork and
make good

walls	m²	0.35	5.95	0.31	0.94	7.20
ceiling	m²	0.40	6.80	0.31	1.07	8.18

Hack off wall tiling and make
good surface | m² | 0.75 | 12.75 | 6.00 | 2.81 | 21.56 |

	Unit	Hours	Hours £	Mat'ls £	O & P £	Total £

Plumbing

Remove sanitary fittings and
supports, seal off supply and
waste pipes and prepare to
receive new fittings

bath	nr	3.50	63.00	13.17	11.43	87.60
sink	nr	3.50	63.00	13.17	11.43	87.60
lavatory basin	nr	3.25	58.50	7.90	9.96	76.36
WC	nr	3.75	67.50	7.90	11.31	86.71
shower cubicle	nr	4.00	72.00	10.54	12.38	94.92
bidet	nr	3.25	58.50	7.90	9.96	76.36

Take down length of existing
gutter, prepare ends and install
new length of gutter to existing
brackets

PVC-U

76mm	nr	0.60	10.80	9.36	3.02	23.18
112mm	nr	0.65	11.70	11.46	3.47	26.63

cast iron

100mm	nr	1.12	20.16	28.36	7.28	55.80

Take down existing brackets from
fascias and replace with galvanised
steel repair brackets at 1m

centres	m	0.30	5.40	2.82	1.23	9.45

Take down length of existing
pipe, prepare ends and install
new length of pipe

PVC-U

68mm diameter	nr	0.75	13.50	10.36	3.58	27.44
68mm square	nr	0.75	13.50	11.38	3.73	28.61

cast iron

75mm	nr	0.90	16.20	41.94	8.72	66.86
100mm	nr	1.05	18.90	57.84	11.51	88.25

	Unit	Hours	Hours £	Mat'ls £	O & P £	Total £
Take down fittings and install new						
PVC-U						
68mm diameter bend	nr	0.50	9.00	3.87	1.93	14.80
68mm offset	nr	0.50	9.00	9.02	2.70	20.72
68mm branch	nr	0.50	9.00	9.22	2.73	20.95
68mm shoe	nr	0.75	13.50	8.33	3.27	25.10
Cast iron						
75mm diameter bend	nr	0.60	10.80	7.76	2.78	21.34
75mm offset	nr	0.60	10.80	19.61	4.56	34.97
75mm branch	nr	0.60	10.80	8.52	2.90	22.22
75mm shoe	nr	0.60	10.80	18.39	4.38	33.57
Cast iron						
100mm diameter bend	nr	0.70	12.60	26.13	5.81	44.54
100mm offset	nr	0.70	12.60	36.16	7.31	56.07
100mm branch	nr	0.70	12.60	10.43	3.45	26.48
100mm shoe	nr	0.70	12.60	22.01	5.19	39.80
Cut out 500mm length of copper pipe, install new pipe with compression fittings at each end						
15mm	nr	0.80	14.40	5.10	2.93	22.43
22mm	nr	0.90	16.20	8.95	3.77	28.92
Take off existing radiator valve and replace with new	nr	0.90	16.20	11.67	4.18	32.05

	Unit	Hours	Hours £	Mat'ls £	O & P £	Total £

Take out existing galvanised
steel water storage tank and install
new plastic tank complete with
ball valve, lid and insulation
including cutting holes, make up
pipework and connectors to
existing pipework

	Unit	Hours	Hours £	Mat'ls £	O & P £	Total £
SC10, 18 litres	nr	3.60	64.80	59.28	18.61	142.69
SC15, 36 litres	nr	3.70	66.60	64.21	19.62	150.43
SC20, 54 litres	nr	3.75	67.50	71.02	20.78	159.30
SC25, 68 litres	nr	3.80	68.40	79.98	22.26	170.64
SC30, 86 litres	nr	4.90	88.20	86.40	26.19	200.79
SC40, 114 litres	nr	5.00	90.00	90.08	27.01	207.09

Glazing

	Unit	Hours	Hours £	Mat'ls £	O & P £	Total £
Hack out glass and remove	m²	0.45	7.65	0.00	1.15	8.80
Clean rebates, remove sprigs or clips and prepare for reglazing	m	0.20	3.40	0.00	0.51	3.91

Painting

Prepare, wash down painted
surfaces, rub down to receive new
paintwork

	Unit	Hours	Hours £	Mat'ls £	O & P £	Total £
brickwork	m²	0.14	2.38	0.00	0.36	2.74
blockwork	m²	0.15	2.55	0.00	0.38	2.93
plasterwork	m²	0.12	2.04	0.00	0.31	2.35

Prepare, wash down previously
painted wood surfaces, rub down
to receive new paintwork surfaces

	Unit	Hours	Hours £	Mat'ls £	O & P £	Total £
over 300mm girth	m²	0.28	4.76	0.00	0.71	5.47
isolated surfaces not exceeding 300mm girth	m	0.10	1.70	0.00	0.26	1.96
isolated surfaces not exceeding 0.5m²	nr	0.12	2.04	0.00	0.31	2.35

	Unit	Hours	Hours £	Mat'ls £	O & P £	Total £

Wallpapering

Strip off one layer of existing
paper, stop cracks and rub down
to receive new paper

	Unit	Hours	Hours £	Mat'ls £	O & P £	Total £
woodchip						
walls	m²	0.18	3.06	0.00	0.46	3.52
walls in staircase areas	m²	0.20	3.40	0.00	0.51	3.91
ceilings	m²	0.22	3.74	0.00	0.56	4.30
ceilings in staircase areas	m²	0.24	4.08	0.00	0.61	4.69
vinyl						
walls	m²	0.23	3.91	0.00	0.59	4.50
walls in staircase areas	m²	0.25	4.25	0.00	0.64	4.89
ceilings	m²	0.27	4.59	0.00	0.69	5.28
ceilings in staircase areas	m²	0.29	4.93	0.00	0.74	5.67
standard patterned						
walls	m²	0.21	3.57	0.00	0.54	4.11
walls in staircase areas	m²	0.23	3.91	0.00	0.59	4.50
ceilings	m²	0.25	4.25	0.00	0.64	4.89
ceilings in staircase areas	m²	0.27	4.59	0.00	0.69	5.28

Part Seven

PLANT AND TOOL HIRE

	24 hours	Additional 24 hours	Week
	£	£	£

TOOL AND EQUIPMENT HIRE

These selected rates are based on
average hire charges made by
hire firms in UK. Check your
local dealer for more information.
These prices exclude VAT.

CONCRETE AND CUTTING EQUIPMENT

Concrete mixers

	24 hours	Additional 24 hours	Week
Petrol, tip up	14.00	7.00	28.00
Electric, tip up	16.00	8.00	32.00
Bulk mixer	36.00	18.00	72.00

Vibrating pokers

Pokers	24 hours	Additional 24 hours	Week
petrol	40.00	20.00	80.00
electric	32.00	16.00	64.00
air poker, 50mm	28.00	14.00	56.00
air poker, 75mm	32.00	16.00	64.00

Power floats

Floats	24 hours	Additional 24 hours	Week
power float, petrol	48.00	24.00	96.00

Beam screeds

Screed units	24 hours	Additional 24 hours	Week
with 3.2 - 7.2 beam	34.00	66.00	128.00

Floor preparation units

Floor saw, petrol	24 hours	Additional 24 hours	Week
350mm	64.00	32.00	128.00
450mm	88.00	44.00	176.00
Scabbler, hand held	26.00	13.00	52.00

	24 hours	Additional 24 hours	Week
	£	£	£
Air needle gun	24.00	12.00	48.00
Concrete planer	72.00	36.00	154.00

Disc cutters

Cutters			
electric, 300mm	24.00	12.00	48.00
two stroke, 300mm	26.00	13.00	52.00
two stroke, 350mm	28.00	14.00	56.00
electric wall chasers	40.00	20.00	80.00

Block and slab splitters

Splitters			
clay	28.00	14.00	56.00
block	24.00	12.00	48.00
slab	40.00	20.00	80.00

ACCESS AND SITE EQUIPMENT

Ladders

Push up double ladder, glass fibre			
2.9m	24.00	12.00	48.00
3.5m	30.00	15.00	60.00
Push up triple ladder, glass fibre			
2.4m	24.00	12.00	48.00
3.0m	32.00	16.00	74.00
Roof ladder			
5.9m single	26.00	13.00	52.00
7.6m double	28.00	14.00	56.00
Rope operated			
6.0m double	40.00	20.00	80.00
5.2m triple	48.00	24.00	96.00

	24 hours	Additional 24 hours	Week
	£	£	£

Props

Shoring props
type 0	8.00	4.00	16.00
type 1	8.00	4.00	16.00
type 2	8.00	4.00	16.00
type 3	8.00	4.00	16.00

Rubbish chutes

Chutes
1m section	14.00	7.00	28.00
y section	22.00	11.00	44.00
hopper	20.00	10.00	40.00

Trestles and staging

Staging
2.4	18.00	9.00	36.00
3.6m	20.00	10.00	40.00
4.8m	26.00	13.00	52.00

Painters' trestle
1.8m	12.00	6.00	24.00
2.4m	14.00	7.00	28.00

Alloy towers

Single width, height
2.20m	50.00	25.00	100.00
3.20m	60.00	30.00	120.00
4.20m	70.00	35.00	140.00
5.20m	80.00	40.00	160.00
6.20m	90.00	45.00	180.00
7.20m	100.00	50.00	200.00
8.20m	110.00	55.00	220.00
9.20m	120.00	60.00	240.00
10.20m	130.00	65.00	260.00

	24 hours	Additional 24 hours	Week
	£	£	£

LIFTING AND MOVING

Box truck	18.00	9.00	36.00
Platform truck	18.00	9.00	36.00
Panel trolley	16.00	8.00	32.00
Stair climber, light duty	16.00	8.00	32.00

COMPACTION

Plate compactors

Compactors			
light	26.00	13.00	52.00
medium	26.00	13.00	52.00

Vibrating rollers

Light	28.00	14.00	56.00
Medium	30.00	15.00	60.00

BREAKING AND DEMOLITION

Breakers

Hydraulic			
petrol	72.00	36.00	144.00
heavy duty, petrol	76.00	38.00	152.00

POWER TOOLS

Drills

Cordless drill	28.00	14.00	56.00
Cordless hammer	26.00	13.00	52.00
Two speed impact	10.00	5.00	20.00
Right angle drill	22.00	11.00	44.00
Combi hammer			
light duty	18.00	9.00	36.00
medium duty	24.00	12.00	48.00
heavy duty	26.00	13.00	52.00

	24 hours	Additional 24 hours	Week
	£	£	£

Grinders

Angle grinder
230mm	14.00	7.00	28.00
300mm	30.00	15.00	60.00

Saws

Reciprocating saw
standard	24.00	12.00	48.00
heavy duty	30.00	15.00	60.00

Circular saw
150mm	18.00	9.00	36.00
230mm	20.00	10.00	40.00

Woodworking

Router	20.00	10.00	40.00
Cordless jig	20.00	10.00	40.00

Fixing equipment

Masonry nailer	28.00	14.00	56.00
Cartridge hammer	28.00	14.00	56.00
Electric screwdriver	18.00	9.00	36.00

Sanders

Belt	20.00	10.00	40.00
Triangle	12.00	6.00	24.00
Orbital	18.00	9.00	36.00

WELDING AND GENERATORS

Generating

Generators
petrol, 2.6kVA	26.00	13.00	52.00
diesel, 2.6kVA	38.00	19.00	76.00
diesel, silenced, 6kVA	94.00	47.00	188.00
diesel, silenced, 20kVA	222.00	111.00	444.00

	24 hours	Additional 24 hours	Week
	£	£	£

PUMPING EQUIPMENT

Pumps

	24 hours	Additional 24 hours	Week
Submersible			
50mm	42.00	21.00	84.00
Centrifugal pump			
50mm	34.00	17.00	68.00
75mm	40.00	20.00	80.00
Diaphragm pump, 75mm	64.00	32.00	128.00